METHODS IN MOLECULAR BIOLOGY

For further volumes:
http://www.springer.com/series/7651

Proteomics

Methods and Protocols

Edited by

Lucio Comai

Keck School of Medicine, University of Southern California, Los Angeles, CA, USA

Jonathan E. Katz

Keck School of Medicine, University of Southern California, Los Angeles, CA, USA

Parag Mallick

Stanford University, Palo Alto, CA, USA

Humana Press

Editors
Lucio Comai
Keck School of Medicine
University of Southern California
Los Angeles, CA, USA

Jonathan E. Katz
Keck School of Medicine
University of Southern California
Los Angeles, CA, USA

Parag Mallick
Stanford University
Palo Alto, CA, USA

ISSN 1064-3745 ISSN 1940-6029 (electronic)
Methods in Molecular Biology
ISBN 978-1-4939-6745-2 ISBN 978-1-4939-6747-6 (eBook)
DOI 10.1007/978-1-4939-6747-6

Library of Congress Control Number: 2017930810

Printed on acid-free paper

This Humana Press imprint is published by Springer Nature
The registered company is Springer Science+Business Media LLC
The registered company address is: 233 Spring Street, New York, NY 10013, U.S.A.

Preface

In the catalog of biochemical techniques, proteomics has barely reached its adolescent stage, and a very immature adolescent at that. Like any teen, potential still overshadows realized accomplishment, but the future is still bright with potential. This particular adolescent has shown quite a level of promise. Indeed, for a number of tasks, proteomics is fully proficient—determining the identity of a small number of proteins, providing absolute quantitation of a similar number of proteins. For others, it is still testing its limits: How many proteins? How many orders of magnitude of sensitivity? And we begin to doubt, but it is not impossible to imagine the realization of the full parental dream—given a sample, what are the concentrations and identity of every protein and every modification on those proteins. And then there are the unexpected questions we can answer—our teenager has shown potential in areas we never first imagined: What is the three-dimensional structure of a protein? How do proteins interact? What is the turnover rate of various post-translational modifications?

We view proteomics as a pipeline with four discrete components: The isolation of material from a biological specimen, sample preprocessing, sample analysis, and data interpretation.

Recognizing proteomic analysis almost always is a collaborative effort and that specialized analyses will always require domain-specific knowledge, our goals with this book are to provide step-by-step protocols on a wide range of biochemical methods, analytical approaches, and bioinformatics tools developed to analyze the proteome. Here are our specific goals for this book:

1. Accessible. Most scientists in the life sciences will be able to employ the methods described in this book. Aside from basic mass spectrometers, we have avoided unusual and/or expensive equipment and reagents. (Specialists do not consult books as a primary reference.)
2. Practical. The techniques herein described are broadly applicable, commonly employed protocols.
3. Current. Mature well-established protocols will be referenced and briefly described. "State-of-the-art" emerging standard protocols will be clearly and completely described—common wisdom included at no extra charge!
4. One stop. Recognizing that proteomics is often a collaborative effort, this book shall describe, as we see it, the complete proteomic pipeline, upfront biology through data analysis. For analysis that has become or is emerging as routine, our hopes are for this to be *the* "go to" reference.

Los Angeles, CA — *Lucio Comai*
Los Angeles, CA — *Jonathan E. Katz*
Palo Alto, CA — *Parag Mallick*

Contents

Contributors

RAVALI ADUSUMILLI • *Department of Radiology, Canary Center at Stanford for Cancer Early Detection, Stanford University, Stanford, CA, USA*
RUEDI AEBERSOLD • *Institute of Molecular Systems Biology, ETH Zurich, Zurich, Switzerland; Faculty of Science, University of Zurich, Zurich, Switzerland*
DANIELE ALBUQUERQUE • *Department of Biochemistry and Immunology, Ribeirão Preto Medical School, University of São Paulo, Ribeirão Preto, SP, Brazil*
MICHELLE ATALLAH • *Canary Center at Stanford for Cancer Early Detection, Stanford University, Palo Alto, CA, USA*
MIRIAM BARRIOS-RODILES • *Center for Systems Biology, Lunenfeld-Tanenbaum Research Institute, Mount Sinai Hospital, Toronto, ON, Canada*
FLORIAN BELL • *Grace Bio-Labs, Bend, OR, USA*
BENJAMIN J. BLENCOWE • *Donnelly Centre, University of Toronto, Toronto, ON, Canada; Department of Molecular Genetics, University of Toronto, Toronto, ON, Canada*
JOSEPH CAPRI • *Department of Pharmacology, David Geffen School of Medicine, Los Angeles, CA, USA*
RICCARDO ZENEZINI CHIOZZI • *Department of Chemistry, Sapienza Università di Roma, Rome, Italy*
CAROLINE S. CHU • *Agilent Technologies, Inc., Santa Clara, CA, USA*
JÉROME DÉJARDIN • *INSERM AVENIR, Institute of Human Genetics CNRS UPR1142, Montpellier, France*
PETRA DOMASINSKA • *Biomedical Research Center, University Hospital Hradec Kralove, Hradec Kralove, Czech Republic; Faculty of Chemical Technology, Department of Biological and Biochemical Sciences, University of Pardubice, Pardubice, Czech Republic*
JOSHUA E. ELIAS • *Department of Chemical & Systems Biology, Stanford University, Stanford, CA, USA*
JONATHAN D. ELLIS • *Donnelly Centre, University of Toronto, Toronto, ON, Canada*
JONATHAN ERDE • *Department of Chemistry and Biochemistry, University of California-Los Angeles, Los Angeles, CA, USA*
VIRGINIA ESPINA • *Center for Applied Proteomics and Molecular Medicine, George Mason University, Manassas, VA, USA*
VITOR M. FAÇA • *Department of Biochemistry and Immunology, Ribeirão Preto Medical School, University of São Paulo, Ribeirão Preto, SP, Brazil; Center for Cell Based Therapy - Hemotherapy Center of Ribeirão Preto, Ribeirão Preto Medical School, University of São Paulo, Ribeirão Preto, SP, Brazil*
STEVE M. FISCHER • *Agilent Technologies, Inc., Santa Clara, CA, USA*
MARK R. FLORY • *Canary Center at Stanford for Cancer Early Detection, Stanford University, Palo Alto, CA, USA*
MARTIN FREJNO • *Chair of Proteomics and Bioanalytics, Technische Universität München, Freising, Germany; Department of Oncology, University of Oxford, Oxford, UK*
BRADFORD W. GIBSON • *The Buck Institute for Research on Aging, Redwood City, CA, USA*

Andy Gieschen • *Agilent Technologies, Inc., Santa Clara, CA, USA*
Anne-Claude Gingras • *Lunenfeld-Tanenbaum Research Institute, Mount Sinai Hospital, Toronto, Canada; Department of Molecular Genetics, University of Toronto, Toronto, Canada*
Alexander Goncearenco • *National Center for Biotechnology Information, National Institutes of Health, Bethesda, MD, USA*
Mariana L. Grassi • *Department of Biochemistry and Immunology, Ribeirão Preto Medical School, University of São Paulo, Ribeirão Preto, SP, Brazil; Center for Cell Based Therapy - Hemotherapy Center of Ribeirão Preto, Ribeirão Preto Medical School, University of São Paulo, Ribeirão Preto, SP, Brazil*
Hannes Hahne • *OmicScouts GmbH, Freising, Germany*
Michael Harpole • *Center for Applied Proteomics and Molecular Medicine, George Mason University, Manassas, VA, USA*
Geoffrey G. Hesketh • *Lunenfeld-Tanenbaum Research Institute, Mount Sinai Hospital, Toronto, Canada*
Christie L. Hunter • *SCIEX, Framingham, MA, USA*
Sophie L. Kan • *INSERM AVENIR, Institute of Human Genetics CNRS UPR1142, Montpellier, France*
Kian Kani • *USC Center for Applied Molecular Medicine, USC Keck School of Medicine, Los Angeles, CA, USA*
Jonathan E. Katz • *USC Center for Applied Molecular Medicine, Los Angeles, CA, USA*
Heiner Koch • *Chair of Proteomics and Bioanalytics, Technische Universität München, Freising, Germany; German Cancer Consortium (DKTK), Heidelberg, Germany; German Cancer Research Center (DKFZ), Heidelberg, Germany*
Majlinda Kullolli • *Department of Radiology, Canary Center at Stanford for Cancer Early Detection, Stanford University School of Medicine, Palo Alto, CA, USA*
Bernhard Kuster • *Chair of Proteomics and Bioanalytics, Technische Universität München, Freising, Germany; Center for Integrated Protein Science Munich (CIPSM), Freising, Germany; German Cancer Consortium (DKTK), Heidelberg, Germany; German Cancer Research Center (DKFZ), Heidelberg, Germany; Bavarian Biomolecular Mass Spectrometry Center, Technische Universität München, Freising, Germany*
Peter Lanzerstorfer • *School of Engineering and Environmental Sciences, University of Applied Sciences Upper Austria, Wels, Austria*
Grace Lawrence • *Center for Applied Proteomics and Molecular Medicine, George Mason University, Manassas, VA, USA*
Simone Lemeer • *Chair of Proteomics and Bioanalytics, Technische Universität München, Freising, Germany; Center for Integrated Protein Science Munich (CIPSM), Freising, Germany; Biomolecular Mass Spectrometry and Proteomics, Bijvoet Center for Biomolecular Research and Utrecht Institute of Pharmaceutical Sciences, Utrecht University, Utrecht, The Netherlands*
Li-Hua Li • *Department of Pathology and Laboratory Medicine, Taipei Veterans General Hospital, Taipei, Taiwan, R.O.C.*
Minghui Li • *National Center for Biotechnology Information, National Institutes of Health, Bethesda, MD, USA*
Joseph A. Loo • *Department of Chemistry and Biochemistry, University of California-Los Angeles, Los Angeles, CA, USA; Department of Biological Chemistry, University of California-Los Angeles, Los Angeles, CA, USA*

PARAG MALLICK • *Department of Radiology, Canary Center at Stanford for Cancer Early Detection, Stanford University, Stanford, CA, USA; School of Medicine, Stanford University, Palo Alto, CA, USA*
ANDREA G. MARIA • *Department of Pediatrics, Ribeirão Preto Medical School, University of São Paulo, Ribeirão Preto, SP, Brazil*
CHRISTINE A. MILLER • *Agilent Technologies, Inc., Santa Clara, CA, USA*
RACHEL R. OGORZALEK LOO • *Department of Biological Chemistry, University of California-Los Angeles, Los Angeles, CA, USA*
CAMILA S. PALMA • *Department of Biochemistry and Immunology, Ribeirão Preto Medical School, University of São Paulo, Ribeirão Preto, SP, Brazil; Center for Cell Based Therapy - Hemotherapy Center of Ribeirão Preto, Ribeirão Preto Medical School, University of São Paulo, Ribeirão Preto, SP, Brazil*
ANNA R. PANCHENKO • *National Center for Biotechnology Information, National Institutes of Health, Bethesda, MD, USA*
SHARON J. PITTERI • *Department of Radiology, Canary Center at Stanford for Cancer Early Detection, Stanford University School of Medicine, Palo Alto, CA, USA*
ALINE POERSCH • *Department of Biochemistry and Immunology, Ribeirão Preto Medical School, University of São Paulo, Ribeirão Preto, SP, Brazil; Center for Cell Based Therapy - Hemotherapy Center of Ribeirão Preto, Ribeirão Preto Medical School, University of São Paulo, Ribeirão Preto, SP, Brazil*
BRIAN RAUGHT • *Princess Margaret Research Institute, Princess Margaret Cancer Centre, University Health Network, Toronto, Canada; Department of Medical Biophysics, University of Toronto, Toronto, Canada*
HANNES L. RÖST • *Institute of Molecular Systems Biology, ETH Zurich, Zurich, Switzerland; Department of Genetics, Stanford University, Stanford, CA, USA*
DANIEL RUDERMAN • *Lawrence J. Ellison Institute for Transformative Medicine of USC, USC Keck School of Medicine, Los Angeles, CA, USA*
BENJAMIN RUPRECHT • *Chair of Proteomics and Bioanalytics, Technische Universität München, Freising, Germany; Center for Integrated Protein Science Munich (CIPSM), Freising, Germany*
NEHMÉ SAKSOUK • *INSERM AVENIR, Institute of Human Genetics CNRS UPR1142, Montpellier, France*
PAYMAN SAMAVARCHI-TEHRANI • *Lunenfeld-Tanenbaum Research Institute, Mount Sinai Hospital, Toronto, Canada*
BIRGIT SCHILLING • *The Buck Institute for Research on Aging, Redwood City, CA, USA*
OLGA T. SCHUBERT • *Department of Human Genetics, University of California, Los Angeles, CA, USA; Institute of Molecular Systems Biology, ETH Zurich, Zurich, Switzerland*
GERHARD J. SCHÜTZ • *Institute of Applied Physics, TU Wien, Vienna, Austria*
EVA SEVCSIK • *Institute of Applied Physics, TU Wien, Vienna, Austria*
MICHAEL SHULTZ • *Grace Bio-Labs, Bend, OR, USA*
CAROLINA H. THOMÉ • *Department of Biochemistry and Immunology, Ribeirão Preto Medical School, University of São Paulo, Ribeirão Preto, SP, Brazil; Center for Cell Based Therapy - Hemotherapy Center of Ribeirão Preto, Ribeirão Preto Medical School, University of São Paulo, Ribeirão Preto, SP, Brazil*
SARAH M. TOTTEN • *Department of Radiology, Canary Center at Stanford for Cancer Early Detection, Stanford University School of Medicine, Palo Alto, CA, USA*

DONGXUE WANG • *Chair of Proteomics and Bioanalytics, Technische Universität München, Freising, Germany*
XIA WANG • *Department of Mathematical Sciences, University of Cincinnati, Cincinnati, OH, USA*
JULIAN WEGHUBER • *School of Engineering and Environmental Sciences, University of Applied Sciences Upper Austria, Wels, Austria*
JULIAN P. WHITELEGGE • *The Pasarow Mass Spectrometry Laboratory, The Jane and Terry Semel Institute for Neuroscience and Human Behavior, David Geffen School of Medicine, UCLA, Los Angeles, CA, USA*
JEFFREY L. WRANA • *Department of Molecular Genetics, University of Toronto, Toronto, ON, Canada; Center for Systems Biology, Lunenfeld-Tanenbaum Research Institute, Mount Sinai Hospital, Toronto, ON, Canada; Breast Cancer Research, Mary Janigan Chair in Molecular Cancer Therapeutics, Toronto, ON, Canada*
SOLOMON YEON • *Center for Applied Proteomics and Molecular Medicine, George Mason University, Manassas, VA, USA*
JI-YOUNG YOUN • *Lunenfeld-Tanenbaum Research Institute, Mount Sinai Hospital, Toronto, Canada*
JANA ZECHA • *Technische Universität München, Freising, Germany; German Cancer Consortium (DKTK), Heidelberg, Germany; German Cancer Research Center (DKFZ), Heidelberg, Germany*
LICHAO ZHANG • *Department of Chemical & Systems Biology, Stanford University, Stanford, CA, USA*
DANIEL P. ZOLG • *Technische Universität München, Freising, Germany*

Chapter 1

A Robust Protocol for Protein Extraction and Digestion

Michelle Atallah, Mark R. Flory, and Parag Mallick

Abstract

Proteins play a key role in all aspects of cellular homeostasis. Proteomics, the large-scale study of proteins, provides in-depth data on protein properties, including abundances and post-translational modification states, and as such provides a rich avenue for the investigation of biological and disease processes. While proteomic tools such as mass spectrometry have enabled exquisitely sensitive sample analysis, sample preparation remains a critical unstandardized variable that can have a significant impact on downstream data readouts. Consistency in sample preparation and handling is therefore paramount in the collection and analysis of proteomic data.

Here we describe methods for performing protein extraction from cell culture or tissues, digesting the isolated protein into peptides via in-solution enzymatic digest, and peptide cleanup with final preparations for analysis via liquid chromatography-mass spectrometry. These protocols have been optimized and standardized for maximum consistency and maintenance of sample integrity.

Key words Proteomics, Protein extraction, Acetone precipitation, Enzymatic solution digest, Liquid chromatography-mass spectrometry

1 Introduction

Analysis of proteins, the key effectors of most cellular processes, is critical for understanding biological processes in healthy and diseased states, and liquid chromatography-mass spectrometry (LC-MS) analysis of peptides has proven to be a workhorse tool to this end. Over the last 10 years, LC-MS instrumentation for biological proteomics has undergone a remarkable evolution with significant gains in analysis speed, precision, consistency, and sensitivity and has in many cases given rise to commoditized instrumentation accessible even to non-experts [1]. As a result, a major factor in the success of proteomics-focused assays has now become the quality of the input material and the consistency of, and care in performing, methods to generate proteomic samples for LC-MS assays.

While the relative simplicity of nucleic acid analytes enabled a reasonably rapid and facile unification of methods in the genomics field, the huge variation in biophysical properties of proteins, and

Lucio Comai et al. (eds.), *Proteomics: Methods and Protocols*, Methods in Molecular Biology, vol. 1550,
DOI 10.1007/978-1-4939-6747-6_1, © Springer Science+Business Media LLC 2017

their several orders of magnitude abundance variation in cells, has proven a significant challenge to the proteomics field [1]. Variations in methods for proteomics sample generation across labs, and even within groups across operators, can significantly affect and even confound the outputs of MS-based proteomics. This will undoubtedly prove to be a challenge going forward for the field in highly specific experimental contexts and in answering unique biological questions, especially given the highly varied biophysical properties exhibited by proteins and protein complexes. However, methods striving for complete solubilization of proteomes, for example for global profiling under different states or conditions, are much more amenable to unification across the field.

However, an operator striving to develop such a sample workup regime for global proteomics analysis faces what is now a daunting variety of choices at each step of the workflow. Often reagents for upstream sample preparation are not compatible with LC-MS analysis downstream, necessitating additional steps that can potentially cause sample loss. For example, while robust detergents are generally required to affect protein solubility, many common detergents used in this role, such as SDS, are not compatible with downstream proteolytic digestion and/or LC-MS. A wide variety of choices to deal with this challenge now face the operator including traditional methods such as protein precipitation for detergent removal [2], commercial innovations such as alternative LC-MS compatible surfactants (e.g., Rapigest SF Surfactant, Waters Corporation), and development of relatively new methods such as FASP (Filter-Aided Sample Preparation), a method facilitating detergent removal and digestion on a solid phase [3], just to name a few. In developing a composite workflow, assessing which of these approaches best balances robustness and consistency while avoiding significant sample loss is a difficult task. Moreover, a similarly daunting set of choices must be addressed at all other steps of a proteomics workflow, including a host of variables for proteolytic digestion and peptide cleanup, among others.

Here we provide a simple, robust method for whole-cell protein extraction from tissue culture cells, including methods for downstream enzymatic digestion and peptide cleanup prior to LC-MS. This method employs a classic ionic detergent at a relatively high percentage to efficiently extract and solubilize proteins. A facile acetone precipitation step is employed to remove this detergent prior to straightforward and proven in-solution proteolytic digestion. A simple solid-phase extraction method is performed at the end of the workflow to clean and concentrate peptides in advance of LC-MS. While no one method is perfect, based on our experiences working in the field we feel the simplicity and consistency of this method will provide new operators to the field with an accessible, robust, and consistent starting point, and in fact we use this routinely in our group for global proteomic analyses of our valuable primary samples. Specific method steps are provided, and critical variables for each method are emphasized so that experimental bias is minimized as much as possible.

2 Materials

1. Cell Lysis and Acetone Precipitation.

Given the sensitivity of mass spectrometry analysis, utmost care should be taken to ensure clean working conditions free of all possible contaminants. All reagents should be prepared using LC-MS grade water (Honeywell brand is preferred). If possible, sequestering a separate set of pipettes along with the use of nitrile examination gloves is extremely useful in avoiding contamination of samples with ubiquitous keratin from the environment.

2.1 Solutions and Reagents

1. 4% w/v sodium dodecyl sulfate (SDS).
2. 1 M Tris hydrochloride, pH 7.5.
3. 1 M dithiothreitol (DTT): Dissolve 87 mg in 5 mL 100% MeOH, store at −20 °C.
4. Thermo HALT protease and phosphatase inhibitor cocktail (100×).
5. 1 M (10×) PMSF (phenyl-methyl-sulfonamide): Dissolve 87 mg in 5 mL LC-MS water, store at −20 °C.
6. LC-MS-grade water.
7. 99.5%+acetone, chilled to −20 °C.

2.2 Equipment

1. Probe tip sonicator (alternatively, a Diagenode or Covaris water bath sonicator, *see* **Note 1**).
2. (for probe tip sonicators) Clear plastic or acrylic box (6×6×6 in. works well).
3. (for probe tip sonicators) Foam tube floater.
4. Pierce Micro BCA Protein Assay Kit.

2.3 Peptide Digestion

2.3.1 Solutions and Reagents

1. 1 M TCEP: Dissolve 287 mg in 1 mL MS water. Store at −20 °C in suitable (e.g., 10 μL) aliquots.
2. 1 M Tris pH 8, *see* **Note 2**.
3. 500 mM iodoacetamide (IAA): Dissolve a single-use pre-weighed tube of 9.3 mg IAA (Pierce/Thermo Fisher) in 200 μL LC/MS water.
4. Protein resuspension solution: 8 M urea, 100 mM Tris pH 8.
5. Protein dilution solution: 100 mM Tris pH 8.
6. 100 mM $CaCl_2$.
7. Sequence-grade trypsin enzyme.

2.4 C18 Cleanup

2.4.1 Solutions and Reagents

1. Honeywell Brand Reverse Phase A (RP-A): LC-MS water with 0.1 % formic acid.
2. Honeywell Brand Reverse Phase B (RP-B): acetonitrile with 0.1 % formic acid.
3. Wetting solution (50 % Honeywell RP-B in RP-A).
4. Equilibration/Wash solution (2 % RP-B in Honeywell RP-A), *see* **Note 3**.
5. Elution solution (90 % Honeywell RP-B in RP-A).

2.5 Equipment

1. Thermo Pierce C18 tips (Thermo product 87784).

3 Methods

3.1 Cell Lysis and Acetone Precipitation

1. Fill an ice bucket with wet ice and fill up to ice level with water.
2. Prepare 500 μL of cell lysis buffer for each 1e7 cells.
   (a) Lysis buffer composition: 3 % SDS, 0.02 M DTT, 0.10 M Tris–HCl pH 7.5, 1× Thermo Protease/Phosphatase inhibitor, 1× PMSF. Make fresh and add protease/phosphatase inhibitors, especially PMSF that is active in aqueous solution for 30 min, just before use.
   (b) Preheat lysis buffer at 95 °C. *See* **Note 4**.
3. Keep cell pellet tubes on dry ice. Add preheated lysis buffer to frozen cell pellet; 500 μL for each ten million cells (*see* **Notes 5** and **6**). Pipette and/or vortex to mix. Critical point: the frozen pellet must not be allowed to thaw until covered with hot SDS lysis solution—thawing then should be rapid and right into the concentrated surfactant.
4. Put cap lock on tube. Place tube in 95 °C heat block for 3 min. Vortex every 15–20 s.
5. If necessary, split contents into tubes of ~500 μL each, *see* **Note 7**.
6. Pour chilled ice water from ice bucket into clear plastic/acrylic box (*see* **Note 8**), avoiding ice as much as possible. Wedge an ice pack to the bottom of the acrylic box to keep the water cool (*see* **Note 9**), and a foam tube holder to float the tubes at the surface.
7. Vent the tubes by opening them briefly (with the opening cap oriented away from you). Place them in the foam holder in the acrylic box.
8. Sonicate samples while the tubes remain in ice water (*see* **Note 10**). There are two critical points in this step, both facilitated by the use of the clear acrylic box: first, the samples must be kept cold to avoid degradation of proteins due to the heat generated by the sonicator. We have found that keeping

the tubes in water is the best way to accomplish this. Secondly, it is necessary to submerge the sonicator probe tip before switching on sonicator, and switch it off before removing the tip. Moving the probe in and out of solution while on will cause foaming of the solution and may damage proteins. Additionally, be consistent with tip depth (2/3 into solution from top is ideal).

(a) Set sonicator amplitude to 40%.

(b) Press the "Set" button to select "Continuous".

(c) Sonicate for 3 cycles. 1 cycle = ON for 30 s, OFF on ice for 2 min (*see* **Notes 11** and **12**). Avoid contact between probe tip and tube walls as much as possible to prevent shedding of polymers into your sample.

9. Centrifuge tubes for 15 min at 15,000 rpm at 20 °C to clarify the lysate of any particulates or insoluble material (*see* **Note 13**).
10. Transfer ~200–250 µL of supernatant to labeled Eppendorf tubes (*see* **Note 14**).

    (a) Pull out 10% of the volume for a separate precipitation to generate a pellet to bring up in 3% SDS for BCA quantitation (*see* **Note 15**).

    (b) Note the exact volume of each sample that goes into each new tube. This will be important later when calculating the amount of protein per tube.

11. Using a glass Pasteur pipette (*see* **Note 16**), add cold (−20 °C) 100% acetone to tubes. Fill to 1 mL line (*see* **Note 17**). Critical point: ensure that least 4× volume acetone is added.
12. Invert tubes several times, vortex well, and place tubes at −20 °C overnight (*see* **Note 18**).
13. The next day, spin tubes at 15,000 rpm for 15 min at 4 °C (*see* **Note 19**).
14. Keep tubes on ice. Without disturbing the pellet or pipetting up and down, remove and discard the acetone supernatant with a glass Pasteur pipette (*see* **Note 20**). Add fresh ice-cold acetone to the 1 mL mark, again without disturbing the pellet.
15. Centrifuge samples at 15,000 rpm for 10 min at 4 °C (*see* **Note 21**).
16. Repeat **steps 14** and **15** for a second wash (*see* **Note 22**).
17. Remove and discard acetone supernatant with a glass Pasteur pipette.
18. Air-dry pellets with the lids open (covered with a Kimwipes) for 1 h while on ice (*see* **Note 23**).
19. Store dry pellets at −80 °C until ready for further use.

20. Resuspend the pellet from the 10 % (v/v) aliquot precipitation from **step 10a** in 100–300 μL 3% SDS and perform BCA (*see* **Note 24**) quantification according to the manufacturer's instructions. From the results determine how much protein is in each acetone-precipitated pellet based on the fraction of the total sample that went into each tube. Critical point: ensure complete resuspension and solubilization of the pellet. In addition to visually inspecting the pellet to confirm complete resuspension, vortexing the sample and/or heating at 37 °C for up to 2 min can aid in resolubilization.

3.2 Peptide Digestion

1. Bring up dried acetone pellet in 50 μL 8 M urea, 100 mM Tris pH 8 for a final concentration of 80 μg in 25 μL (*see* **Note 25**).
   (a) Use manual pipetting and if necessary heat and/or sonication to break up pellet. Avoid any heating over 37 °C to avoid carbamylation by urea.
2. Add TCEP to final concentration of 5 mM and incubate for RT for 30 min (*see* **Note 26**). Mix well by vortexing, and then knock down by pulse spinning in microfuge.
3. Add IAA to final concentration of 10 mM and incubate for RT for 30 min in the dark (covered in foil, *see* **Note 27**). Mix well by vortexing, and then knock down by pulse spinning in microfuge.
4. Bring up to 250 μL with 100 mM Tris pH 8, reducing the urea concentration to <1 M (*see* **Note 28**). Mix well by vortexing, and then knock down by pulse spinning in microfuge.
5. Add 100 mM $CaCl_2$ to a final concentration of 1 mM. Mix well by vortexing, and then knock down by pulse spinning in microfuge.
6. Add 8 μg trypsin (1:10 mass:mass) for a final concentration of 30 ng/μL. Mix well by vortexing, and then knock down by pulse spinning in microfuge.
7. Incubate overnight at 37 °C in static or shaking incubator.

3.3 C18 Cleanup

The following is a modified protocol is based on the manufacturer's recommendations for Pierce C18 tips (100 μL bed, Catalog No. 87784).

Preloading tubes, or well plates, with each solution to be used can increase throughput speed and minimizes downtime in which the tip can dry. Each 100 μL aliquot of sample should be loaded separately, and should be separately aliquoted into wells or tubes.

1. Wet C18 filter tip by aspirating 100 μL of wetting solution and then discarding solvent (*see* **Note 29**).
2. Repeat wash step and discard solvent.
3. Equilibrate tip by aspirating 100 μL of equilibration/wash solution and discarding the solvent.

4. Repeat equilibration step and discard solvent.
5. Aspirate 100 μL of sample and, using a clean Eppendorf tube, slowly aspirate 100 μL up and down 10× (*see* **Note 30**).
   (a) Repeat as many times as necessary until the entire sample volume has been absorbed onto the column.
6. Wash tip by aspirating 100 μL of equilibration/wash solution and discarding the solvent.
7. Repeat the wash step and discard the solvent.
8. Elute the sample by slowly aspirating 100 μL of elution solution into the tip, and slowly depositing into a new elution tube.
9. Repeat the elution step for a total final sample volume of 200 μL. Deposit the second elution into the same tube as the first.
10. Dry the sample down to 10% of its elution volume in a SpeedVac (*see* **Note 31**).
11. Resuspend peptides in 0.1% formic acid in 2% RP-B in RP-A. Resuspend to a final volume of 1 μg/ul (or other suitable concentration and volume depending on your intended LC-MS injection amount).
12. Spin sample for 3 min at maximum speed in a centrifuge before transferring the supernatant to an autosampler vial for downstream mass spectrometry analysis (*see* **Note 32**).

4 Notes

1. The sonicator is used to shear the chromatin in the sample, decreasing its viscosity to allow for efficient protein recovery. Branson water baths are not powerful enough for this; only brands such as Diagenode and Covaris that are typically used for shearing fixed chromatin (e.g., for Chip-SEQ) will work, although the sonication efficiencies of water bath type instruments in our experience are much more sensitive to input amounts (e.g. input overloading) versus probe-tip sonicators.
2. Tris pH 8 is optimal when using trypsin as the digestive enzyme. If using LysC, substitute Tris pH 9 throughout the protocol.
3. Including a small amount of acetonitrile in the solution aids in solubilization of the peptides.
4. When heating solutions to near-boiling temperatures, it is helpful to add cap locks to the Eppendorf tubes to ensure that they stay closed. Furthermore, after boiling tubes should be opened facing away from you to vent and avoid splashing up of expanded air/liquid in the tube.
5. Scale volume of lysis buffer up for pellets with more cells.

6. Volume of lysis buffer stated is for probe tip sonicator; volumes should be adjusted for different sonicators according to manufacturer's recommendations. Additionally, appropriate tubes should be used if necessary —for example, older Diagenode models require their own commercial tubes which have a harder consistency for sufficient sonication efficiency.
7. 500 μL per tube is the ideal volume for sonication on a probe tip sonicator. Smaller volumes can be used with suitably powerful water bath sonicators.
8. For probe tip sonicators, it is essential to visually monitor the depth of the probe tip during sonication while keeping the sample cold. We recommend using a clear enclosure so that the placement depth of the probe tip can be visually ascertained for as much consistency as possible.
9. It is important to keep the samples as cold as possible to counteract the heating induced by sonication.
10. Invert tubes before and after sonication to check for viscosity—viscosity should be much reduced after sonication.
11. Different brands of sonicators have different strengths, so optimal conditions (where minimal sonication is applied to sufficiently disrupt chromatin) should be determined empirically for each sonicator. Slowly increasing strength and/or time of sonication and monitoring viscosity of the resulting lysate with test lysates (e.g., by inverting tubes and observing fluidity of the solution during a pilot timecourse) is useful for quickly determining an optimal sonication regime. Over-sonication should be avoided to minimize heating and/or physical compromise of proteins in samples.
12. In the case of probe tip sonicators it is absolutely critical is to submerge the sonicator probe tip before switching on the sonicator, and switch it off before removing the tip.
13. Orient tubes inside centrifuge so hinge faces outward to consistently position the pellet.
14. Avoid touching pellet with pipette tip—do not hesitate to leave some supernatant behind if necessary. Split as necessary as you wan to be able to add at least 4–5 volumes of ice-cold acetone for a good precipitation.
15. A 10% precipitation is done to remove DTT and so that the lysate can be resuspended in a strong denaturant compatible with the protein quantification BCA assay. Interferences can significantly bias results of the assay.
16. It is important to always use glass when working with acetone—the alcohol can potentially solubilize plastic on pipette tips, causing polymer interference in downstream LC-MS analysis.

17. Again, make sure this is at least a fourfold volume addition of neat ice-cold acetone to ensure the dielectric constant of the solution is sufficiently reduced for efficient precipitation. Keep acetone on ice or put it back in the freezer when not in use.
18. Ideally should see some flecks of particulate matter starting to precipitate even before putting tubes into freezer, although visible precipitating material may not be immediately observable with inputs of less than 300 μg.
19. Again, orienting the tubes inside the centrifuge so that the hinges face outward will ensure you know where the pellet is.
20. Leaving behind a small amount (~10 μL) of acetone is preferable to disturbing the pellet.
21. The pellet will usually appear lightly opaque or white on the backside of conical bottom.
22. Here it is fine to simply add cold acetone and re-spin. No vortexing or release of pellet from tube required, or recommended.
23. Pellets may shrink during the drying process. Do not over-dry pellets, as this will make them more difficult to resuspend. Do not use the Speed-Vac at this step.
24. Run triplicates of each sample. May also run sample undiluted in addition to 1:5–1:20 dilutions to ensure accurate range of the BCA assay.
25. If sample amounts permit, reserve a small amount (equivalent of ~5 μg) for protein gel analysis pre- vs post-digest.
26. Dilute TCEP 1:10 in MS-grade water (e.g., dilute 1:10 and add 5 μL to each 100 μL digest).
27. Dilute single use tube (9.3 mg per tube) in 200 μL MS-grade water for 200 mM stock (e.g., add 5 μL to each 100 μL digest)—use of the pre-weighed tubes minimizes weighing errors and saves on reagent, which is not reusable once resuspended.
28. This dilution step is very important—trypsin is only active in up to 1 M urea. Less trypsin (to 1:50 w/w) can be used to reduce cost, but care should be taken to avoid incomplete digestion - trypsin concentration in the final volume should not drop below 20 ng/μL.
29. Once wet, do not introduce air into the tip resin—this could dry the column and cause any bound material to become permanently attached. This is important throughout the protocol, but especially after the methanol wetting step as methanol will evaporate quickly if delays occur. Likewise, proceed continuously through the protocol as any stopping or pauses could cause the tip to dry out.

30. Aspirations up and down into the tip should be done slowly (each single up/down cycle over 3–5 s, and multiple aspirations up and down into the tip can be done to improve efficiency. Recommended are 10 aspirations for each 100 μL aliquot of sample loaded, and 5 for each 100 μL elution, and 3 aspirations at every other step. With experience and care, implementation an 8-channel multichannel pipette and well plates can improve throughput speed.
31. This requires careful monitoring of the drying, which can be achieved with a strobe device (e.g. Labconco Centrizap) that allows visualization of the sample without stopping the Speed-Vac.
32. Avoid any pelleted material at bottom of tube. This final spin step helps remove any particulates or debris that may clog the mass spectrometer.

References

1. Mallick P, Kuster B (2010) Proteomics: a pragmatic perspective. Nat Biotechnol 28: 695–709
2. Kachuk C, Stephen K, Doucette A (2015) Comparison of sodium dodecyl sulfate depletion techniques for proteome analysis by mass spectrometry. J Chromatogr A 1418:158–166
3. Wiśniewski JR, Zougman A, Nagaraj N, Mann M (2009) Universal sample preparation method for proteome analysis. Nat Methods 6:359–362

Chapter 2

Improving Proteome Coverage and Sample Recovery with Enhanced FASP (eFASP) for Quantitative Proteomic Experiments

Jonathan Erde, Rachel R. Ogorzalek Loo, and Joseph A. Loo

Abstract

Enhanced Filter Aided Sample Preparation (eFASP) incorporates plastics passivation and digestion-enhancing surfactants into the traditional FASP workflow to reduce sample loss and increase hydrophobic protein representation in qualitative and quantitative proteomics experiments. Resulting protein digests are free of contaminants and can be analyzed directly by LC-MS.

Key words Enhanced filter aided sample preparation, Quantitative proteomics, Detergent, Ammonium deoxycholate

1 Introduction

The integrity of proteomic experiments hinges on consistent and robust protein extraction, solubilization, and digestion. Protocols using anionic detergents and/or chaotropes to extract and solubilize cellular and matrix proteins efficiently provide samples that must be purified prior to digestion and analysis. Methods such as organic precipitation to remove contaminants, denaturants, and other undesired species (e.g., salts, nucleic acids, lipids, and alkylating reagents), are subject to poor recoveries, re-solubilization problems, and protein-to-protein variation. Enhanced Filter Aided Sample Preparation (eFASP) provides efficient protein extraction, purification, and digestion for a variety of samples [1, 2].

Traditional Filter Aided Sample Preparation (FASP) circumvents many protein purification challenges by exchanging buffers in spin filter units (ultrafiltration assemblies) that can remove sodium dodecyl sulfate (SDS) and sample contaminants completely [3, 4]. Proteins are reduced, alkylated, washed, and digested in the filter unit, releasing product free of detergent, reductant, and alkylating agent. Nevertheless, when applied to very small sample sizes,

Lucio Comai et al. (eds.), *Proteomics: Methods and Protocols*, Methods in Molecular Biology, vol. 1550,
DOI 10.1007/978-1-4939-6747-6_2, © Springer Science+Business Media LLC 2017

this method can suffer sample losses near 50% [5]. Enhanced FASP (eFASP) addresses this challenge by incorporating deoxycholic acid and, optionally, TWEEN® 20 into the FASP workflow.

Deoxycholic acid (DCA) is a secondary bile acid that, amongst many other uses, is employed as a mild detergent for membrane proteins. It increases the efficiency with which trypsin digests cytosolic and membrane proteins and is easily removed by acidification and phase transfer (PT) to peptide-immiscible ethyl acetate (EA) in liquid–liquid extraction [1, 6–11]. PT decreases EA-soluble contaminants, including SDS, *n*-octylglucoside, NP-40, and Triton X-100 [12].

The eFASP protocol optionally uses the surfactant TWEEN®-20 to passivate surfaces of Microcon® filter units and collection tubes. TWEEN®-20, as well as SDS, are recognized choices for minimizing protein binding to surfaces and have been recommended by Amicon Centricon for use in filter units [13]. Passivation of the filter units and collection tubes used in eFASP can reduce peptide and protein loss due to nonspecific surface binding, but care is needed to prevent TWEEN®-related ions from contaminating mass spectra.

Presented here is the eFASP approach, which utilizes 0.2% DCA and (optionally) TWEEN®-20 to quantitatively increase recovery and proteomic coverage of hydrophilic and hydrophobic proteins. An express eFASP method variant is also included, which uses a one-step reduction/alkylation employing tris(2-carboxyethyl)phosphine (TCEP) and 4-vinylpyridine (4-VP) prior to deposition on the Microcon® filter, increasing alkylation specificity and speeding processing [1, 14–17].

2 Materials

Prepare all solutions fresh using ultrapure water and MS-grade reagents. Follow all waste disposal regulations and chemical safety guidelines.

2.1 Solutions and Reagents for eFASP (See Subheading 3.1)

1. Passivation Solution: 5% (v/v) TWEEN® 20.
2. Lysis Buffer: 4% SDS, 0.2% deoxycholic acid (DCA, Sigma, D2510), 50 mM TCEP, 100 mM ammonium bicarbonate (ABC), pH 8 (*see* **Notes 1** and **2**).
3. Exchange Buffer: 8 M urea, 0.2% DCA, 100 mM ABC, pH 8 (*see* **Note 1**).
4. Alkylation Buffer: 50 mM iodoacetamide, 8 M urea, 0.2% DCA, 100 mM ABC, pH 8 (*see* **Notes 1** and **2**).
5. Digestion Buffer: 0.2% DCA, 50 mM ABC, pH 8 (*see* **Note 1**).
6. Trypsin Buffer: 0.5 μg/μl trypsin, 50 mM ABC, pH 8.

7. Peptide Recovery Buffer: 50 mM ABC, pH 8.
8. Ethyl acetate.
9. 50% methanol.
10. Trifluoroacetic acid.
11. MS-grade H_2O.

2.2 Solutions and Reagents for Express eFASP (See Subheading 3.2)

1. Passivation Solution: 5% (v/v) TWEEN® 20.
2. Lysis Buffer: 4% SDS, 0.2% DCA (Sigma, D2510), 50 mM TCEP, 100 mM ammonium bicarbonate (ABC), pH 8 (*see* **Notes 1** and **2**).
3. Exchange Buffer: 8 M urea, 0.2% DCA, 100 mM ABC, pH 8 (*see* **Note 1**).
4. Alkylation Stock: 500 mM 4-vinylpyridine (4-VP) in ethanol (*see* **Note 3**).
5. Quench Buffer: 1 M dithiothreitol (DTT), 100 mM ABC, pH 8.
6. Digestion Buffer: 0.2% DCA, 50 mM ABC, pH 8 (*see* **Note 1**).
7. Trypsin Buffer: 0.5 μg/μl trypsin, 50 mM ABC, pH 8.
8. Peptide Recovery Buffer: 50 mM ABC, pH 8.
9. Ethyl acetate.
10. 50% methanol.
11. Trifluoroacetic acid.
12. MS-grade H_2O.

2.3 Equipment

1. Microcon® ultrafiltration units (YM-30 30 kDa cutoff limit; Millipore, Billerica, MA).
2. Bench-top centrifuge.
3. Thermo-mixer (initially set to 90 °C).
4. SpeedVac®.
5. Squeeze bottle containing MS-grade H_2O.
6. Sonicator/homogenizer for disrupting cells.
7. Eppendorf LoBind® tube, 2 ml.
8. Ultrasonic bath.

3 Methods

Two eFASP protocols are described. The first is the standard procedure that utilizes in-filter alkylation with iodoacetamide (*see* Subheading 3.1). The second is an express procedure that utilizes in-solution alkylation with 4-vinylpyridine to increase speed, eliminating some buffer exchange steps (*see* Subheading 3.2).

Carry out all procedures at room temperature or as specified and follow instrument and chemical safety guidelines. The passivation steps may be omitted to save time or if TWEEN®-related background ions cannot be minimized in mass spectra.

3.1 eFASP: Standard

3.1.1 Surface Passivation (Optional)

1. On a shaker, incubate filter units and collection tubes overnight in Passivation Solution. Small batches of items may be incubated in 50 ml Falcon centrifuge tubes.
2. With clean tweezers, remove each item and rinse its outer and inner surfaces with MS-grade H_2O dispensed from a squeeze bottle.
3. Transfer items to a clean beaker containing a large volume of MS-grade H_2O; e.g., 250 ml or more. Incubate items for 30 min at room temperature, shaking at low speed.
4. Repeat **step 3** two additional times with fresh MS-grade H_2O.
5. Reserve the passivated collection tubes for peptide recovery from ultrafiltration devices.

3.1.2 Sample Lysis

1. Wash and pellet cells according to established guidelines for cell type.
2. Add sufficient Lysis Buffer to the pelleted cells such that a 25 μl aliquot of lysate will provide the quantity desired for processing by eFASP (or a maximum protein concentration of 10 μg/μl). Ensure that the selected volume and tube size can accommodate the sonicator probe.
3. Place the lysate into a 90 °C thermo-mixer and incubate for 10 min, shaking at 600 rpm. Remove lysate and decrease thermo-mixer temperature to 37 °C for later use.
4. Sonicate lysate (employing sonicator/homogenization probe) three times for 10 s each.
5. Centrifuge the lysate at 14,000 × *g* for 10 min.
6. Repeat sonication and centrifugation once (**steps 4** and **5**).
7. Sonicate the lysate (including any pelleted material), for 10 s. Cool to 37 °C.

3.1.3 Sample Processing

1. Transfer 25 μl of lysate to a sample tube containing 200 μl of Exchange Buffer. Vortex briefly to mix.
2. Place a (passivated) filter unit atop a non-passivated collection tube.
3. Dispense the 225 μl lysate/Exchange Buffer sample to the filter unit and centrifuge at 14,000 × *g* for 10 min. Discard filtrate.
4. Add 200 μl Exchange Buffer to the filter unit and centrifuge at 14,000 × *g* for 10 min. Discard filtrate.
5. Repeat **step 4** two more times.

6. Dispense 100 μl Alkylation Buffer to the filter unit and transfer it to a 37 °C thermo-mixer for 1 h, shaking at 300 rpm.
7. Centrifuge the filter unit at 14,000×*g* for 10 min. Discard filtrate.
8. Add 200 μl Exchange Buffer to the filter unit and centrifuge at 14,000×*g* for 10 min. Discard filtrate.
9. Add 200 μl eFASP Digestion Buffer to the filter unit and centrifuge at 14,000×*g* for 10 min. Discard filtrate.
10. Repeat **step 9** two more times.
11. Transfer the filter unit to a passivated collection tube.
12. Add 100 μl eFASP Digestion Buffer to the filter unit.
13. Calculate the volume of Trypsin Buffer to dispense in order to achieve the desired enzyme-to-substrate ratio; e.g., 1:50 w:w.
14. Deposit the calculated volume of Trypsin Buffer to the filter unit, and place in a 37 °C thermo-mixer for 12 h, shaking at low speed. Secure the filter unit cap to minimize evaporation.
15. Remove the filter unit/collection tube assembly from the thermo-mixer and centrifuge at 14,000×*g* for 10 min. *Retain peptide-containing filtrate.*
16. Deposit 50 μl of Peptide Recovery Buffer onto the filter unit and centrifuge at 14,000×*g* for 10 min.
17. Repeat **step 16** step once. *Retain peptide-containing filtrate.*

3.1.4 Phase Transfer

1. To the collection tube with peptide-containing filtrate, add 200 μl of ethyl acetate and transfer to a 2 ml Eppendorf LoBind® tube.
2. Add 2.5 μl TFA and quickly vortex. A white, thread-like precipitate may be visible if a large quantity of peptide is present.
3. Add ethyl acetate to nearly fill the tube, leaving only enough space to agitate without losing liquid.
4. Agitate the mixture for 10 s in an ultrasonic bath and centrifuge at 16,000×*g* for 10 min.
5. Carefully pipet most of the upper (organic) layer into a tube for discard. Do not disturb the organic/aqueous boundary layer.
6. Repeat **steps 3–5** two times.
7. Place the uncapped sample tube in a 60 °C thermo-mixer, in a fume hood, for 5 min to remove residual ethyl acetate.
8. Remove residual organic solvent and volatile salts by vacuum drying in a SpeedVac®.
9. Resuspend the dried sample in 50% methanol and vacuum-dry.
10. Repeat **step 9** two times.

3.2 Express eFASP

3.2.1 Surface Passivation (Optional)

3.2.2 Sample Lysis

1. Wash and pellet cells according to established guidelines for cell type.
2. Add sufficient Lysis Buffer to the pelleted cells such that a 25 μl aliquot of lysate will provide the quantity desired for processing by eFASP (or a maximum protein concentration of 10 μg/μl). Ensure that the selected volume and tube size can accommodate the sonicator probe.
3. Place the lysate into a 90 °C thermo-mixer and incubate for 10 min, shaking at 600 rpm. Remove lysate and decrease thermo-mixer temperature to 37 °C for later use.
4. Sonicate lysate (employing sonicator/homogenization probe) three times for 10 s each.
5. Centrifuge the lysate at 14,000 × *g* for 10 min.
6. Repeat sonication and centrifugation once (**steps 4** and **5**).
7. Sonicate the lysate (including any pelleted material), for 10 s. Cool to 37 °C.
8. Add Alkylation Stock to the lysate to a final concentration of 25 mM 4-VP, and place the sample tube into a 37 °C thermo-mixer for 1 h at 300 rpm.
9. Add Quench Buffer to the lysate to a final concentration of 40 mM DTT.

3.2.3 Sample Processing

1. Transfer 25 μl of lysate to a sample tube containing 200 μl of Exchange Buffer. Vortex briefly to mix.
2. Place a (passivated) filter unit atop a non-passivated collection tube.
3. Dispense the 225 μl lysate/Exchange Buffer sample to the filter unit and centrifuge at 14,000 × *g* for 10 min. Discard filtrate.
4. Add 200 μl eFASP Digestion Buffer to the filter unit and centrifuge at 14,000 × *g* for 10 min. Discard filtrate.
5. Repeat **step 4** two more times.
6. Detach the filter unit from the non-passivated collection tube, and place it on top of a passivated tube.
7. Add 100 μl eFASP Digestion Buffer to the filter unit.
8. Calculate the volume of Trypsin Buffer to dispense in order to achieve the desired enzyme-to-substrate ratio; e.g., 1:50 w:w.
9. Deposit the calculated volume of Trypsin Buffer to the filter unit, and move the filter/collection tube assembly to a 37 °C thermo-mixer for 12 h of shaking at low speed. Cap the filter unit to reduce evaporation.
10. Remove the filter/collection tube assembly from the thermo-mixer and centrifuge at 14,000 × *g* for 10 min. *Retain the peptide-containing filtrate.*

11. Dispense 50 μl of Peptide Recovery Buffer to the filter unit and centrifuge at 14,000×*g* for 10 min.
12. Repeat **step 11** step once.

3.2.4 Phase Transfer

1. To the collection tube with peptide-containing filtrate, add 200 μl of ethyl acetate and transfer to a 2 ml Eppendorf LoBind® tube.
2. Add 2.5 μl TFA and quickly vortex. A white, thread-like precipitate may be visible if a large quantity of peptide is present.
3. Add ethyl acetate to nearly fill the tube, leaving only enough space to agitate without losing liquid.
4. Agitate the mixture for 10 s in an ultrasonic bath and centrifuge at 16,000 × g for 10 min.
5. Carefully pipet most of the upper (organic) layer into a tube for discard. Do not disturb the organic/aqueous boundary layer.
6. Repeat steps 3–5 two times.
7. Place the uncapped sample tube in a 60 °C thermo-mixer, in a fume hood, for 5 min to remove residual ethyl acetate.
8. Remove residual organic solvent and volatile salts by vacuum drying in a SpeedVac®.
9. Resuspend the dried sample in 50 % methanol and vacuum-dry.
10. Repeat step 9 two times.

4 Notes

1. This protocol employs deoxycholic acid rather than sodium deoxycholate, in order to minimize analyte exposure to sodium, which can degrade mass spectrometry analyses. Even with online liquid chromatography to remove much of the Na^+, acidic peptides may appear sodium-adducted. A trade-off in substituting deoxycholic acid for sodium deoxycholate is the former's low solubility; 0.2 % DCA is about the maximum solubility achievable in ABC buffer. Dissolving deoxycholic acid in a small volume of ethanol prior to mixing with buffer can facilitate dissolution. Alternatively, ABC buffer may be added to the appropriate quantity of solid DCA immediately before use and vortexed extensively; even if slightly cloudy, the freshly prepared DCA solution can be employed in eFASP. DCA solutions should not be refrigerated; irreversible precipitation may occur. We have found that 0.1 % DCA performs almost as well as 0.2 % in eFASP; thus, solubility problems may be eased by using the lower concentration. Finally, sodium deoxycholate

could substitute for deoxycholic acid, with STAGE tip cleanup prior to injection onto the LC column.

2. Alternatively, Lysis Buffer can employ 5 mM TCEP and/or Alkylation Buffer can be formulated with 5 mM iodoacetamide.
3. We assume that neat 4-vinylpyridine is 8.8 M in concentration.

References

1. Erde J, Loo RRO, Loo JA (2014) Enhanced FASP (eFASP) to increase proteome coverage and sample recovery for quantitative proteomic experiments. J Proteome Res 13(4):1885–95
2. Erde J (2012) High throughput analysis of proteome perturbations induced by radiation, radiomitigators and chemotherapeutics. University of California, Los Angeles
3. Manza LL, Stamer SL, Ham A-JL, Codreanu SG, Liebler DC (2005) Sample preparation and digestion for proteomic analyses using spin filters. Proteomics 5(7):1742–5
4. Wisniewski JR, Mann M (2009) Spin filter–based sample preparation for shotgun proteomics. Nat Methods 6(11):785–6
5. Wisniewski JR, Zielinska DF, Mann M (2011) Comparison of ultrafiltration units for proteomic and N-glycoproteomic analysis by the filter-aided sample preparation method. Anal Biochem 410(2):307–9
6. Masuda T, Tomita M, Ishihama Y (2008 Feb) Phase transfer surfactant-aided trypsin digestion for membrane proteome analysis. J Proteome Res 7(2):731–40
7. Masuda T, Sugiyama N, Tomita M, Ishihama Y (2011) Microscale phosphoproteome analysis of 10,000 cells from human cancer cell lines. Anal Chem 83(20):7698–703
8. Masuda T, Saito N, Tomita M, Ishihama Y (2009) Unbiased quantitation of escherichia coli membrane proteome using phase transfer surfactants. Mol Cell Proteomics 8(12):2770–7
9. Zhou J, Zhou T, Cao R, Liu Z, Shen J, Chen P et al (2006) Evaluation of the application of sodium deoxycholate to proteomic analysis of rat hippocampal plasma membrane. J Proteome Res 5(10):2547–53
10. Lin Y, Zhou J, Bi D, Chen P, Wang X, Liang S (2008) Sodium-deoxycholate-assisted tryptic digestion and identification of proteolytically resistant proteins. Anal Biochem 377(2):259–66
11. Lin Y, Liu Y, Li J, Zhao Y, He Q, Han W et al (2010) Evaluation and optimization of removal of an acid-insoluble surfactant for shotgun analysis of membrane proteome. Electrophoresis 31(16):2705–13
12. Yeung Y-G, Nieves E, Angeletti RH, Stanley ER (2008) Removal of detergents from protein digests for mass spectrometry analysis. Anal Biochem 382(2):135–7
13. Passivation of Amicon Microcon Concentrators for Improved Recovery (1999). Bedford, MA: Millipore Corporation, Technical Note PC1001EN00
14. Sebastiano R, Citterio A, Lapadula M, Righetti PG (2003) A new deuterated alkylating agent for quantitative proteomics. Rapid Commun Mass Spectrom 17(21):2380–6
15. Bai F, Liu S, Witzmann FA (2005) A "de-streaking" method for two-dimensional electrophoresis using the reducing agent tris(2-carboxyethyl)-phosphine hydrochloride and alkylating agent vinylpyridine. Proteomics 5(8):2043–7
16. Liu S, Bai F, Witzmann F (2006) Destreaking strategies for two-dimensional electrophoresis. In: Separation methods in proteomics. Eds.: Smejkal GB, Lazarev A, CRC Press/Taylor & Francis, Boca Raton, pp. 207–17
17. Righetti PG (2006 Sep) Real and imaginary artefacts in proteome analysis via two-dimensional maps. J Chromatogr B 841(1–2):14–22

Chapter 3

Proteome Characterization of a Chromatin Locus Using the Proteomics of Isolated Chromatin Segments Approach

Sophie L. Kan, Nehmé Saksouk, and Jérome Déjardin

Abstract

The biological functions of given genomic regions are ruled by the local chromatin composition. The Proteomics of Isolated Chromatin segments approach (PICh) is a powerful and unbiased method to analyze the composition of chosen chromatin segments, provided they are abundant (repeated) or that the organism studied has a small genome. PICh can be used to identify novel and unexpected regulatory factors, or when combined with quantitative mass spectrometric approaches, to characterize the function of a defined factor at the chosen locus, by quantifying composition changes at the locus upon removal/addition of that factor.

Key words Proteomics, Chromatin, Protein Purification, Repeated Sequences, Heterochromatin, Telomeres

1 Introduction

Each cell of an organism harbors the same genetic information. Yet cells can express distinct genetic programs according to their differentiation state or to environmental factors. These features control the chromatin composition, which in turn regulates the biological features of a locus of interest. A major tool to study the composition of chromatin is the chromatin immunoprecipitation approach (ChIP), which allows to interrogate using various analytical approaches, whether a DNA sequence is associated with a factor of interest [1]. This method is widely used because it is sensitive and it has become relatively easy to implement. However, ChIP is limited by the knowledge of the factor to test and the availability of suitable antibodies against that factor. We have developed a method which we named PICh (Proteomics of Isolated Chromatin segments) which allows to purify a given genomic region together with the factors associated in vivo, without the need for any prior knowledge about the identity of bound factors [2]. Briefly PICh is based on the hybridization capture of a specific chromatin fragment.

Lucio Comai et al. (eds.), *Proteomics: Methods and Protocols*, Methods in Molecular Biology, vol. 1550,
DOI 10.1007/978-1-4939-6747-6_3, © Springer Science+Business Media LLC 2017

The PICh method is powerful because it provides a comprehensive knowledge about the steady state protein composition of the target locus. However it must be kept in mind that PICh has one important technical limitation: the signal to noise ratio. We estimate the PICh affinity purification enrichment to reach ~10,000 fold, and thus PICh is not suitable to purify single copy targets from human cells. A single 1 kb locus of interest represents only ~0.00003 % of the human genome. To obtain a 50 % pure mixture for a relatively straightforward downstream mass spectrometric analysis of such low abundance species, it would require an enrichment factor of 1,500,000 fold, which to our knowledge is challenging with a single affinity step. The nonspecific capture of minute amounts of non-relevant chromatin fragments is also an issue that needs to be carefully eliminated from the PICh material. For instance, the nonspecific capture of 0.003 % of the starting material would be acceptable in any ChIP experiment because enrichment of a specific target is compared to any other similarly abundant "negative control." Yet in PICh the entire nonspecific material contributes proteins in the eluate, 0.003 % of input represents a 100 times more material than proteins specifically contributed by a unique target. Thus PICh is applicable only to abundant targets from mammalian genomes and importantly, entails the use of dedicated reagents. The protocol described here is optimized for the purification of proteins associated to abundant sequences such as mammalian telomeres or pericentromeric repeats (0.01–3 % of the genomes) (Fig. 1). For regions with higher complexity such as ribosomal RNA gene promoters, refer to the end-targeting PICh approach (ePICh) [3]. In PICh, chromatin is reversibly cross-linked with formaldehyde and sheared by sonication. The locus of interest is targeted with desthiobiotinylated oligonucleotide probes containing 50 % of locked nucleic acids (LNA). LNA–DNA hybrids have high melting temperatures ensuring strong and stable hybridization to the end sequence of the fragment of interest [4]. In our hands, DNA or RNA capture probes do not hybridize stably enough to allow any specific enrichment in PICh, presumably because short D-loops or R-loops are notoriously highly unstable in aqueous solution [5, 6]. The probe–chromatin hybrids are captured by streptavidin coupled magnetic beads. Desthiobiotin is a biotin analog that has weaker affinity for streptavidin than biotin (Kd ~ 10–12 M) which allows a competitive gentle elution from streptavidin beads with biotin (Kd ~ 10–14 M) [7]. This gentle condition allows maintaining nonspecific proteins bound to the beads. Finally the chromatin mixture is precipitated by trichloroacetic acid, and protein–protein and protein–DNA cross-links are reversed. The preparation can be used for western-blot assays or analyzed by mass spectrometry to identify new regulatory factors. The quantitative version of the PICh technique (qPICh) is less suited to identify low abundance factors (the sensitivity is 2-fold lower), but is used to quantify differences in protein composition at the studied locus in two distinct samples [8]

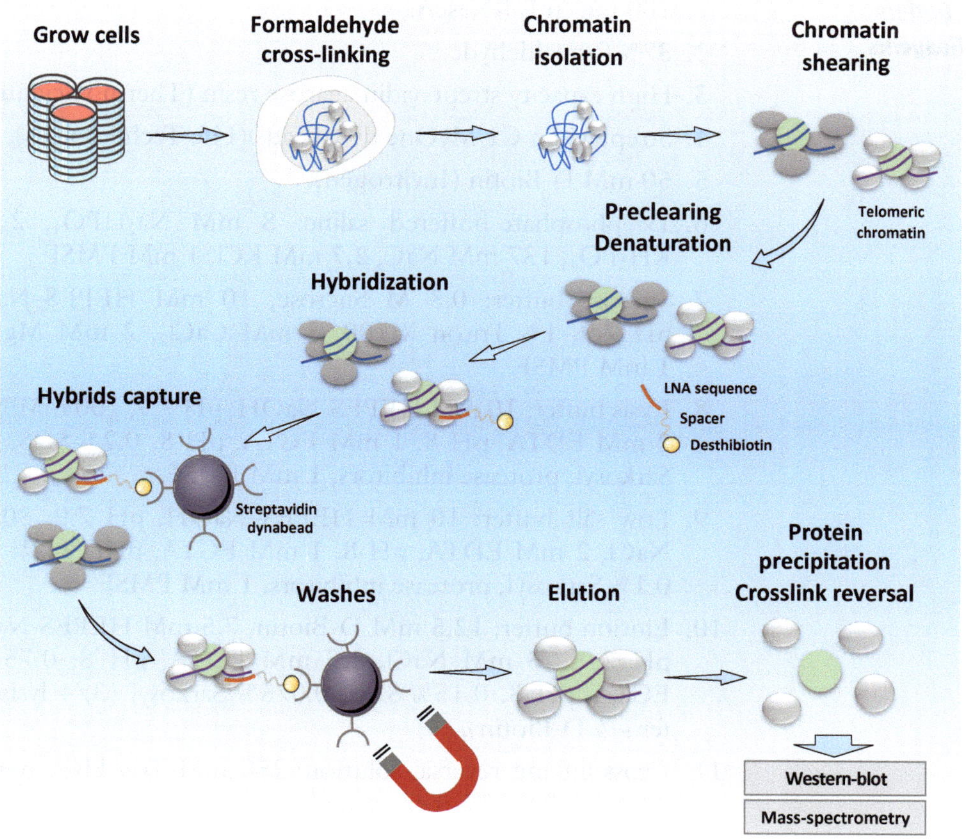

Fig. 1 Outline of the PICh protocol

provided the abundance of that locus is comparable in the two samples. qPICh combines SILAC with PICh and is appropriate to address the function of a given factor at a region of interest. PICh and qPICh are described in the following protocol.

2 Materials

Prepare all solutions using ultrapure water and analytical grade reagents. Store all buffers at 2–8 °C. Add phenylmethanesulfonylfluoride (PMSF) just prior to use.

2.1 Special Equipment

1. Sonicator with high power sonication probe (Qsonica).
2. DynaMag™-2 magnet or equivalent for 1.5 mL microcentrifuge tubes (Life Technologies).
3. DynaMag™-15 magnet or equivalent for 15 mL conical tubes (Life Technologies).
4. 10 mL chromatography columns (Pierce).

2.2 Buffers and Reagents

1. 100 mg/mL RNaseA.
2. 37% formaldehyde.
3. High capacity streptavidin agarose resin (Thermo Scientific).
4. Streptavidin C1 MyOne dynabeads (Life Technologies).
5. 50 mM D-Biotin (Invitrogen).
6. 1× phosphate-buffered saline: 8 mM Na_2HPO_4, 2 mM KH_2PO_4, 137 mM NaC, 2.7 mM KCl, 1 mM PMSF.
7. Sucrose buffer: 0.3 M Sucrose, 10 mM HEPES–NaOH, pH 7.9, 1% Triton X-100, 3 mM $CaCl_2$, 2 mM MgOAc, 1 mM PMSF.
8. Lysis buffer: 10 mM HEPES-NaOH, pH 7.9, 100 mM NaCl, 2 mM EDTA, pH 8, 1 mM EGTA, pH 8, 0.2% SDS, 0,1% Sarkosyl, protease inhibitors, 1 mM PMSF.
9. Low salt buffer: 10 mM HEPES–NaOH, pH 7.9; 30 mM NaCl, 2 mM EDTA, pH 8, 1 mM EGTA, pH8, 0.2% SDS, 0.1% Sarkosyl, protease inhibitors, 1 mM PMSF.
10. Elution buffer: 12.5 mM D-Biotin, 7.5 mM HEPES-NaOH; pH 7.9, 75 mM NaCl, 1.5 mM EDTA, pH 8; 0.75 mM EGTA, pH 8, 0.15% SDS, 0.075% Sarkosyl (3/4 lysis buffer + ¼ D-Biotin).
11. Cross-linking reversal solution: 250 mM Tris–HCl, pH 8.8, 2% SDS, 0.9 M 2-mercaptoethanol.
12. Sephacryl S-400 high resolution gel filtration resin (GE Healthcare).
13. SilverQuest Staining Kit (Invitrogen).
14. LNA probes: LNA probe synthesis is performed in-house using an H-6 K&A synthesizer, on a 1 μ-mole scale, following manufacturer's recommendations and standard phosphoramidite synthesis guidelines. Alternatively LNA probes can be ordered from Exiqon or any company offering to make LNA containing probes. Capital letters represent LNA residues while small letters are DNA residues:

    Telomere: Desthiobiotin-TEG-4XC18 spacers-5′TtAgGg TtAgGgTtAgGgTtAgGgt-3′

    Major satellites: Equimolar mixture of 4 distinct capture probes:

Desthiobiotin-TEG-4XC18 spacers– 5′ CaCtTTaGgaCGTgAa ATaTGg-3′

Desthiobiotin-TEG-4XC18 spacers– 5′ TgTaGgAcAtGgAaTa TgGcAaGaAaAc-3′ Desthiobiotin-TEG-4XC18 spacers-5′CgAgGaAaAcTgAaAaAgGttgAaAaTt-3′ Desthiobiotin-TEG-4XC18spacers-5′AaTCaCGgAaAaGgAGaaATac-3′

Scramble: Desthiobiotin-TEG-4XC18 spacers – 5′GaTgTgT gGaTgTggAtGtGgAtgTgg-3′

15. Media for stable isotope labeling by amino acids in cell culture:

    Arginine and lysine free cell growth medium.

    Dialyzed FBS.

    Light arginine/lysine.

    Heavy arginine/lysine (Cambridge Isotope Labeling).

3 Methods

3.1 Tissue Culture

1. Plate cells on enough 150 mm plates to harvest 10^9 cells per condition (*see* **Note 1**).
2. Incorporation of isotopically stable amino acids into cellular proteins for qPICh. Grow the two cell lines (e.g., WT and mutant for a gene of interest) in culture media with isotopically distinct amino acids (*see* **Note 2**).
3. Count cells of one or two representative plate for each condition. For qPICh this step is extremely important.
4. FACS analyze the cell cycle profile of the cell populations to account for potential differences cell cycle distributions among the distinct cell populations.

3.2 Formaldehyde Cross-Linking

3.2.1 Adherent Cells

1. 1. Add formaldehyde directly to the growth medium to a 3% final concentration for 30 min at room temperature (*see* **Note 3**). Gently shake the plates on an orbital shaker.
2. Remove the growth medium and dispose of it appropriately.
3. Wash twice with PBS-PMSF to remove the formaldehyde of the cross-linking solution (*see* **Note 4**).
4. Scrap cells in 3 mL PBS containing 0,05% Tween-20 per 150 mm plate with a cell lifter and collect them in a 50 mL conical tube (10–15 plates should fit into one 50 mL conical tube) (*see* **Note 5**).
5. Centrifuge the cross-linked cells at 3200 × *g* for 10 min at 4 °C.
6. Pool the pellets in one 50 mL conical tube (*see* **Note 6**).
7. Wash the pellet three times in PBS-PMSF. Centrifuge each time at 3200 × *g* for10 min at 4 °C.

3.2.2 Suspension Cells

1. Cross-link the protein–DNA complexes by incubating the cells with formaldehyde diluted to a 3% final concentration for 30 min at room temperature on a nutator (volume of the cross-linking solution must not be below 50 times the volume of pelleted cells: for a pellet of 4 mL cells, use at least 200 mL of cross-linking solution).

2. Centrifuge the cross-linked cells at 3200 × *g* for 10 min at 4 °C.
3. Remove the growth medium and dispose of it appropriately.
4. Wash the pellet twice in PBS-PMSF (3200 × g, 10 min) (*see* **Notes 5** and **6**).
5. The pellet can be stored at −80 °C up to 1 month (*see* **Note** 7).
6. For qPICh mix pelleted cells in a one-to-one ratio (*see* **Note 8**).

3.3 Chromatin Preparation

1. Resuspend the pellet in 10 mL of sucrose solution by vortexing and bring volume to 50 mL (*see* **Note 9**).
2. Centrifuge at 3200 × *g* for 10 min at 4 °C.
3. Remove and discard the supernatant, resuspend the pellet in 10 mL of sucrose solution and bring the volume to 20 mL with sucrose solution.
4. Transfer the solution to a 40 mL dounce homogenizer.
5. Dounce on ice approximately 20 times with a tight pestle.
6. Transfer the lysed cells into a new 50 mL conical tube, wash the homogenizer twice with 15 mL of sucrose solution and pool these washes with the sample to minimize material loss.
7. Centrifuge at 3200 × *g* for 10 min at 4 °C.
8. Discard the supernatant.
9. Resuspend the pellet in the same pelleted volume of PBS.
10. Add Triton-X100 (from a 20% stock solution) to a final concentration of 0.5% Triton X-100.
11. Add RNaseA to a final concentration of 10 μg/μL (10 μL of concentrated RNaseA (100 mg/mL) per 1 mL of chromatin).
12. Incubate overnight on a nutator at 4 °C (*see* **Note 10**).
13. Transfer the mixture into a 50 mL conical tube and bring the volume to 50 mL with PBS—PMSF.
14. Centrifuge the sample at 3200 × *g* for 10 min at 4 °C and discard the supernatant.
15. Wash three times in PBS-PMSF to dilute RNase A.
16. Resuspend the pellet in 50 mL of freshly prepared lysis buffer.
17. Centrifuge the sample at 3200 × *g* for 10 min and discard the supernatant.
18. Resuspend the pellet in 2.5 volume of pelleted volume of lysis buffer by pipetting with 1 mL micropipette and split into 3 mL aliquots in 15 mL conical tubes (*see* **Note 11**).
19. Sonicate the sample to shear the chromatin to an average length of about 1 kb. Sonicate on ice and ensure that the samples do not foam during the process. Sonication parameters for a Misonix S-4000 sonicator with a high power probe (*see* **Note 12**): power setting amplitude 70% (36–45 W), 15 s constant pulse, 45 s pause, 7 min total process time.

3.4 Chromatin Pre-clearing

1. Incubate the sonicated chromatin at 58 °C for 5 min (*see* **Note 13**).
2. Aliquot the chromatin into 1.5 mL microcentrifuge tubes and pellet insoluble material by centrifugation at 16,000 × *g* for 10 min at room temperature.
3. Transfer the supernatants into a fresh 15 mL conical tube.
4. Save 20 μL to check the chromatin shearing efficiency (*see* Subheading 3.11).
5. In the meantime equilibrate the high capacity streptavidin agarose resin with lysis buffer for 5 min at room temperature. Use 200 μL of slurry Ultralink streptavidin resin for every 700×10^6 cells. Centrifuge at 500 × *g* for 2 min at room temperature. Repeat this step once.
6. Add streptavidin beads to the chromatin sample for the pre-clearing step to remove any endogenously biotinylated protein (e.g., carboxylases) that might interfere with the subsequent capture.
7. Incubate overnight at 4 °C on a nutator (*see* **Note 14**).
8. Prepare the desalting columns by placing a 10 mL PIERCE empty chromatography column into a 50 mL conical tube. Add the same volume of Sephacryl S-400 HR gel filtration media as that of the chromatin sample (*see* **Note 15**). Do not add the chromatin at this step.
9. Centrifuge at 750 × *g* for 3 min to dry the resin in the column.
10. Place the column in a fresh 50 mL conical tube and add chromatin (together with the streptavidin agarose beads used for pre-clearing) to the dried column.
11. Pre-clear the chromatin by centrifugation at 750 × *g* for 5 min. Recover the flow through, it is the pre-cleared chromatin fraction.
12. Measure the optical density (OD) of the pre-cleared chromatin. Use the low salt buffer as a blank solution. A high quality PICh chromatin sample has the following characteristics: OD260: 2–2.5 mg/mL, OD260/OD280: 1.3–1.45 (*see* **Note 16**).
13. Bring the pre-cleared chromatin to a final concentration of 0.02 % of SDS (1/100 th of the final volume of 20 % SDS).
14. Spare 10 μL for the input.

3.5 Hybridization and Chromatin Capture

1. Split the chromatin in two equal volumes, one for the experiment and the other for the scrambled control (*see* **Note 17**).
2. Add the LNA probe to a final concentration of 0.25 μM (*see* **Note 18**).
3. Aliquot 150 μL of the chromatin-LNA mixture into PCR tubes.

4. Hybridize in a standard thermocycler using the following program (*see* **Note 19**):
   (a) 25 °C for 3 min
   (b) 71 °C for 9 min.
   (c) 38 °C for 1 h.
   (d) 60 °C for 3 min.
   (e) 38 °C for 30 min.
   (f) 60 °C for 3 min.
   (g) 38 °C for 30 min.
   (h) 25°C final temperature.
5. Pool the samples from the PCR tubes into 1.5 mL microcentrifuge tubes and centrifuge at 16,000 × g for 15 min at room temperature in order to pellet any precipitate that could have formed during the hybridization step.
6. Pool the supernatants into 15 mL conical tubes.
7. In the meanwhile equilibrate the MyONE C1 magnetic streptavidin beads twice 10 mL of low salt buffer. Use 900 μL of MyONE C1 magnetic streptavidin beads for every 24 μL of LNA probe.
8. Immobilize the beads on the magnetic stand and discard the supernatant.
9. Keep the beads on the stand while adding the chromatin.
10. Still keeping beads on the stand dilute the chromatin in the same volume of Milli-Q water as the volume of chromatin (*see* **Note 20**).
11. Resuspend the beads very gently and slowly. Incubate for 1 h and a half on a nutator at room temperature.
12. Bring the volume to 10 mL with lysis buffer.
13. Immobilize on magnetic stand.

3.6 Washes

1. Wash the beads five times with lysis buffer. Resuspend the beads gently between the washes, avoid vortexing (*see* **Note 21**). Perform the additional wash with low salt buffer.
2. Resuspend the beads in 1.2 mL lysis buffer per tube and transfer to low binding 1.5 mL microtubes (*see* **Note 22**).
3. Immobilize the beads on the magnetic stand.
4. Discard the supernatant.
5. Resuspend in 1 mL of lysis buffer.
6. Immobilize the beads on the magnetic stand. A "halo" should be apparent on the sides of the wall of the microcentrifuge tube (*see* **Note 23**).
7. Discard the supernatant.

8. Resuspend in 1 mL of low salt buffer.
9. Incubate for 5 min at 42 °C in a thermomixer shaking at 500 rpm.
10. Immobilize the beads on the magnetic stand and discard the supernatant.

3.7 Elution

1. Resuspend the beads in 900 μL of elution buffer.
2. Incubate for 30 min at 42 °C and an additional 10 min at 65 °C in the Thermomixer shaking at 500 rpm.
3. Immobilize the beads on the magnetic stand.
4. Transfer the eluates to fresh 1.5 mL microcentrifuge tubes.
5. Immobilize the beads on the magnetic stand.
6. Transfer the eluates again to fresh 1.5 mL microcentrifuge tubes. This step is important to eliminate all the beads from the eluates (*see* **Note 24**).
7. Verify that the eluates OD260 is around 15–25 ng/μL (*see* **Note 25**).

3.8 Protein Precipitation

1. Prepare the inputs: add 890 μL of lysis buffer to the 10 μL of input saved at **step 37**.
2. Precipitate proteins by adding 200 μL of 100% trichloroacetic acid to the 900 μL of eluates (the final concentration of trichloroacetic acid should be in the range of 15–20%).
3. Incubate on ice for 10 min.
4. Centrifuge at 16,000 × *g* for 10 min at 4 °C.
5. Carefully remove the supernatant with a pipette and leave about 50 μL (*see* **Note 26**).
6. Wash the pellet twice with 1 mL of ice-cold acetone 100%.
7. Vortex for 10 s.
8. Centrifuge at 16,000 × *g* for 10 min at 4 °C.
9. Carefully remove the supernatant and leave about 50 μL.
10. Evaporate the acetone by heating the tube at 100 °C for 10 s. Repeat this step if the acetone is not entirely evaporated.
11. Resuspend in 70 μL of cross-linking reversal solution if the sample will be analyzed by mass spectrometry or in 100 μL of reversal cross-linking solution if the sample is used for western-blot analysis. In both cases, resuspend the input in 400 μL of cross-linking reversal solution.
12. Incubate for 24 min at 99 °C, after 12 min vortex briefly and centrifuge to retrieve condensation from the cap back to the bottom of the tube and incubate another 12 min at 99 °C.

3.9 Control for PICh Efficiency

1. Separate proteins on a 12 % Bis-tris gel run with MOPS buffer. Load 2 μL of input and 5–10 μL of scrambled and PICh extracts (*see* **Note 27**).
2. Silver stain the gel to reveal the proteins (*see* **Note 28**). If 1/6th of eluted material was used, there should be visible proteins bands after 3 min of developing. If the bands do not appear after 5 min it means there are not enough proteins in the remaining PICh extracts for analysis by mass spectrometry (Fig. 2).
3. The PICh extracts can be used for western-blot analysis (Fig. 3), mass spectrometry analysis or stored at −80 °C.

3.10 Protein Analysis by Mass Spectrometric Approaches

1. Resolve proteins (usually up to 40 % of the total, ~30 μL) on a 12 % Bis-Tris gel.
2. Cut each lane into 5–10 bands.
3. Process each band using standard mass spectrometric identification protocols.

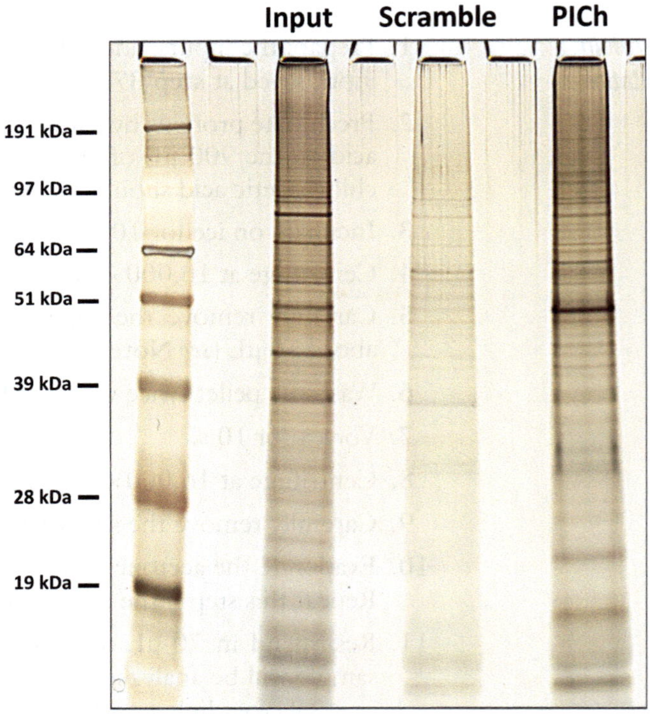

Fig. 2 Silver staining of a PICh experiment. The protein complexes associated with mouse ESCs telomeres were specifically enriched in the PICh extract where a telomere-specific probe was used. Note that the banding pattern in the PICh lane is different from the input lane. The scrambled lane shows nonspecific protein interactions eluted from the beads. Proteins were separated by SDS-PAGE and visualized by silver staining. Molecular weight marker is indicated on the left. 0.0125 % of input and 4 % of scramble and PICh extracts were loaded

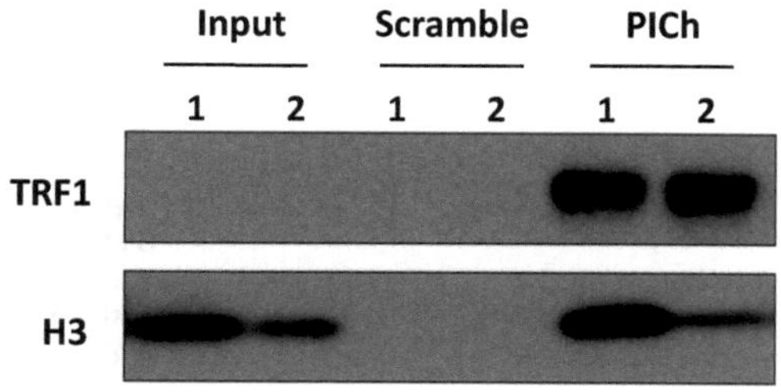

Fig. 3 Immuno-blot analyses of a telomere PICh experiment. Telomeric PICh extracts were separated on a 12% Bis-Tris gel. 1 and 2 are two independent experiments 2 starting from less material than 1. PICh efficiency is shown by the specific enrichment of the telomere-associated protein TRF1. Note that TRF1 is undetectable in similar amounts of input chromatin, as this factor is specific to telomeres

4. Protein quantifications in the SILAC experiments are performed using the MaxQuant software using the default parameters (*see* **Note 29**).

3.11 Chromatin Shearing Analysis

1. Reverse cross-link 20 μL of chromatin input by adding 200 μL of reverse cross-linking solution and add 10 μL of NaCl 5 M (final concentration > 200 mM).
2. Incubate overnight at 65 °C by shaking 600 rpm.
3. The next day, add 24 μg of RNase A for at least 1 h at 37 °C.
4. Add 10 μL of Proteinase K (20 μg/μL) and digest proteins for 2 h at 65 °C.
5. Purify DNA with phenol–chloroform and ethanol-precipitate the DNA.
6. Run 10 μg of DNA on a 0,8% agarose gel (Fig. 4) (*see* **Note 30**).

4 Notes

1. To harvest 10^9 mouse embryonic stem cells, more or less 50 plates of 150 mm gelatin coated dishes should be foreseen. Lowering the scale of the experiment is not advised, as this usually results in higher background contaminations.
2. SILAC is an approach based on the protein incorporation of stable isotope-labeled amino acids into newly synthetized proteins in vivo. The compared cell populations should be grown with the "light" isotope of the used amino acids in the control condition and with the "heavy" version in the experiment. Use a drop-out medium and also use dialyzed serum. It is highly recommended to perform the reverse experiment in which the amino acid isotopes are swapped between the two conditions. It is also advised to perform an incorporation on a test plate to make sure there is a negligible conversion of arginine into proline, which would complicate the analyses.

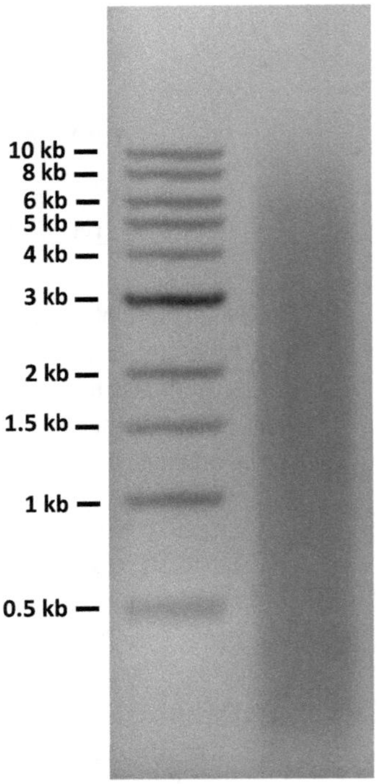

Fig. 4 Agarose gel electrophoresis of DNA isolated from PICh extracts. Size distribution of purified mouse embryonic stem cell DNA after sonication. 2 μg of DNA is analyzed on a 0.8 % agarose gel, post-stained with ethidium bromide

The time needed for complete incorporation into proteins depends on the cell line. In general six cell cycles are considered sufficient for full incorporation and replacement of "long" lived proteins. For stem cells that cycle every 8 h this corresponds to 3 days in culture. Also it is good to check that the different media formulations are not affecting the cell cycle distribution by FACS analyses.

3. The cross-linking step usually involves much higher formaldehyde concentrations and volumes than in standard chromatin immunoprecipitation. Follow appropriate working practices and dispose of solutions properly.
4. Do not quench formaldehyde with glycine solutions as usually performed during classical chromatin immunoprecipitation experiments. It may result in nonspecific cross-linking of glycine to proteins and prevent peptide mass attribution during the mass spectrometry analysis. Instead, dilute and wash out unreacted formaldehyde with PBS washes.
5. If the samples are used for mass spectrometry analysis, it is important to avoid any keratin or other source of contamination that could interfere with the analysis. Therefore it is advisable to work with powder free gloves, filter tips, disposable pipettes and clean but non autoclaved tubes. Before starting the SDS PAGE,

extensively wash the migration tank with MilliQ water, use commercial pre-cast gels, premade buffers, loading dye etc., that have limited chances to be contaminated by keratins or other unwanted protein. Stain your gel in a brand new plastic dish (diam 150 mm) and open the cover only to change buffers.

6. A 4–5 mL volume of cell pellet is sufficient for the sequence-specific pull-down and the scramble pull-down.
7. Flash freeze the cross-linked material in liquid nitrogen and store it at −80 °C to minimize cross-linking reversal of proteins from the chromatin. It is not recommended to use material stored longer than one month as we observed significant de-cross-linking and much lower protein retrieval. The best results are obtained when fresh material is used and the storage steps are skipped.
8. It is extremely important that the same number of cells is mixed for quantitative mass spectrometry analysis.
9. Resuspend pellets by vortexing until the sonication **step 22**. To avoid foam vortex very gently and slowly increase to the maximal speed. Using a pipette to resuspend results in losing a significant amount of chromatin that sticks to the plastic. Pay attention to minimize material loss at each step, as the sample sticks to plastic ware.
10. This RNase A step is to optimize the subsequent LNA probe–chromatin hybridization step and to avoid with the nonspecific capture of RNA–protein complexes.
11. Do not vortex, as vortexing leads to excessive foaming of the sample which inhibits sonication efficiency. Resuspend by pipetting with 1 mL micropipette.
12. At this level of cross-linking, indirect ultrasonication in water baths does not solubilize the chromatin. We have recently developed an alternative solubilization method involving restriction enzyme digestion and high pressure solubilization with a *French* high pressure system [3].
13. Heating at 58 °C favors the unmasking of endogenously biotinylated proteins which have to be cleared from the sample.
14. Pre-clearing is necessary to remove most of endogenously biotinylated proteins that might compete with desthiobiotinylated probe during the streptavidin binding step. If the signal is weak on silver-stained gel, reduce the pre-clear to 2 h at room temperature instead of overnight.
15. This step is very important for the quality of the purification. The ultra-fast gel filtration reduces the salt concentration of the sample by about two-thirds leaving roughly 30 mM of NaCl. Low salt concentrations make the hybridization step more stringent. Also, it will prevent nonspecific precipitation of chromatin to the dynabeads during the capture. The use of

DNA denaturing agents in buffers to increase hybrid specificity (formamide or urea) is not advised at any step, as this will result in non specific precipitation on the magnetic beads during the capture step.

16. Higher values of 260/280 ratio would mean that RNaseA treatment was not effective and that chromatin is contaminated with RNA–protein complexes. This will strongly affect the outcome of the PICh procedure as this usually results in the nonspecific capture of those RNP complexes. In such a case both scrambled and telomere pull downs contain many hnRNPs ribosomal proteins.
17. The scrambled probes are nonspecific probes used as a negative control.
18. An LNA containing probe offers increased affinity for its complementary strand and the LNA–DNA hybrids have high melting temperatures ensuring strong and stable interactions compared to traditional DNA or RNA oligonucleotides.
19. The hybridization program was optimized so that LNA probe can invade DNA sequence into chromatin without observable protein de-cross-linking. Denaturation temperatures higher than 75 °C should be avoided as they result in significant cross-linking reversal and proteins loss from the target chromatin. Denaturation temperature below 68 °C should also be avoided because no capture was observed.
20. The addition of water prevents chromatin precipitation on the beads.
21. Do not vortex but resuspend the beads by gently inverting the tubes until all the beads are totally resuspended.
22. To avoid material loss in the pipette tip add 400 μL of lysis buffer with one pipette and resuspend the beads and transfer the mixture with another pipette (keep the tip on this pipet). Repeat this **step 2** additional times (3×400 μL = 1.2 mL).
23. The "halo" indicates that the PICh is efficient. The chromatin bound dynabeads need more space when immobilized on the magnetic stand thus forming a "halo" on the wall of the microcentrifuge tube.
24. Any bead remaining in the eluate will contribute, together with nonspecifically bound proteins, to the nonspecific background. As the target is usually of low abundance, the nonspecific background can significantly alter the outcome of the experiment as discussed in the introduction.
25. This value is a good indication that the LNA probe has been eluted from the beads. Usually at the same concentration, the scramble LNA probe absorbs more than the telomere LNA probe in the eluates. The reason for this discrepancy is unknown.

26. Put the 1.5 mL microcentrifuge tube with the hinge outwards. The protein pellet will end up all along on this side of the wall of the tube which explains why it is sometimes difficult to see.
27. If the PICh assay is very clean the scrambled extract should be devoid of any proteins. Most frequent contaminants are histones. The PICh extract should show different banding patterns that input meaning the probe used specifically purified proteins bound to the genomic region of interest.
28. Silver staining is about 50–100 times more sensitive than Coomassie blue staining. Although some Silver staining methods are compatible with protein identification by mass spectrometry, we found that the analysis from Silver stained bands strongly decreased the sensitivity of the analysis. Thus, while the identification of proteins from such silver stained gels is technically feasible, we do not recommend this.
29. The direct mass spectrometric analysis of liquid samples is doable. However, it reproducibly resulted in a much smaller and less complex proteome and therefore is not recommended.
30. The DNA smear ranges from below 0.5 to more than 3 kb. This is higher than the average fragment size distribution obtained after the typical lower cross-linking conditions used in ChIP (1 % HCHO for 10 min). Do not pre-stain, but post-stain the gel for accurate size estimation.

Acknowledgments

We would like to thank Titia de Lange for the kind gift of the TRF1 antibody.

References

1. Carey MF, Peterson CL, Smale ST (2009) Chromatin immunoprecipitation (ChIP). Cold Spring Harb Protoc 9:pdb prot5279
2. Dejardin J, Kingston RE (2009) Purification of proteins associated with specific genomic Loci. Cell 136:175–186
3. Ide S, Dejardin J (2015) End-targeting proteomics of isolated chromatin segments of a mammalian ribosomal RNA gene promoter. Nat Commun 6:6674
4. Braasch DA, Corey DR (2001) Locked nucleic acid (LNA): fine-tuning the recognition of DNA and RNA. Chem Biol 8:1–7
5. Belotserkovskii BP, Reddy G, Zarling DA (1999) DNA hybrids stabilized by heterologies. Biochemistry 38:10785–10792
6. Landgraf R, Chen CH, Sigman DS (1995) R-loop stability as a function of RNA structure and size. Nucleic Acids Res 23:3516–3523
7. Hirsch JD, Eslamizar L, Filanoski BJ, Malekzadeh N, Haugland RP, Beechem JM, Haugland RP (2002) Easily reversible desthiobiotin binding to streptavidin, avidin, and other biotin-binding proteins: uses for protein labeling, detection, and isolation. Anal Biochem 308:343–357
8. Saksouk N, Barth TK, Ziegler-Birling C, Olava N, Nowak A, Rey E, Mateos-Langerak J, Urbach S, Reik W, Torres-Pedilla ME, Imhof A, Dejardin J, Simboeck E (2014) Redundant mechanisms to form silent chromatin at pericentromeric regions rely on BEND3 and DNA methylation. Mol Cell 56:580–594

Chapter 4

Profiling Cell Lines Nuclear Sub-proteome

Aline Poersch, Andrea G. Maria, Camila S. Palma, Mariana L. Grassi, Daniele Albuquerque, Carolina H. Thomé, and Vitor M. Faça

Abstract

Proteins are very dynamic within the cell and their localization and trafficking between subcellular compartments are critical for their correct function. Indeed, the abnormal localization of a protein might lead to the pathogenesis of several diseases. The association of cell fractionation methods and mass spectrometry based proteomic methods allow both the localization and quantification of proteins in different subcompartments. Here we present a detailed protocol for enrichment, identification, and quantitation of the nuclear proteome in cell lines combining nuclear subproteome enrichment by differential centrifugation and high-throughput proteomics.

Key words Nuclear fractionation, Subcellular proteomics, Protein localization, Cell Line, Mass spectrometry

1 Introduction

Eukaryotic cells are arranged in compartments characterized by distinct biochemical process and particular sets of proteins [1]. As each compartment provides specific physiological conditions, the presence in a subcellular compartment and the traffic into different organelles are crucial to determine the function of a protein as well as the function of the compartment [2]. Moreover, the subcellular localization of a protein defines whether it will go through post-translational modifications, interact with other molecules and integrate with different biological networks [1]. Because proteins are very dynamic within the cell, the localization of a protein into a specific cellular compartment defines its correct function [3] and, therefore, the abnormal localization of a protein might lead to pathologies that include cardiovascular, neurodegenerative, cancer, and metabolic diseases [1]. In addition, modulation of some protein subcellular localization has been proposed as promising therapeutic strategies [4]. Altogether, these factors lead to an increased interest in performing studies that enable the identification and quantification of proteins in specific subcellular localization.

Lucio Comai et al. (eds.), *Proteomics: Methods and Protocols*, Methods in Molecular Biology, vol. 1550,
DOI 10.1007/978-1-4939-6747-6_4, © Springer Science+Business Media LLC 2017

The nucleus has important roles in several cellular functions such as gene expression, cell cycle control, cell growth and signal transduction [5]. The shuttling of specific proteins out of the nucleus is essential for the regulation of important functions in the cell [6, 7]. A variety of tumor suppressors demand to localize in the nucleus for their correct function and aberrant protein localization or failure on the dynamics of spatio-temporal cell signaling are involved in tumor cancer cell survival, tumorigenesis, tumor progression, and drug resistance [1, 8, 9]. Consequently, for particular sets of proteins, cytoplasmic localization can be explored as a marker of cell growth disorder and tumor development risk [6, 10, 11]. For instance, nuclear export machinery through XPO1 (Exportin-1/Chromosome Region Maintenance 1/CRM1) is upregulated in different cancer types and may be used as a prognostic indicator [7]. Abnormal XPO1 intracellular localization might affect the correct function of important suppressors and oncogenic proteins such as p53, Rb, FOXO, p21, and others, contributing to cancer development and progression [12–16]. Therefore, identification of the correct location of proteins is a potential strategy for therapeutic intervention in many types of cancer.

A subcellular proteome may be identified by applying specific enrichment strategies that concentrate organelles and cellular compartments [17]. These strategies are usually based on differences in sedimentation coefficients and organelles density [18]. Thus, the subcellular compartments can be separated by centrifugation, taking advantage of the density gradient and buffers that maintain the compartment integrity. Moreover, detergents might be used to enrich the subcellular fractions, depending on the compartment solubility [19, 20]. Once the subcellular fraction of interest is isolated and proteins are enriched, analytical methods can be applied to identify and quantify the proteins present in each compartment. Several studies have applied mass spectrometry for this purpose [17] and large-scale identification of proteins with proteomic methods is showing to be an important approach to elucidate protein function and regulation [21]. Here, we present a protocol to study the nuclear proteome using centrifugation to enrich for cellular nucleus, combined to in-depth proteomics based on intact protein fractionation and highthroughput LC-MS/MS protein identification. Further, we show data obtained by our group applying the protocol described here, illustrating the quality and importance to analyze the protein content in the nucleus.

2 Materials

2.1 Cell Culture

1. Cell culture media: Mammary Epithelial Cell Growth Medium (MEBM) supplemented with 100 ng/mL cholera toxin (Sigma), MEGM SingleQuots kit (bovine pituitary extract—BPE, human epidermal growth factor—hEGF,

insulin, hydrocortisone, and gentamicin amphotericin-b—GA-1000; Lonza), 10–0 % (v/v) Fetal Bovine Serum (FBS) and 1 % (v/v) penicillin/streptomycin (*see* **Note 1**).

2. Pre-warmed (37 °C) phosphate buffered saline (PBS), pH 7.2.
3. 100 mm treated culture dishes for adherent cell culture.

2.2 Cell Lysis Components

1. Lysis buffer: 50 mM HEPES pH 7.4, 10 mM NaCl, 5 mM $MgCl_2$, 0,1 mM EDTA, 1 mM Na_3VO_4 (fosfatase inhibitor), 1 mM NaF and 1 mM $Na_4P_2O_7.10dH_2O$, with protease inhibitor cocktail 5 % (v/v) (Sigma-Aldrich) (*see* **Note 2**).
2. Cell scraper.
3. Ice-cold phosphate buffered saline (PBS), pH 7.2.
4. 3 mL syringe with needle
5. Tissue homogenizer model D-130 (Biosystems) or equivalent model.
6. 2 mL microcentrifuge tubes.
7. Refrigerated microcentrifuge.

2.3 Nuclear Fractionation Components

1. Nuclear protein extraction buffer: 8 M urea and 2% (v/v) CHAPS containing 5 % (v/v) protease inhibitor cocktail and 1 mM Na_3VO_4 (*see* **Note 3**).
2. Vortex mixer.
3. Refrigerated microcentrifuge.

2.4 Sample Preparation for Proteomic Analysis

2.4.1 Protein Fractionation by SDS-PAGE

1. Bradford quantification kit.
2. 10% Mini-PROTEAN® TGX™ Precast Gel (Bio-Rad) or equivalent, and gel electrophoresis equipment (Bio-Rad).
3. 2× Laemmli sample buffer (Bio-Rad).
4. Reduction solution: Dissolve 10 mg of dithiothreitol (DTT) in 1 mL of ultrapure water (10 μg/μL) (*see* **Note 4**).
5. Alkylation solution: Dissolve 5 mg of iodoacetamide (IAM) in 0.1 mL of ultrapure water (50 μg/μL) (*see* **Note 5**).
6. Fixing solution: 40% ethanol, 10% acetic acid, and 50% ultrapure water.
7. Staining solution: 0.1 g of Coomassie Brilliant Blue G-250, 1.175 mL of phosphoric acid, 10 g of ammonium sulfate, and q.s.p. 100 mL of ultrapure water (*see* **Note 6**).
8. Staining solution 20% methanol: Dilute 40 mL of staining solution in 10 mL of methanol.
9. Washing solution: 25% methanol–ultrapure water.

2.4.2 In-Gel Trypsin Digestion

1. Clean glass plate and sterile scalpel (*see* **Note 7**).
2. Ammonium bicarbonate 100 mM pH 8 solution: Dissolve 1.58 g of ammonium bicarbonate in 200 mL of ultrapure water (*see* **Note 8**).
3. Destain solution: 50 mL of ammonium bicarbonate 100 mM pH 8 with 50 mL of 100% acetonitrile.
4. Trypsin solution: Resuspend one vial containing 20 μg of sequencing-grade modified trypsin (Promega) in 100 μL of ammonium bicarbonate 100 mM pH 8 solution (*see* **Note 9**).
5. Thermomixer.

2.5 Peptide Extraction and Sample Desalting Components

1. Peptide extraction buffer I (50% acetonitrile–50% ultrapure water–0.1% formic acid solution).
2. Peptide extraction buffer II (70% acetonitrile–30% ultrapure water–0.1% formic acid solution).
3. Peptide extraction buffer III (100% acetonitrile–0.1% formic acid solution).
4. ZipTip® C_{18} pipette tips (Merck Millipore).
5. Mass spectrometry compatible injection vials.
6. Thermomixer.
7. SpeedVac concentrator.

2.6 High-Throughput Mass Spectrometry Components

1. Equipment: LTQ-Orbitrap Velos mass spectrometer (Thermo Fisher) coupled to a nanoflow chromatography system (Eksigent).
2. Chromatographic column: 25 cm long column (Picofrit 75 mm ID, New Objectives, packed in-house with MagicC18 resin).
3. Solvents for reversed-phase chromatography: aqueous solvent (A)—5% acetonitrile–95% water–0.1% formic acid; organic solvent (B)—95% acetonitrile–5% water–0.1% formic acid. Bottled water, acetonitrile, and formic acid are obtained from Fisher Scientific.

3 Methods

3.1 Cell Lysis

1. Cultivate MCF10A cell line in 75 cm^2 flasks in described culture media. For the fractionation, plate 5×10^6–1×10^7 cells in a 100 mm culture dish.
2. After 24 h and/or cellular confluence >80%, remove the media, wash the cells twice with pre-warmed PBS, substitute with new culture media and proceed with the treatments of interest.

3.2 Preparation of Cell Lysate

1. After treatments of interest, remove the cell media and wash the cells twice with ice-cold PBS. Remove any solution excess that remains on the dish. Add 150 μL of lysis buffer containing protease inhibitors in each dish and scrape cells using a cell scraper. Collect cell lysate and transfer to 2 mL microtubes. Keep samples on ice.
2. Pass the cell lysate 20 times through a thin 25-gauge needle using a 3 mL syringe and homogenize in a tissue homogenizer (dounce homogenizer) for 1 min. Keep samples on ice.
3. Centrifuge samples at 500×*g* for 20 min at 4 °C to pellet nuclei (no brake applied during the deceleration of the centrifuge). Carefully transfer all of the supernatant to a 1.5 mL microtube (*see* **Note 10**).

3.3 Nuclear Fractionation

1. Add to the dry pellet obtained previously (*see* Subheading 3.2) 100 μL of nuclear extraction buffer. Nuclear proteins extraction are obtained by sonication of samples using an ultrasound probe for 5 min followed by cycles of vortexing for 20 s and ice bath for 5 min. Repeat this cycle three times.
2. Centrifuge samples at 20,000×*g* for 30 min at 4 °C and transfer the supernatant enriched in nuclear proteins to a 1.5 mL microtube. Store samples at −80 °C until use.

3.4 Sample Preparation for Proteomic Analysis

3.4.1 SDS-Polyacrylamide Gel Electrophoresis

1. Quantify samples containing nuclear proteins using Bradford Assay kit according to the manufacturer's instructions.
2. Aliquot 50 μg of nuclear proteins sample, mix with 2× Laemmli sample buffer and perform protein reduction by adding 5 μL of reduction solution and maintaining the reaction at 95 °C for 5 min. Centrifuge samples for 5 min at 8500×*g* at room temperature.
3. After cooling, perform protein alkylation by adding 3 μL of alkylating solution and maintain the reaction at room temperature for 20 min protected from light.
4. Load each sample onto a 10% 1.0 mm precast gel and perform SDS-polyacrylamide gel electrophoresis according to the manufacturer's instructions.
5. Fix the gel with 25 mL of fixing solution, shaking for 30 min at room temperature.
6. Remove fixing solution and add 25 mL of staining solution. Incubate overnight shaking at room temperature.
7. Remove staining solution and wash several times with 25% methanol washing solution (*see* **Note 11**).

3.4.2 In-Gel Trypsin Digestion

1. Place the gel on a clean glass plate and excise each lane containing proteins with a clean scalpel. Cut each lane into approximately 1 cm square pieces and transfer to a clean 1.5 mL microtube pre-rinsed with methanol.

2. Add 200 μL of destaining solution and incubate for 10 min, shaking at room temperature. Repeat this step four times. Finally, add 200 μL of destaining solution and incubate overnight at 4 °C (*see* **Note 12**). Add 200 μL of acetonitrile for 10 min and dry with SpeedVac concentrator.
3. Perform in-gel trypsin digestion of gel slices adding 20 μL (0.5 μg) of trypsin solution and agitate vigorously. Carry out digestion for 30 min at 37 °C constantly shaking in a thermomixer (450 rpm) (*see* **Note 13**).
4. Add 200 μL of ammonium bicarbonate 100 mM pH 8 and incubate overnight at 37 °C constantly shaking in a thermomixer (450 rpm).

3.5 Peptide Extraction and Sample Desalting

1. After digestion, transfer to a 1.5 mL microtube the trypsin/ammonium bicarbonate 100 mM pH 8 solution and dry in SpeedVac.
2. Add 200 μL of peptide extraction buffer to the gel pieces and incubate for 90 min at room temperature, constantly shaking in a thermomixer (450 rpm).
3. Combine the supernatant containing the peptide extract I with the initial supernatant in the same 1.5 mL microtube and dry in SpeedVac.
4. Again, add 200 μL of peptide extraction buffer II to the gel pieces and incubate for 90 min at room temperature, constantly shaking in a thermomixer (450 rpm).
5. Combine the supernatant containing the peptide extract II supernatant with previous peptide extracts and dry in SpeedVac.
6. Add 200 μL of peptide extraction buffer to the gel slices and incubate for 90 min at room temperature, constantly shaking in a thermomixer (450 rpm).
7. Finally, combine supernatant containing the peptide extract III and dry in SpeedVac.
8. Dissolve peptide mixture in 10 μL 5% acetonitrile–0.1% formic acid.
9. Desalt samples using ZipTip® C_{18} pipette tips as follows:
10. Condition tips with 20 μL of 100% acetonitrile–0.1% formic acid. Repeat three times.
11. Equilibrate tips with 20 μL 5% acetonitrile–0.1% formic acid. Repeat three times.
12. Apply sample through the tip.
13. Wash with 20 μL 5% acetonitrile–0.1% formic; repeat three times.

14. Elute peptides with 20 μL of 50% acetonitrile–water in 0.1% formic acid. Repeat three times.
15. Dry eluted peptides in SpeedVac. Samples can be stored dried at −80 °C until ready for LC-MS analysis.
16. Reconstitue samples in 15 μL 5% acetonitrile–0.1% formic acid. Centrifugate at 12,000 × *g* for 15 min and transfer to mass spectrometry compatible injection vials.

3.6 High-Throughput Mass Spectrometry

1. Carry out the high throughput LC-MS/MS data collection for each individual fraction. Inject 10 μL of peptide extract and analyze samples over a 90 min linear gradient from 5 to 35% of organic solvent at 350 nL/min in the system described in section 2.6 (*see* **Note 14**).
2. Process LC-MS/MS files through data bank search, protein inference and quantitative analysis (*see* **Note 15**).
3. Match the lists of proteins identified in the nuclear enriched fraction with proteins identified in the total cell extract, cytoplasmic or membrane enriched fractions (*see* **Note 10**). Select for nuclear proteins based on the higher value of enrichment obtained from the ratio of spectral counts observed in enriched nuclear fractions/total cell extract cytoplasmic or membrane enriched fraction profile (*see* **Note 16**). A Sample dataset obtained for MCF10A cell line is presented in Table 1 and Fig. 1 (*see* **Note 17**).

4 Notes

1. Cultivate cell lineage according to the culture method specified in ATCC. Cultivate cells in 75 cm^2 flasks or 100 mm dishes and maintain at 37 °C in a 5% CO_2 humidified incubator. Cultivate cells in SILAC media, using standard SILAC protocols, if relative quantification is required [22].
2. Prepare a fresh lysis buffer solution and add protease inhibitor cocktail (P8340, Sigma Aldrich) and Na_3VO_4 (phosphatase inhibitor) right before proceeding with the cell lysis.
3. Prepare a fresh nuclear protein extraction buffer solution and add protease inhibitor cocktail and Na_3VO_4 right before performing the protein extraction and solubilization.
4. Dissolve DTT in water right before proceeding protein reduction.
5. Dissolve IAM in water right before proceeding protein alkylation and protect from light. Protein alkylation before SDS-PAGE run improves peptide recovery after in situ protein digestion.

Table 1
Top 20 proteins enriched in the nuclear fraction of MCF10A breast epithelial cells.

				MCF10A		
	Uniprot entry name	Gene name	Description	Cytoplasm counts	Nucleus counts	"Enrichment" nucleus/cytoplasm
1	FBRL_HUMAN	FBL	rRNA 2′-O-methyltransferase fibrillarin	1	54	54
2	IF16_HUMAN	IFI16	Gamma-interferon-inducible protein 16	N/D	45	45
3	H2AX_HUMAN	H2AFX	Histone H2AX	1	44	44
4	HMGA1_HUMAN	HMGA1	High mobility group protein HMG-I/HMG-Y	N/D	35	35
5	DKC1_HUMAN	DKC1	H/ACA ribonucleoprotein complex subunit 4	N/D	35	35
6	CSK21_HUMAN	CSNK2A1	Casein kinase II subunit alpha	N/D	34	34
7	HBA_HUMAN	HBA1	Hemoglobin subunit alpha	N/D	34	34
8	NEP1_HUMAN	EMG1	Ribosomal RNA small subunit methyltransferase NEP1	N/D	31	31
9	FUS_HUMAN	FUS	RNA-binding protein FUS	N/D	31	31
10	SNUT1_HUMAN	SART1	U4/U6.U5 tri-snRNP-associated protein 1	N/D	27	27
11	HNRDL_HUMAN	HNRPDL	Heterogeneous nuclear ribonucleoprotein D-like	1	25	25
12	SRSF2_HUMAN	SRSF2	Serine/arginine-rich splicing factor 2	N/D	25	25
13	TMOD3_HUMAN	TMOD3	Tropomodulin-3	1	24	24
14	DDX21_HUMAN	DDX21	Nucleolar RNA helicase 2	1	24	24
15	HNRPM_HUMAN	HNRNPM	Heterogeneous nuclear ribonucleoprotein M	2	48	24
16	SP16H_HUMAN	SUPT16H	FACT complex subunit SPT16	N/D	24	24
17	NOP56_HUMAN	NOP56	Nucleolar protein 56	N/D	24	24
18	TOP1_HUMAN	TOP1	DNA topoisomerase 1	N/D	22	22
19	CBX5_HUMAN	CBX5	Chromobox protein homolog 5	N/D	21	21
20	SRRT_HUMAN	SRRT	Serrate RNA effector molecule homolog	N/D	21	21

Breast epithelial cell line MCF10A was profiled by high-throughput proteomics according the method here described. The ratio of spectral counts of proteins detected in the nuclear enriched fraction over spectral counts detected in the cytoplasmic enriched fraction was used as an "enrichment factor" for nuclear proteins. The list presents the top 20 proteins of MCF10A cell line detected in the nuclear fraction. For calculation purposes, proteins not detected (N/D) in the cytoplasmic fraction were considered as having one spectral count.

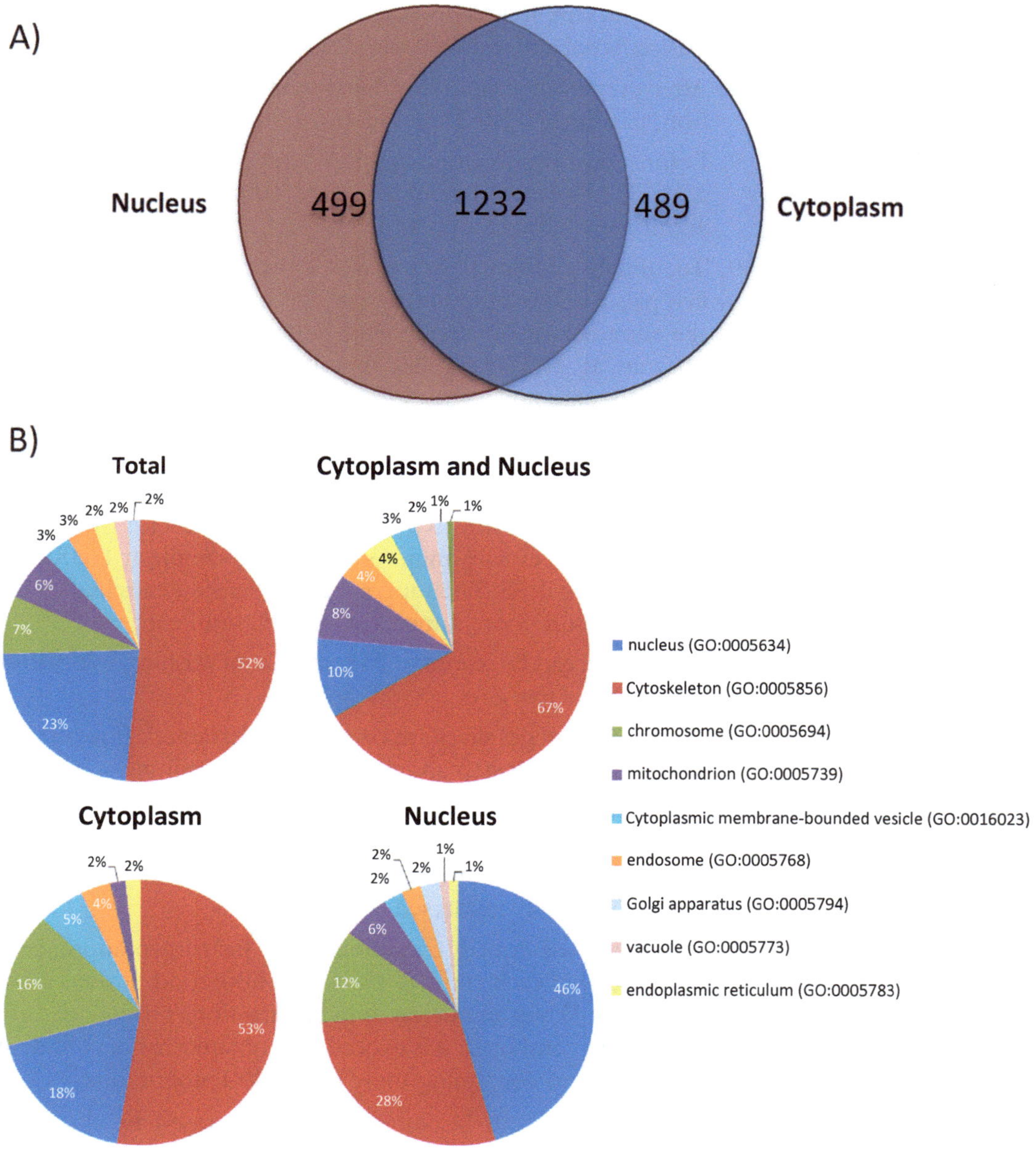

Fig. 1 Protein dataset for nuclear profile of MCF10A breast epithelial cell line. (**a**) Venn diagram of proteins identified by LC-MS/MS in the nuclear and cytoplasmic enriched fractions of MCF10A cells. The fractionation methodology allowed confident overall protein identification of 2220 proteins with less than 1 % FDR. The overlap between nuclear enriched and cytoplasmic enriched fraction (55 %) as well as the number of proteins identified in only one of those fractions (45 %) shows the complementarity of both subcellular proteomic profiles. (**b**) Gene Ontology classification of proteins enriched in the nuclear and cytoplasmic compartments also supports the complementarity of the proteomic profiles

6. Staining solution can be prepared before use and stored in a clean amber or brown glass bottle at room temperature. Prepare a fresh staining solution in 20 % methanol right before proceeding gel staining.

7. To avoid sample keratin contamination wash glass plate with 70% ethanol and always wear gloves.
8. After preparing ammonium bicarbonate 100 mM pH 8 solution, pass it thru a 0.22 μm filter.
9. Before starting in-gel digestion, resuspend trypsin vial in 100 μL of ammonium bicarbonate 100 mM pH 8 solution and incubate for 15 min at 37 °C.
10. To obtain additional subcellular enriched fractions, such as cytoplasm and membrane, follow the next steps: (1) centrifuge the supernatant obtained (Subheading 3.2) at 16,000 × *g* for 20 min at 4 °C to pellet membranes. The supernatant is denoted enriched cytosol fraction; (2) resuspend pellet in 50 μL of membrane extraction buffer (25 mM MES (2-(N-morpholino)-ethanesulfonic acid), pH 6.5, 150 mM NaCl, 2% Triton X-100) containing 5% (v/v) protease inhibitor cocktail and 1 mM $N_{a3}VO_4$ and incubate on ice for 60 min with agitation every 5 min; (3) centrifuge sample at 16,000 × *g* for 20 min at 4 °C. The supernatant is denoted enriched membrane fraction. Store samples at −80 °C until use.
11. Stained gel can be stored in 25% methanol washing solution at 4–6 °C for a few days.
12. Ensure that all gel slices are completely destained. If necessary, repeat destaining step adding 200 μL of destaining solution.
13. Incubate the gel pieces for few minutes (30 min) until the trypsin solution is completely absorbed. Only after that complete with ammonium bicarbonate solution to keep the gel piece completely immersed in solution.
14. Using the described conditions and instrumentation, it is expected that each run will yield 200–500 good protein identifications. When all nuclear enriched fractions protein identifications are combined, it is expected to obtain a list of confident identifications containing more than 2000 protein hits.
15. Acquired data can be automatically processed by the open source Labkey Server (www.labkey.org) platform, which employs the TransProteomic Pipeline, developed at the Institute of Systems Biology [23]. Search data against the most recent version of the human proteome database (Uniprot) or other appropriate human protein database of your choice. A fixed modification of 57.021464 is added to cysteine residues and a variable modification of 15.994915 is added to methionine residues for database search. Optionally, when SILAC strategy is used, account for incorporation of the light and heavy amino acid isotopes. To estimate the significance of peptide and protein matches, we apply the tools PeptideProphet [24] and ProteinProphet [25]. Identifications with a PeptideProphet probability score greater than 0.9 are selected

and submitted to ProteinProphet to account for the protein inference problem. Overall, false discovery rates for this procedure are less than 1 %. Protein quantification can also be performed automatically with the Q3 tool available in the distribution of labkey server [23].

16. The spectral counting method can be used to estimate nuclear protein enrichment as previously described [26]. Briefly, the normalized spectral counts for each protein group output by ProteinProphet for nuclear protein profile divided by the normalized spectral counts output for the total cell extract or cytoplasmic/membrane profile can be considered as a semiquantitative enrichment analysis. The total number of counts in the entire experiment is used as a normalization parameter for each profile.
17. Several proteins were identified only in the nuclear enriched fraction dataset. As expected, protein markers for nucleus were detected among the top enriched protein in our analysis, such as histones and DNA binding proteins (*see* Table 1). Gene ontology analysis supports the enrichment of nuclear protein based on annotation for cellular compartments (*see* Fig. 1). A comparison for annotations observed in a cytoplasmic profile for the same cell line highlights the enrichments obtainded by the methodology described here.

Acknowledgments

This research was supported by FAPESP (Young Scientist Grant—Proc.No. 2011/0947-1), CNPq, Center for Cell Based Thereapy—CTC-CEPID (Proc.FAPESP 2013/08135-2) and CISBi-NAP. A.G.M. C.S.P, M.L.G., C.H.T., and D.A. received fellowships from FAPESP Proc. No., 2014/16839-2, 2012/09682-4, 2013/08755-0, 2013/07675-3, and 2012/02518-4, respectively. A.P receives a PNPD fellowship from CAPES. V.M.F. receives a fellowship from CNPq, Proc.No. (308561/2014-7). We thank Profs. Emanuel Carrilho and Daniel Cardoso for allowing our data collection with the LTQ-Orbitrap Velos at the Analytical Central – Chemistry Institute of São Carlos—University of São Paulo.

References

1. Hung MC, Link W (2011) Protein localization in disease and therapy. J Cell Sci 124(Pt 20):3381–3392. doi:10.1242/jcs.089110
2. Simha R, Briesemeister S, Kohlbacher O, Shatkay H (2015) Protein (multi-)location prediction: utilizing interdependencies via a generative model. Bioinformatics 31(12):i365–i374. doi:10.1093/bioinformatics/btv264
3. Butler GS, Overall CM (2009) Proteomic identification of multitasking proteins in unexpected locations complicates drug targeting. Nat Rev Drug Discov 8(12):935–948. doi:10.1038/nrd2945

4. Tomas A, Futter CE, Eden ER (2014) EGF receptor trafficking: consequences for signaling and cancer. Trends Cell Biol 24(1):26–34. doi:10.1016/j.tcb.2013.11.002
5. Chahine MN, Pierce GN (2009) Therapeutic targeting of nuclear protein import in pathological cell conditions. Pharmacol Rev 61(3):358–372. doi:10.1124/pr.108.000620
6. Turner JG, Sullivan DM (2008) CRM1-mediated nuclear export of proteins and drug resistance in cancer. Curr Med Chem 15(26):2648–2655
7. Takeda A, Yaseen NR (2014) Nucleoporins and nucleocytoplasmic transport in hematologic malignancies. Semin Cancer Biol 27:3–10. doi:10.1016/j.semcancer.2014.02.009
8. Kau TR, Way JC, Silver PA (2004) Nuclear transport and cancer: from mechanism to intervention. Nat Rev Cancer 4(2):106–117. doi:10.1038/nrc1274
9. Wang SC, Hung MC (2005) Cytoplasmic/nuclear shuttling and tumor progression. Ann N Y Acad Sci 1059:11–15. doi:10.1196/annals.1339.002
10. Fabbro M, Henderson BR (2003) Regulation of tumor suppressors by nuclear-cytoplasmic shuttling. Exp Cell Res 282(2):59–69
11. Salmena L, Pandolfi PP (2007) Changing venues for tumour suppression: balancing destruction and localization by monoubiquitylation. Nat Rev Cancer 7(6):409–413. doi:10.1038/nrc2145
12. Santiago A, Li D, Zhao LY, Godsey A, Liao D (2013) p53 SUMOylation promotes its nuclear export by facilitating its release from the nuclear export receptor CRM1. Mol Biol Cell 24(17):2739–2752. doi:10.1091/mbc.E12-10-0771
13. Ohtani N, Brennan P, Gaubatz S, Sanij E, Hertzog P, Wolvetang E, Ghysdael J, Rowe M, Hara E (2003) Epstein-Barr virus LMP1 blocks p16INK4a-RB pathway by promoting nuclear export of E2F4/5. J Cell Biol 162(2):173–183. doi:10.1083/jcb.200302085
14. Yang H, Zhao R, Yang HY, Lee MH (2005) Constitutively active FOXO4 inhibits Akt activity, regulates p27 Kip1 stability, and suppresses HER2-mediated tumorigenicity. Oncogene 24(11):1924–1935. doi:10.1038/sj.onc.1208352
15. Hu MC, Lee DF, Xia W, Golfman LS, Ou-Yang F, Yang JY, Zou Y, Bao S, Hanada N, Saso H, Kobayashi R, Hung MC (2004) IkappaB kinase promotes tumorigenesis through inhibition of forkhead FOXO3a. Cell 117(2):225–237
16. Alt JR, Gladden AB, Diehl JA (2002) p21(Cip1) Promotes cyclin D1 nuclear accumulation via direct inhibition of nuclear export. J Biol Chem 277(10):8517–8523. doi:10.1074/jbc.M108867200
17. Drissi R, Dubois ML, Boisvert FM (2013) Proteomics methods for subcellular proteome analysis. FEBS J 280(22):5626–5634. doi:10.1111/febs.12502
18. Lee YH, Tan HT, Chung MC (2010) Subcellular fractionation methods and strategies for proteomics. Proteomics 10(22):3935–3956. doi:10.1002/pmic.201000289
19. Ramsby ML, Makowski GS, Khairallah EA (1994) Differential detergent fractionation of isolated hepatocytes: biochemical, immunochemical and two-dimensional gel electrophoresis characterization of cytoskeletal and noncytoskeletal compartments. Electrophoresis 15(2):265–277
20. Sawhney S, Stubbs R, Hood K (2009) Reproducibility, sensitivity and compatibility of the ProteoExtract subcellular fractionation kit with saturation labeling of laser microdissected tissues. Proteomics 9(16):4087–4092. doi:10.1002/pmic.200800949
21. Walther TC, Mann M (2010) Mass spectrometry-based proteomics in cell biology. J Cell Biol 190(4):491–500. doi:10.1083/jcb.201004052
22. Ong SE, Mann M (2006) A practical recipe for stable isotope labeling by amino acids in cell culture (SILAC). Nat Protoc 1(6):2650–2660. doi:10.1038/nprot.2006.427
23. Rauch A, Bellew M, Eng J, Fitzgibbon M, Holzman T, Hussey P, Igra M, Maclean B, Lin CW, Detter A, Fang R, Faca V, Gafken P, Zhang H, Whiteaker J, States D, Hanash S, Paulovich A, McIntosh MW (2006) Computational proteomics analysis system (CPAS): an extensible, open-source analytic system for evaluating and publishing proteomic data and high throughput biological experiments. J Proteome Res 5(1):112–121. doi:10.1021/pr0503533
24. Keller A, Nesvizhskii AI, Kolker E, Aebersold R (2002) Empirical statistical model to estimate the accuracy of peptide identifications made by MS/MS and database search. Anal Chem 74(20):5383–5392
25. Nesvizhskii AI, Keller A, Kolker E, Aebersold R (2003) A statistical model for identifying proteins by tandem mass spectrometry. Anal Chem 75(17):4646–4658
26. Faca VM, Ventura AP, Fitzgibbon MP, Pereira-Faca SR, Pitteri SJ, Green AE, Ireton RC, Zhang Q, Wang H, O'Briant KC, Drescher CW, Schummer M, McIntosh MW, Knudsen BS, Hanash SM (2008) Proteomic analysis of ovarian cancer cells reveals dynamic processes of protein secretion and shedding of extra-cellular domains. PLoS One 3(6):e2425. doi:10.1371/journal.pone.0002425

Chapter 5

Optimized Enrichment of Phosphoproteomes by Fe-IMAC Column Chromatography

Benjamin Ruprecht, Heiner Koch, Petra Domasinska, Martin Frejno, Bernhard Kuster, and Simone Lemeer

Abstract

Phosphorylation is among the most important post-translational modifications of proteins and has numerous regulatory functions across all domains of life. However, phosphorylation is often substoichiometric, requiring selective and sensitive methods to enrich phosphorylated peptides from complex cellular digests. Various methods have been devised for this purpose and we have recently described a Fe-IMAC HPLC column chromatography setup which is capable of comprehensive, reproducible, and selective enrichment of phosphopeptides out of complex peptide mixtures. In contrast to other formats such as StageTips or batch incubations using TiO_2 or Ti-IMAC beads, Fe-IMAC HPLC columns do not suffer from issues regarding incomplete phosphopeptide binding or elution and enrichment efficiency scales linearly with the amount of starting material. Here, we provide a step-by-step protocol for the entire phosphopeptide enrichment procedure including sample preparation (lysis, digestion, desalting), Fe-IMAC column chromatography (column setup, operation, charging), measurement by LC-MS/MS (nHPLC gradient, MS parameters) and data analysis (MaxQuant). To increase throughput, we have optimized several key steps such as the gradient time of the Fe-IMAC separation (15 min per enrichment), the number of consecutive enrichments possible between two chargings (>20) and the column recharging itself (<1 h). We show that the application of this protocol enables the selective (>90%) identification of more than 10,000 unique phosphopeptides from 1 mg of HeLa digest within 2 h of measurement time (Q Exactive Plus).

Key words Phosphorylation, Proteomics, Phosphocapture, LC-MS

Abbreviations

ACN	Acetonitrile
AGC	Acquisition gain control
CAA	Chloroacetamide
DTT	Dithiothreitol
FA	Formic acid
FCS	Fetal calf serum
HCD	Higher energy collision induced dissociation
HCl	Hydrochloride
HPLC	High-performance liquid chromatography

Lucio Comai et al. (eds.), *Proteomics: Methods and Protocols*, Methods in Molecular Biology, vol. 1550,
DOI 10.1007/978-1-4939-6747-6_5, © Springer Science+Business Media LLC 2017

I.D.	Inner diameter
IMAC	Immobilized metal ion affinity chromatography
MeOH	Methanol
MS	Mass spectrometry
MS/MS	Tandem mass spectrometry
PBS	Phosphate buffered saline
Ppm	parts per million
PSM	Peptide spectrum match
pY/pS/Pt	Phosphotyrosine, -serine, -threonine
TFA	Trifluoro acetic acid
TiO_2	Titanium dioxide
Tris	Tris(hydroxymethyl)aminomethane
v/v	Volume/volume
w/w	Weight/weight
ZrO_2	Zirconium dioxide

1 Introduction

Reversible protein phosphorylation is a posttranslational modification that plays a key role in signal transduction and aberrant regulation has been implicated in a number of diseases [1]. As a consequence, mass spectrometry-based large-scale identification of phosphorylation events has received considerable attention over the last years. Due to the low abundance and sub stoichiometric levels of the phosphorylation events, enrichment of phosphopeptides or phosphoproteins is required prior to mass spectrometric detection [2, 3]. In the field of phosphoproteomics, various large-scale enrichment strategies have been developed over the last years, all with their own strengths and weaknesses. The most widely used enrichment strategies utilize the affinity of phosphate groups to metal ions (immobilized on a solid support). Examples of such strategies are metal oxide affinity chromatography (TiO_2, ZrO_2) [4, 5], immobilized metal ion affinity chromatography (IMAC) with different metal ions (Fe^{3+}, Ga^{3+}, or Zr^{4+}) [6–8] and Ti-IMAC [9]. It is thought that there is a high degree of complementarity between these different enrichment materials [9–12]; however, we have recently shown that such complementarity can be attributed to the format of the enrichment, the inefficient elution from the material, as well as the insufficient acquisition speed of the mass spectrometer, rather than the material [13]. To date, most phosphopeptide enrichments are still performed in either batch mode or in micro-column format with the material packed in gel-loader tips. These formats suffer from a high degree of variability due to the multitude of manual handling steps in these protocols. In addition, variability is further increased by the use of, for example, different loading solvents, different washing procedures and incubation times [14, 15]. More importantly, the enrichment efficiency and selectivity is largely dependent on the so called bead-to-sample ratio [16, 17]. Consequently, batch- and tip-based enrichment strategies require optimization for each sample. Here, we describe the workflow for the reproducible and comprehensive enrichment of phosphopeptides using Fe-IMAC HPLC columns that overcomes most of these issues. As reported previously, the Fe-IMAC column does not suffer from bead-to-sample ratio issues and allows for the comprehensive depletion of phosphopeptides from digests without showing any bias in the type of phosphopeptides that are enriched [13]. The protocol describes the entire workflow starting from sample preparation to data analysis. Moreover, it includes several improvements over the published method such as shortened gradient length and improved column recharging, ultimately resulting in considerably increased throughput. We provide a detailed description of the column setup and operation (including charging of the columns and gradients) and describe guidelines for monitoring column performance. Finally, we apply the protocol to the enrichment phosphopeptides from 1 mg cell line digest which led to the identification of >10,000 unique phosphopeptides in 2 h of measurement time.

2 Material

Unless stated otherwise, all solvents and solutions are prepared fresh, using ultrapure water and analytical grade reagents. Devices such as centrifuges, vacuum centrifuges/lyophilizer, thermoshaker, or refrigerators (−20 °C/−80 °C) are not explicitly listed.

2.1 Preparation of Proteome Digests for Phosphopeptide Analysis

1. Cell culture: RPMI 1640 medium supplemented with 10% fetal calve serum (FCS). Add 55 ml FCS to 500 ml RPMI 1640 medium. Sterile phosphate buffered saline (PBS) without calcium and magnesium. 150 × 20 mm cell culture dishes. Cell scraper. HeLa S3 cervix carcinoma cells (DMSZ, Braunschweig, Germany).
2. Lysis buffer: Prepare a 40 mM Tris(hydroxymethyl)aminomethane (Tris)–HCl, pH = 7.6 solution, containing 8 M urea, protease and phosphatase inhibitors. Prepare a stock solution of 2 M Tris–HCl by dissolving 2.42 g Tris in 5 ml water, adjust the pH to 7.6 using a 5 M HCl solution and fill up to 10 ml with water. A 100 fold stock solution of phosphatase inhibitor cocktail 1, 2, and 3 is commercially available (Sigma Aldrich, Munich, Germany). Add 4.8 g of urea to a 15 ml falcon tube, add 200 µl Tris–HCl stock solution, add one protease inhibitor tablet complete mini EDTA-free (Roche, Mannheim, Germany) and 100 µl of each phosphatase inhibitor stock solution and fill up to 10 ml with water. Store the lysis buffer on ice.
3. Reducing agent: 1 M stock solution of dithiothreitol (DTT) in water. Dissolve 1.54 g DTT in a falcon tube and fill up to 10 ml

with water. Prepare 200 µl aliquots and store the reducing agent at −20 °C.

4. Alkylating agent: 550 mM stock solution of CAA (CAA) in water. Dissolve 514 mg CAA in a falcon tube and fill up to 10 ml with water. Prepare 200 µl aliquots and store the alkylating agent at −20 °C.
5. 40 mM Tris–HCl solution, pH 7.6: Prepare a stock solution of 2 M Tris–HCl by dissolving 2.42 g Tris in 5 ml water. Adjust the pH to 7.6 using a 5 M HCl solution and fill up to 10 ml with water. Take 200 µl of the 2 M stock solution and fill up to 10 ml with water.
6. Trypsin stock solution: Prepare a stock solution of 1 µg/µl trypsin (sequencing grade modified trypsin, Promega) in 50 mM acetic acid, store at −80 °C.
7. Sep-Pak C-18 peptide purification: 50 mg Sep-Pak cartridges (Waters Corp., Eschborn, Germany). Solvent A: 0.07% (v/v) TFA in water. Dilute 70 µl of 100% TFA in 99.93 ml water. Solvent B: 50% (v/v) ACN, 0.07% (v/v) TFA in water. Prepare 50 ml ACN and 70 µl of TFA and dilute in 49.93 ml water. Store at 4 °C.
8. Vacuum manifold for Sep-Pak desalting.

2.2 Phosphopeptide Enrichment by Fe-IMAC Column Chromatography

To avoid column clogging, solvents should be degassed and vacuum-filtered prior to use.

1. Formic acid solvent: 100 ml of 0.1% FA (v/v) in water.
2. IMAC charging solvent: 250 ml of 25 mM $FeCl_3$ (reagent grade, Sigma-Aldrich, Product No. 157740) in 100 mM acetic acid. Put 200 ml water into a cylinder, add 1.43 ml acetic acid and fill up to 250 ml with water. Add 1.014 g $FeCl_3$ and use a magnetic stirrer to dissolve $FeCl_3$. Leave the solution stirring for 30 min and vacuum filter it afterwards to remove insoluble $FeCl_3$ remnants. Store the filtered solution at 4 °C.
3. IMAC stripping solvent: 250 ml of 50 mM EDTA in water, pH 8. Use a magnetic stirrer to dissolve 3.653 g EDTA in 200 ml water. Add 5 M NaOH solution until the EDTA is dissolved, adjust the pH to 8 and fill up to 250 ml with water. Vacuum filter the solvent to remove insoluble EDTA remnants and store at 4 °C.
4. IMAC loading solvent: 1 l of 0.07% (v/v) TFA in 30% (v/v) ACN. Always prepare fresh.
5. IMAC elution solvent: 100 ml of 0.3% (v/v) NH_4OH in water (caution, *see* **Note 1**).
6. ProPack IMAC-10 column: 10 µm, nonporous, polymeric beads; 4 mm inner diameter × 50 mm length (Thermo Scientific, Product No. 063276).

7. HPLC system with the following requirements: flow rates ranging from 0.1 ml/min to 4 ml/min; 0.5–1 ml sample loop; UV detector set to read fixed wavelengths of 214 nm and 280 nm; stable at pH 2–12.

2.3 Desalting of the Fe-IMAC Eluate

1. Preparation of the C-18 StageTips: A detailed instruction on how to construct micro-column tips is provided in Chapter 8. Prepare a StageTip containing five C-18 disks (Empore Octadecyl C-18 47 mm Solid Phase Extraction Disks #2215, 3 M Purification, Eagan, MN, USA).
2. Desalting solvents: Solvent A: prepare 5 ml of 0.07% (v/v) TFA in water. Solvent B: prepare 5 ml of 0.07% (v/v) TFA and 60% (v/v) ACN in water.

2.4 LC-MS/MS and Data Analysis

1. 50 mM citric acid and 1% (v/v) FA in water. Dissolve 2.1 g of citric acid in 9.9 ml of water and add 100 μl of FA.
2. LC-MS/MS: nano-HPLC setup coupled to a high resolution mass spectrometer. Here, we use an Eksigent NanoLC-Ultra 1D+ (Eksigent, Dublin, CA) coupled to a Q Exactive Plus mass spectrometer (Thermo Scientific, Bremen). LC-trap column: 75 μm×2 cm, packed with 5 μm Reprosil-Pur ODS-3 C-18 material (Dr. Maisch, Ammerbuch, Germany). Analytical column: 75 μm×42 cm, packed with 3 μm Reprosil-Gold C-18 material (Dr. Maisch, Ammerbuch, Germany).
3. Nano-HPLC solvents: Loading solvent: 0.1% (v/v) FA in water. Solvent A: 0.1% (v/v) FA and 5% (v/v) DMSO [18] in water. Solvent B: 0.1% (v/v) FA and 5% (v/v) DMSO in ACN.
4. Data analysis: Freely available MaxQuant [19] software package (e.g., version 1.5.2.8) with the integrated search engine Andromeda [20]. Protein sequence database in FASTA format (e.g., UniprotKB).
5. Spreadsheet editor or the freely available Perseus software package.

3 Methods

3.1 Preparation of Proteome Digests for Phosphopeptide Analysis

A schematic overview of the experimental steps covered in this protocol is provided in Fig. 1.

1. Seed HeLa cells under sterile conditions in RPMI 1640 medium supplemented with 10% FCS. Use 30 ml medium for 150 mm cell culture dishes. Grow the cells to 80% confluency under humidified atmosphere, 5% CO_2 at 37 °C. For lysis, place the cell culture dishes on ice or work at 4 °C. Wash cells two times with cold PBS. Use a pipette to aspirate residual PBS from cell culture plates after the final washing step (*see* **Note 2**).

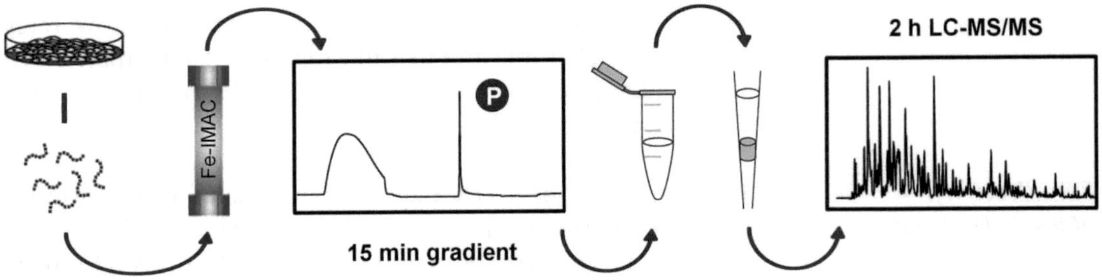

Fig. 1 Experimental workflow for comprehensive phosphopeptide enrichment depicting cell culture and lysis, phosphopeptide enrichment using the Fe-IMAC column, desalting of enriched phosphopeptides, and LC-MS/MS analysis

2. Add 550 μl of precooled lysis buffer to cell culture dishes. Carefully pan the dish to distribute the lysis buffer evenly over all cells and incubate the dishes for 10 min on ice. Use a cell scraper to mix cell lysate in the cell culture plates. Transfer cell lysates to 1.5 ml reaction vessels and spin down insoluble debris for 20 min at 21,000 × *g* and 4 °C. Transfer supernatant to a new tube. Use a Bradford assay (or similar photometric assay) to determine the protein concentration. Store lysates at −80 °C or continue directly.
3. Use 1 mg of protein lysate. To reduce disulfide bonds, 1 M DTT stock solution is added to a final concentration of 10 mM (1:100 dilution). Incubate in a thermoshaker for 40 min at 37 °C and 700 rpm.
4. For alkylation of cysteine residues, add 550 mM CAA to a final concentration of 50 mM (1 : 10 dilution). Carefully invert the sample once and incubate for 30 min at room temperature in the dark.
5. Dilute sample with four volumes of 40 mM Tris–HCl (pH 7.6) to decrease urea concentration to 1.2 M (*see* **Note 3**). Add trypsin in a protease-to-protein ratio of 1:100 (w/w) and pre-digest 4 h in a thermo-shaker at 37 °C and 700 rpm. Add another 1:100 (w/w) trypsin and incubate the digestion mixture over night in a thermo-shaker at 37 °C and 700 rpm.
6. Cool samples down to room temperature and acidify the sample by addition of 0.5 % (v/v) TFA. Centrifuge acidified peptides at 5000 × *g* to precipitate insoluble matter. Use 50 mg Sep-Pak columns and place them into a vacuum manifold (*see* **Note 4**). Prime Sep-Pak columns by adding 1 ml of solvent B. Equilibrate column by adding 2 × 1 ml of solvent A. Transfer the supernatant of the acidified sample to the column and load slowly (*see* **Note 5**). Reapply flow-through a second time and discard it afterwards. Wash the column with 3 × 1 ml solvent A. Elute peptides with 2 × 150 μl solvent B into a 1.5 ml reaction vessel. Adjust the volume to 0.5 ml by

addition of solvent A (*see* **Note 6**). The Sep-Pak eluate has a final concentration of 30% ACN and can thus be directly applied to phosphopeptide enrichment (*see* **Note 7**) or alternatively stored at −80 °C.

3.2 Phosphopeptide Enrichment by Fe-IMAC Column Chromatography

For first time use, the column can be directly charged with $FeCl_3$ solvent (it does not have to be stripped) (*see* **Note 8**). The column is usually operated below 1000 psi. Column stripping and charging should be repeated after 20 enrichments or in case the column has not been used for more than one week (*see* **Note 9**).

1. Column stripping: Connect the IMAC column to your HPLC system and rinse it with ultrapure water (1 ml/min, 10 ml). Inject 1 ml of IMAC stripping solvent into the sample loop and let it run through the column (1 ml/min). After 1 min, inject another 1 ml IMAC stripping solvent. Repeat this step eight more times. Make sure that the sample loop is flushed with water afterwards. Rinse the column with ultrapure water (2 ml/min, 5 ml).
2. Column charging: Inject 1 ml of IMAC charging solvent into the sample loop and let it run over the column (0.2 ml/min). After 5 min, inject another 1 ml of IMAC charging solvent. Repeat this step four more times. Wash the sample loop with 3 × 1 ml EDTA to scavenge remaining Fe^{3+} ions (from both syringe and sample loop) and 10 × 1 ml water to get rid of residual EDTA. Rinse the column with 50 ml FA solvent to wash away unbound Fe^{3+} ions (2 ml/min).
3. Fe-IMAC enrichment: Connect the IMAC loading solvent and IMAC elution solvent to the HPLC system. Flush the column with 5 ml of 50% IMAC elution solvent (3 ml/min). Re-equilibrate the column with 20 ml IMAC loading solvent (3 ml/min) (*see* **Note 10**). Perform a standard enrichment to ensure proper charging (15 min gradient, *see* Table 1, *see* **Note 10**).

Table 1
Settings for a 15 min Fe-IMAC column enrichment, displaying the programmed time, the flow (in ml/min), and percentages IMAC elution solvent used

Time [min]	Flow [ml/min]	IMAC elution solvent [%]
0–0.1	1	0
0.1–5.1	0.2	0
5.1–6.72	3	0 - 16
6.72–11.68	0.55	16 - 26.25
11.68–12.35	3	26.25 - 50
12.35	0	50 - 0
12.35–15.02	3	0

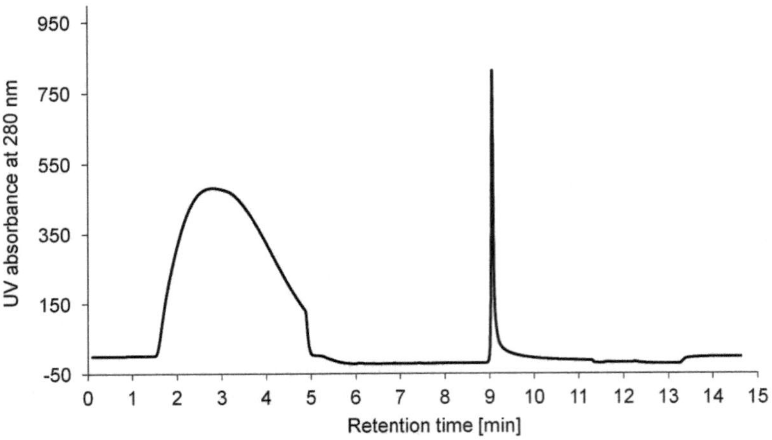

Fig. 2 Typical chromatogram of a 15 min Fe-IMAC column enrichment using 1 mg of HeLa digest. The first peak (retention time between 2 and 5 min.) contains the non-phosphorylated peptides. The second peak (retention time around 9 min) contains the phosphorylated peptides

4. Inject the desalted sample (*see* **Note 11**) dissolved in 0.5 ml IMAC loading solvent. Prepare two 1.5 ml reaction vessels for collecting flow-through and phosphopeptide eluate. Start the 15 min gradient (the gradient setup is shown in Table 1). Monitor the absorption at 280 nm (*see* Fig. 2) and collect 1.3 ml of the flow-through in a 1.5 ml eppendorf tube. The IMAC eluate peak containing the phosphopeptides is collected in another 1.5 ml eppendorf tube (total volume of 1 ml, *see* Fig. 2 and Table 1). Freeze both eppendorf tubes at −80 °C and subsequently dry the samples down using a vacuum centrifuge or a lyophilizer. A typical 15 min Fe-IMAC enrichment chromatogram is depicted in Fig. 2.
5. There is no need to run blanks in between consecutive enrichments. The carryover is minimal. You can reinject the Fe-IMAC column flow-through in order to monitor (absorption at 214 and 280 nm) if all the phosphopeptides were properly depleted.

3.3 Desalting of the Fe-IMAC Eluate

Although most of the ammonia will evaporate during the vacuum centrifugation/lyophilization step, residual ammonia salts might remain. Hence, it is recommended to desalt Fe-IMAC eluates using C-18 StageTips [21]. Pass all liquids through the tips by centrifugation (~800 × *g*, room temperature; *see* **Note 12**).

1. Dissolve the dried sample in 250 μl of solvent A and keep the sample on ice while the StageTips are prepared. Check the pH of the dissolved peptide solution and, if required, adjust it to pH 2 using FA.
2. Sequentially activate the tips using 250 μl of MeOH, 250 μl of solvent B and equilibrate with 250 μl of solvent A. Empty the eppendorf tube in between.

3. Load the dissolved sample and reapply the flow-through. Discard the flow-through afterwards and wash the column with 250 μl solvent A.
4. Use 40 μl of solvent B to elute the peptides of the C-18 material. Transfer the eluate into a 96-well plate and dry the sample down using a vacuum centrifuge/lyophilizer. At this point, the plate can be stored at −20 °C.

3.4 LC-MS/MS and Data Analysis

1. Reconstitute desalted IMAC eluate in 20 μl of 1% FA in 50 mM citrate (*see* **Note 13**).
2. Perform LC-MS/MS measurements by coupling an Eksigent NanoLC-Ultra 1D+ to a Q Exactive Plus instrument. 5 μl of IMAC enriched phosphopeptides corresponding to the enrichment from 250 μg peptide digest are delivered to the trap column at a flow rate of 5 μl/min in loading solvent (0.1% FA in water). During 10 min of sample loading chelated iron is washed out while phosphopeptides are retained.
3. Transfer peptides to the analytical column and separate at a flow rate of 300 nl/min using a 110 min gradient from 0 to 27% solvent B (0–2 min: 0% B; 2–100 min: 0–27% C; 100–101 min 27–80% B, 101–105 min: 80% B, 105–106 min: 80–0% B, 106–110 min: 0% B) (*see* **Note 14**).
4. Operate the Q Exactive Plus in data-dependent mode, automatically switching between MS1 and MS2. Acquire full-scan MS spectra at 360–1300 m/z, 70,000 resolution with acquisition gain control (AGC) target value of 3×10^6 charges and maximum injection time of 100 ms for MS1. Allow up to 20 precursor ions for HCD fragmentation in tandem mass spectra. Acquire MS2 spectra at 17,500 resolution, AGC target value of 1×10^5 charges and max injection time of 50 ms (*see* **Note 15**). Set precursor ion isolation width to 1.7 Th and dynamic exclusion to 20 s. Figure 3 shows an expected MS1 base peak intensity chromatogram of the LC-MS/MS measurement.

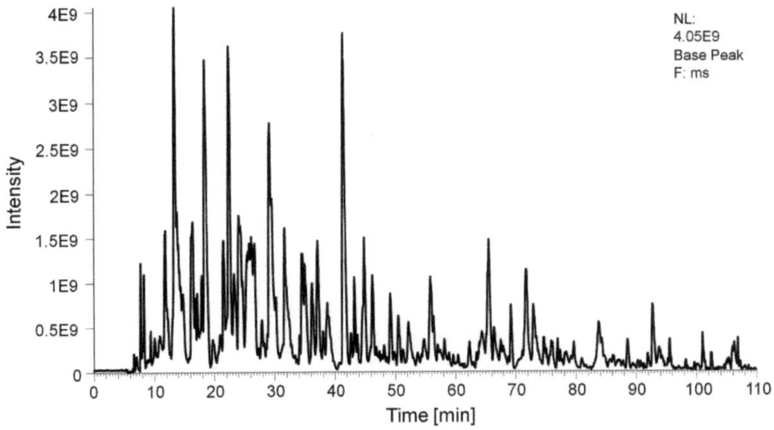

Fig. 3 Typical base peak intensity chromatogram of a desalted Fe-IMAC phosphopeptide eluate (1/4 of a 1 mg HeLa enrichment) measured on a Q Exactive Plus

5. Analyze data using a proteomics software capable of label-free quantification. All results shown in this chapter are based on peptide identifications by search of raw data against the UniProtKB human database, version July 2013 (88,354 sequences) using the freely available MaxQuant version 1.5.2.8 and its built-in Andromeda search engine. Parameters used are specified in Table 2.
6. To facilitate any kind of data analysis, filter the MaxQuant evidence.txt or the phospho(STY)sites.txt output file to remove reverse sequences and potential contaminants. To determine the selectivity of the phosphopeptide enrichment, the reported number of peptides annotated with a phosphorylation sites is divided by the total number of identified sequences. The intensity-based selectivity is acquired similarly by dividing the summed intensity of phosphorylated peptides by the total intensity. Filter the "Modified Sequence" column for duplicates to remove redundancies and obtain the number of

Table 2
Group-specific and global parameters for data analysis using MaxQuant version 1.5.2.8

Group-specific parameters	
Type	Standard
Label	No
Variable modifications	Acetyl (Protein N-term), Oxidation (M), Phospho (STY)
Digestion mode	Specific (Trypsin/P)
Max. missed cleavages	2
Main search peptide tolerance	5 parts per million (ppm)
Max. number of modifications per peptide	5
Global parameters	
Database	UniProtKB
Fixed modifications	Carbamidomethyl
PSM FDR	0.01
Protein FDR	0.01
Site decoy fraction	0.01
Min. peptide length	7
Min. score for (un)modified peptides	0
Min. delta score for (un)modified peptides	0
MS/MS match tolerance	20 ppm
Second peptide search	Enabled

Table 3
Overview of results typically expected from a single Fe-IMAC enrichment, measured on a 2 h LC-MS/MS gradient on a Q Exactive Plus. The Fe-IMAC column eluate was reconstituted in 20 μl of 50 mM Citrate, 1 % FA and 5 μl were injected

Phosphopeptides (MaxQuant—evidence.txt)	
Identified unique phosphopeptides	10089
Quantified unique phosphopeptides	9392
Mono phosphorylated	8392 (83%)
Multiply phosphorylated	1697 (17%)
Identification-based phosphopeptide selectivity	81%
Intensity-based phosphopeptide selectivity	94%
Phosphorylation sites (MaxQuant—Phospho(STY)sites.txt)	
Identified phosphorylation sites	8973
Quantified phosphorylation sites	7451
Class I sites (Loc. prob. > 0.75)	6566
pS sites (class I)	5674 (86%)
pT sites (class I)	727 (11%)
pY sites (class I)	165 (3%)

unique phosphopeptides. Similarly, the phospho (STY)sites. txt is used to determine the number of unique and quantifiable sites. Filter for "Localization probability" ≥ 0.75 to obtain the number of class I sites [22].

7. Table 3 shows expected results in terms of unique phosphopeptides and phosphorylation sites obtained from processing 1 mg of HeLa digest (1/4 of the enrichment was subjected to MS measurement) according to the procedures described in this protocol.

4 Notes

1. We noticed that ammonia is evaporating if the NH_4OH bottle is not tightly sealed or if it has been opened and closed repeatedly. Thus, the 0.3 % (v/v) in the IMAC elution solvent refers to a freshly opened bottle of NH_4OH and might have to be adjusted upon prolonged use. This can be delayed by ensuring proper sealing of the bottle or alternatively by working at 4 °C.

2. Aspiration of remaining PBS is important to avoid dilution of the lysis buffer. Low concentrations of the chaotropic reagent urea might result in insufficient protein denaturation.
3. Considering the additional volume of the lysed cells and the remaining PBS, the concentration of urea is reduced from 8 M to roughly 6 M. Hence, a 1:4 dilution is sufficient to reduce the urea concentration down to 1.2 M.
4. Sep-Pak sorbent weight has to be chosen according to the amount of digest you intend to load. As a rule of thumb, the capacity of Sep-Pack cartridges equals 5% of the sorbent weight (e.g., 2.5 mg for the 50 mg sorbent weight cartridges and 10 mg for the 200 mg sorbent weight cartridges).
5. Load the sample slowly onto the Sep-Pak column. Lower the flow rate by adjusting the vacuum at the vacuum manifold. Loading should take at least 10 min to ensure proper binding of the phosphopeptides. Reapplying the flow-through increases recovery.
6. Avoid letting the columns run dry.
7. Adjusting the volume to 0.5 ml using solvent A enables direct sample loading onto the IMAC column without the need to intermediately dry the sample down.
8. Upon first time use, the column should be thoroughly flushed with water followed by 0.1% (v/v) FA solvent. It is advisable to note down the column backpressure for different types of solvents. This facilitates to monitor column performance over time. If the pressure increase is too severe, the column should be exchanged.
9. From our experience, 20 enrichments can be performed without any performance decrease. However, if the column is not used for a longer period of time (conservatively more than one week), column performance seems to be decreasing. If you want to verify or monitor column performance, run a standard before and after your enrichment set.
10. Make sure to properly equilibrate the column after charging. The absorption at 280 nm has to have reached a stable baseline. We recommend to always run a standard or a blank run before you enrich your first sample.
11. The 4 mm I.D. column can be applied to sample amounts ranging between 0.5 and 3 mg. Please be aware that the enrichment efficiency is also cell line dependent as the degree of cellular phosphorylation is highly dynamic and may therefore vary considerably.
12. Using volumes of 250 μl prevents columns from running dry even upon prolonged centrifugation. This is especially beneficial when parallelized fractionation is intended as not all col-

umns run at the same speed. If only one sample is intended to be desalted, the procedure can be accelerated by manually pushing the liquids through the tips using a 5 ml Eppendorf CombiTip. The volumes can be scaled down accordingly (~40 μl for each step).

13. Citric acid acts as a chelating agent for residual Fe^{3+} ions that might co-elute from the Fe-IMAC column. Remaining Fe^{3+} ions can stick to the trap/analytical nano-HPLC columns and deplete phosphopeptides. Ever since we use citrate we have not detected any iron contamination [23]. If you are in doubt, specify iron as a variable modification during data processing and check if any iron-bound peptides are identified.
14. Phosphopeptides are generally more hydrophilic than non-phosphopeptides. Compared to full proteome separations, we use a shallow gradient which leads to a more efficient use of gradient and MS time.
15. Phosphopeptides often show a pronounced loss of the phosphate group upon fragmentation. Because this neutral loss peak may constitute a big part of the fragment ion intensity, backbone fragment ions might get lost. Therefore you may want to evaluate if increased MS2 injection time and or increased MS2 target values (by a factor of 2) leads to an increase in phosphopeptide identifications. The substantially higher identification rate compensates for decreased scan numbers. Moreover, we found that this AGC/injection time increase is beneficial for phosphosite localization.

References

1. Lu Z, Jiang G, Blume-Jensen P et al (2001) Epidermal growth factor-induced tumor cell invasion and metastasis initiated by dephosphorylation and downregulation of focal adhesion kinase. Mol Cell Biol 21:4016–4031
2. Ruprecht B, Lemeer S (2014) Proteomic analysis of phosphorylation in cancer. Expert Rev Proteomics 11:259–267
3. Lemeer S, Heck AJ (2009) The phosphoproteomics data explosion. Curr Opin Chem Biol 13:414–420
4. Pinkse MWH, Uitto PM, Hilhorst MJ et al (2004) Selective Isolation at the femtomole level of phosphopeptides from proteolytic digests using 2D-nanoLC-ESI-MS/MS and titanium oxide precolumns. Anal Chem 76: 3935–3943
5. Kweon HK, Håkansson K (2006) Selective zirconium dioxide-based enrichment of phosphorylated peptides for mass spectrometric analysis. Anal Chem 78:1743–1749
6. Andersson L, Porath J (1986) Isolation of phosphoproteins by immobilized metal (Fe3+) affinity chromatography. Anal Biochem 154: 250–254
7. Posewitz MC, Tempst P (1999) Immobilized gallium(III) affinity chromatography of phosphopeptides. Anal Chem 71:2883–2892
8. Zhou H, Xu S, Ye M et al (2006) Zirconium phosphonate-modified porous silicon for highly specific capture of phosphopeptides and MALDI-TOF MS analysis. J Proteome Res 5:2431–2437
9. Zhou H, Low TY, Hennrich ML et al (2011) Enhancing the identification of phosphopeptides from putative basophilic kinase substrates using Ti (IV) based IMAC enrichment. Mol Cell Proteomics 10:M110.006452
10. Bodenmiller B, Mueller LN, Mueller M et al (2007) Reproducible isolation of distinct, overlapping segments of the phosphoproteome. Nat Methods 4:231–237

11. Tsai C-F, Hsu C-C, Hung J-N et al (2014) Sequential phosphoproteomic enrichment through complementary metal-directed immobilized metal ion affinity chromatography. Anal Chem 86:685–693
12. Thingholm TE, Jensen ON, Robinson PJ et al (2008) SIMAC (sequential elution from IMAC), a phosphoproteomics strategy for the rapid separation of monophosphorylated from multiply phosphorylated peptides. Mol Cell Proteomics 7:661–671
13. Ruprecht B, Koch H, Medard G et al (2015) Comprehensive and reproducible phosphopeptide enrichment using iron immobilized metal ion affinity chromatography (Fe-IMAC) columns. Mol Cell Proteomics 14:205–215
14. Larsen MR, Thingholm TE, Jensen ON et al (2005) Highly selective enrichment of phosphorylated peptides from peptide mixtures using titanium dioxide microcolumns. Mol Cell Proteomics 4:873–886
15. Kettenbach AN, Gerber SA (2011) Rapid and reproducible single-stage phosphopeptide enrichment of complex peptide mixtures: application to general and phosphotyrosine-specific phosphoproteomics experiments. Anal Chem 83:7635–7644
16. Li Q, Ning Z, Tang J et al (2009) Effect of peptide-to-TiO2 beads ratio on phosphopeptide enrichment selectivity. J Proteome Res 8:5375–5381
17. Zhou H, Di Palma S, Preisinger C et al (2013) Toward a comprehensive characterization of a human cancer cell phosphoproteome. J Proteome Res 12:260–271
18. Hahne H, Pachl F, Ruprecht B et al (2013) DMSO enhances electrospray response, boosting sensitivity of proteomic experiments. Nat Methods 10:989–991
19. Cox J, Mann M (2008) MaxQuant enables high peptide identification rates, individualized p.p.b.-range mass accuracies and proteome-wide protein quantification. Nat Biotechnol 26:1367–1372
20. Cox J, Neuhauser N, Michalski A et al (2011) Andromeda: a peptide search engine integrated into the MaxQuant environment. J Proteome Res 10:1794–1805
21. Rappsilber J, Mann M, Ishihama Y (2007) Protocol for micro-purification, enrichment, pre-fractionation and storage of peptides for proteomics using StageTips. Nat Protoc 2:1896–1906
22. Olsen JV, Blagoev B, Gnad F et al (2006) Global, in vivo, and site-specific phosphorylation dynamics in signaling networks. Cell 127:635–648
23. Winter D, Seidler J, Ziv Y et al (2009) Citrate boosts the performance of phosphopeptide analysis by UPLC-ESI-MS/MS. J Proteome Res 8:418–424

Chapter 6

Full Membrane Protein Coverage Digestion and Quantitative Bottom-Up Mass Spectrometry Proteomics

Joseph Capri and Julian P. Whitelegge

Abstract

A true and accurate bottom-up global proteomic measurement will only be achieved when all proteins in a sample can be digested efficiently and at least some peptides recovered on which to base an estimate of abundance. Integral membrane proteins make up around one-third of the proteome and require specialized protocols if they are to be successfully solubilized for efficient digestion by the enzymes used in bottom-up proteomics. The protocol described relies upon solubilization using the detergents sodium deoxycholate and lauryl sarcosine with heating to 95 °C. A subset of peptides is purified by reverse-phase solid-phase extraction and fractionated by strong-cation exchange prior to nano-liquid chromatography with data-dependent tandem mass spectrometry. For quantitative proteomics experiments a protocol is described for stable-isotope coding of peptides using dimethylation of primary amines allowing for three-way sample multiplexing.

Key words Trypsin, Electrospray ionization, Proteome, StageTip, Dimethylation, Phase transfer

1 Introduction

Integral membrane proteins make up around one-third of the global proteome and a larger proportion of drug targets because of their wide-ranging critical functions in cell biology. A global proteomics method for measuring changes in protein expression level in different conditions must fully cover the membrane proteome. Researchers have battled this challenge with a variety of approaches over the years and detergent solubilization prior to proteolytic digestion has emerged as the most efficacious. Central to the success of the method lies efficient digestion of integral membrane proteins enabling recovery of loop-region peptides for quantitative bottom-up proteomics with transmembrane domains typically ignored since they are lost during peptide work up prior to analysis as a result of their hydrophobic properties. Analysis of membrane protein posttranslational modifications (PTMs) is accommodated

Lucio Comai et al. (eds.), *Proteomics: Methods and Protocols*, Methods in Molecular Biology, vol. 1550,
DOI 10.1007/978-1-4939-6747-6_6, © Springer Science+Business Media LLC 2017

when they are localized to peptides recovered from loop regions. Modifications in transmembrane domains require alternative approaches such as top-down mass spectrometry [1, 2].

Two approaches to detergent solubilization have yielded the most complete coverage of the integral membrane proteome; filter-assisted sample preparation (FASP) was described by the Liebler group [3] and refined by Mann's group [4] relying upon sodium dodecylsulfate (SDS) for solubilization with subsequent removal of detergent through a filter using urea followed by enzymatic digestion. In-solution digestion (ISD) relies upon detergent solubilization using detergents that are tolerated by the proteolytic enzyme used for digestion with subsequent removal of detergent via acid precipitation or phase-transfer after digestion [5]. Masuda's work first described the use of sodium deoxycholate–lauryl sarcosine mixtures with heating to achieve efficient digestion of membrane proteins [5]. Loo's group combined FASP with ISD replacing urea with deoxycholate to remove SDS [6]. It is generally agreed that use of FASP results in some selective sample loss on the filter used and the ISD technique is gaining prevalence [7, 8].

The ISD protocol described herein is based upon the efficiency of membrane protein solubilization with 0.5% sodium deoxycholate and 12 mM lauryl sarcosine, with heating to 95 °C [5, 8]. This mixture is highly denaturing at high-temperature but becomes non-denaturing at 37 °C for trypsin digestion.

Occasionally a protein's amino acid sequence results in failure to detect a membrane protein. Acidic proteins may lack suitable basic residues for trypsin cleavage, while basic proteins may have too many. In such cases a second proteolytic enzyme with different specificity must be used for complete coverage proteomics.

2 Materials

Prepare all solutions using ultrapure water (prepared by purifying deionized water to attain a conductivity of 18 MΩ cm at 25 °C) and analytical grade reagents. All reagents should be stored at room temperature unless otherwise noted.

1. Cell lysis buffer: 0.5% sodium deoxycholate, 12 mM N-laurylsarcosinate sodium, and 50 mM ammonium bicarbonate pH 8.5 with 89 μg/mL of Sigma Protease Inhibitor Cocktail powder (catalog number: P2714) Store at 4 °C. Only store for a maximum of 2 weeks. Lysis buffer that does not contain protease inhibitor and can be stored at room temperature for several months.
2. Reduction buffer stock solution: 1 M tris(2-carboxyethyl) phosphine in water.
3. Alkylation stock solution: 1 mM iodoacetamide in water.

4. Bicinchoninic acid protein assay (Pierce).
5. Digestion buffer: 50 mM ammonium bicarbonate, pH 8.5.
6. Sequencing grade trypsin.
7. Trifluoroacetic acid.
8. 200 mg tC18 Sep-Pak cartridges (Waters).
9. HPLC grade methanol.
10. Sep-Pak elution buffer: 80% acetonitrile with 0.1% trifluoroacetic acid.
11. Sep-Pak loading buffer: 2% acetonitrile with 0.1% trifluoroacetic acid.
12. Dimethyl labeling solutions:

    Prewash: 250 mM 2-(*N*-morpholino)ethanesulfonic acid pH 5.5

    Light: 60 mM sodium cyanoborohydride, 0.4% formaldehyde, and 250 mM 2-(*N*-morpholino)ethanesulfonic acid pH 5.5

    Intermediate: 60 mM sodium cyanoborohydride, 0.4% formaldehyde (CD_2O), and 250 mM 2-(*N*-morpholino)ethanesulfonic acid pH 5.5

    Heavy: 60 mM sodium cyanoborodeuteride, 0.4% formaldehyde ($^{13}CD_2O$), and 250 mM 2-(*N*-morpholino)ethanesulfonic acid pH 5.5.
13. 15 mL Conical vials.
14. Pierce Quantitative Colorimetric Peptide Assay (Thermo, catalog #23275).
15. StageTip loading buffer: 2% acetonitrile with 0.5% acetic acid.
16. StageTip elution buffer: 80% acetonitrile with 0.5% acetic acid.
17. Strong cation exchange (SCX) elution buffers: 30% acetonitrile with 0.5% acetic acid and increasing amounts of ammonium acetate (NH_4AcO): (1) 25 mM NH_4AcO, (2) 35 mM NH_4AcO, (3) 50 mM NH_4AcO, (4) 70 mM NH_4AcO, (5) 100 mM NH_4AcO, (6) 150 mM NH_4AcO, (7) 350 mM NH_4AcO, and (8) 750 mM NH_4AcO.
18. C18-SCX StageTips are made according to Rappsilber et al [9]. Briefly, a Hamilton 16 gauge blunt-ended needle is used to puncture chromatographic filters and seated into a P200 pipet tip. Two SCX frits (Empore™ Cation 47 mm Extraction Disc, Model 2251, Millipore) are seated first followed by two C18 frits (Empore™ C18 47 mm Extraction Disc, Model 2215, Millipore) seated above.
19. For every C18-SCX StageTip, eight C18 StageTips need to be made in order to desalt the SCX fractions. These C18 StageTips are made similarily to **step 18**, except only one C18 frit is seated in the P200 pipet tip.

20. DDA capable mass spectrometer with associated nano flow HPLC, we use a Thermo Orbitrap XL with an Eksigent 2D nanoLC and Spark autosampler.
21. Nano capillary columns, 25 cm × 75 μm, packed with C18 (300Å, 3 μm particle size) resin.
22. NanoLC, mobile phase A: 3 % acetonitrile, 3 % dimethylsulfoxide, and 0.1 % formic acid.
23. NanoLC, mobile phase B: 97 % acetonitrile, 3 % dimethylsulfoxide, and 0.1 % formic acid.
24. MaxQuant analysis software.

3 Methods

3.1 Cell Lysis and Proteolytic Digest

1. Adherent cells are washed twice directly on plate with ice-cold PBS pH 7.6 (*see* **Note 1**).
2. 0.5 mL per 1×10^7 cells of cell lysis buffer is added directly to plate, cells are scraped with a cell scraper, and lysates are triturated with P1000 pipettor.
3. Cell lysates are transferred to 1.5 mL Eppendorf lo-bind microcentrifuge tubes, water bath sonicated at RT for 5 min, and heated at 95 °C for 5 min.
4. Bicinchoninic acid protein assay (Pierce) is performed to determine protein concentration.
5. Disulfide bridges are reduced with 5 mM tris(2-carboxyethyl) phosphine (final concentration) at RT for 30 min and subsequently alkylated with 10 mM iodoacetamide (final concentration) at RT in the dark for 30 min.
6. Cell lysates are transferred to 15 mL Falcon tubes and diluted 1:5 (v:v) with 50 mM ammonium bicarbonate pH 8.5.
7. Proteins are digested with sequencing grade trypsin 1:100 (enzyme:protein by mass) for 4 h at 37 °C under gentle agitation followed by a second aliquot of trypsin 1:100 (enzyme:protein) overnight at 37 °C under gentle agitation.
8. Samples are acidified with 0.5 % trifluoroacetic acid (final concentration), vortexed rapidly for 5 min, and centrifuged at 16,000 × *g* for 5 min at RT to pellet sodium deoxycholate.
9. Transfer supernatant to a new tube and proceed to peptide desalting. If needing to store for up to a week, keep peptide samples at 4 °C, otherwise freeze at −80 °C.

3.2 Peptide Desalting and Reductive Dimethylation

1. 200 mg tC18 Sep-Pak cartridges (Waters) are wetted with 2 mL of 100 % methanol, with solvent pulled through the cartridge using a vacuum manifold. It is critical to stop the flow before all solvent has passed through to prevent any air from

entering the packing material (this applies for all subsequent steps for Sep-Paks). Leaving ~100 μL of solvent above the packing material is ideal.

2. 1 mL of Sep-Pak elution buffer is passed under vacuum and repeated 1×.
3. 1 mL of Sep-Pak loading buffer is passed under vacuum and repeated 2×.
4. Peptide digests are loaded onto Sep-Paks via gravity.
5. 1 mL of 250 mM 2-(*N*-morpholino)ethanesulfonic acid pH 5.5 is passed under vacuum [10].
6. 3 mL of the respective dimethyl labeling solution is passed through. This process needs to take at least 10 min to ensure complete labeling. This can be accomplished by passing solution by gravity.
7. 1 mL of Sep-Pak loading buffer is passed under vacuum and repeated 1×.
8. 1 mL of Sep-Pak elution buffer is passed under gravity, collected in 15 mL conical vial, and repeated 1×.
9. Dimethyl-labeled peptide samples are lyophilized to dryness.

3.3 Strong Cation Exchange Fractionation

For each of the following steps, C18-SCX StageTips will be denoted (S) and C18 StageTips will be denoted (C) if the following step is to be performed on that particular StageTip. Unless otherwise noted, all solvent can be discarded properly (*see* **Note 2**).

1. Pierce Quantitative Colorimetric Peptide Assay is performed.
2. Peptide samples are reconstituted in StageTip loading buffer at a concentration of 0.2 mg/mL.
3. Light, medium, and heavy labeled peptides are mixed 1:1:1.
4. (S,C) StageTips are wetted with 20 μL of 100% methanol, pushing solvent through by applying pressure with hand syringe. It is critical to prevent air from entering frits, leaving ~1–2 μL above frits.
5. (S,C) 20 μL of StageTip elution buffer is passed through using pressure from hand syringe.
6. (S,C) 20 μL of StageTip loading buffer is passed through using pressure from hand syringe.
7. (C) 100 μL of 0.5% acetic acid is deposited into C18 StageTip and set aside for later use.
8. (S) 20 μL of SCX elution buffer 8 is passed through using pressure from hand syringe.
9. (S) 20 μL of StageTip loading buffer is passed through using pressure from hand syringe.
10. (S) 32 μg of differential-labeled peptides are loaded to C18-SCX StageTip using pressure from hand syringe.

11. (S) 20 μL of StageTip loading buffer are passed through using pressure from hand syringe.
12. (S) 20 μL of StageTip elution buffer is passed through using pressure from hand syringe.
13. (S) 20 μL of 30% acetonitrile with 0.5% acetic acid is passed through using pressure from hand syringe.
14. (S) 20 μL of SCX elution buffer 1 is passed through using pressure from hand syringe and collected in 100 μL of pre-deposited 0.5% acetic acid above pre-conditioned C18 StageTip from **step 7**. This is repeated for SCX elution buffers 2–8 and collected into separate pre-conditioned C18 StageTips from **step 7**.
15. (C) SCX fractions are pipetted up and down to mix with pre-deposited 100 μL of 0.5% acetic acid and then passed through C18 using pressure from hand syringe.
16. (C) 20 μL of StageTip loading buffer is passed through using pressure from hand syringe.
17. (C) 20 μL of StageTip elution buffer is passed through using pressure from hand syringe and collected in 1.5 mL microcentrifuge tubes.
18. Peptide fractions are concentrated in vacuum centrifuge to ~2 μL, typically 4 min.
19. Concentrated peptide fractions are reconstituted with 10 uL of 2% acetonitrile with 0.1% formic acid and transferred to autosampler injection vials.

3.4 NanoLC-RP-MS/MS

5 μL of each peptide fraction is analyzed using 180 min data-dependent reverse-phase nLC-MS/MS on Thermo Orbitrap XL equipped with Eksigent Spark autosampler, Eksigent 2D nanoLC, and Thermo nano-ESI source (*see* **Note 3**).

1. Samples are loaded onto a laser-pulled reverse-phase nanocapillary (75 μm I.D., 360 μm O.D. × 25 cm length) with C18 (300Å, 3 μm particle size) for 30 min with mobile phase A (3% acetonitrile, 3% dimethylsulfoxide, and 0.1% formic acid) at 600 nL/min.
2. Peptides are analyzed over 180 min linear gradient to 100% mobile phase B (97% acetonitrile, 3% dimethylsulfoxide, and 0.1% formic acid) at 300 nL/min.
3. Electrospray ionization and source parameters are as follows: spray voltage of 2.2 kV, capillary temperature of 200 °C, capillary voltage at 35 V, and tube lens at 90 V.
4. Data-dependent MS/MS is operated using the following parameters: full MS from 400 to 1700 m/z with 60,000 resolution at 400 m/z and target ion count of 3×10^5 or fill time of 700 ms with lock-mass at 401.922718 m/z, and 12 MS/MS with charge-state screening excluding +1 and unassigned

charge states for ions surpassing 6000 counts, target ion count of 5000 or fill time of 50 ms, CID collision energy of 35, and dynamic exclusion of 30 s.

5. Raw data is searched against respective species Uniprot fasta database using MaxQuant 1.5.3.30 with standard preset search parameters. The search parameters are as follows: 3-plex dimethyl labeling to lysine and peptide N-terminus, trypsin cleavage allowing up to two missed cleavages, fixed modification of carbamidomethyl to cysteines, variable modifications of acetylation to protein N-terminus and methionine oxidation, 10 ppm and 0.5 Da mass errors for Full MS and MS/MS, respectively, 1 % false-discovery rate on peptide and protein identifications, and peptide match between run feature with 1.5 min time window.

4 Notes

1. If proteomic samples are derived from tissue organs or organisms with cell walls, a bead beater should be utilized to homogenize the sample in cell lysis buffer. After homogenization, proceed with Subheading 3.1, **step 3** and centrifuge at 16,000 × *g* for 5 min at room temperature to pellet insoluble material.
2. Depending on sample complexity, more or less strong cation exchange fractions can be performed.
3. Online reverse-phase chromatography, in terms of time and gradient, will need to be optimized depending on sample being analyzed and type of mass spectrometer acquiring data.

References

1. Ryan CM, Souda P et al (2010) Post-translational modifications of integral membrane proteins resolved by top-down Fourier transform mass spectrometry with collisionally activated dissociation. Mol Cell Proteomics 9:791–803
2. Whitelegge JP (2013) Integral membrane proteins and bilayer proteomics. Anal Chem 85:2558–2568
3. Manza LL, Stamer SL et al (2005) Sample preparation and digestion for proteomic analyses using spin filters. Proteomics 5: 1742–1745
4. Wiśniewski JR, Zougman A, Mann M (2009) Combination of FASP and StageTip-based fractionation allows in-depth analysis of the hippocampal membrane proteome. J Proteome Res 8:5674–5678
5. Masuda T, Tomita M, Ishihama Y (2008) Phase transfer surfactant-aided trypsin digestion for membrane proteome analysis. J Proteome Res 7:731–740
6. Erde J, Loo RR, Loo JA (2014) Enhanced FASP (eFASP) to increase proteome coverage and sample recovery for quantitative proteomic experiments. J Proteome Res 13:1885–1895
7. Kulak NA, Pichler G et al (2014) Minimal, encapsulated proteomic-sample processing applied to copy-number estimation in eukaryotic cells. Nat Methods 11:319–324
8. León IR, Schwämmle V et al (2013) Quantitative assessment of in-solution digestion efficiency identifies optimal protocols for unbiased protein analysis. Mol Cell Proteomics 12:2992–3005
9. Rappsilber J, Mann M, Ishihama Y (2007) Protocol for micro-purification, enrichment, pre-fractionation and storage of peptides for proteomics using StageTips. Nat Protoc 2:1896–1906
10. Wilson-Grady JT, Haas W, Gygi SP (2013) Quantitative comparison of the fasted and re-fed mouse liver phosphoproteomes using lower pH reductive dimethylation. Methods 61:277–286

Chapter 7

Hydrophilic Strong Anion Exchange (hSAX) Chromatography Enables Deep Fractionation of Tissue Proteomes

Benjamin Ruprecht, Dongxue Wang, Riccardo Zenezini Chiozzi, Li-Hua Li, Hannes Hahne, and Bernhard Kuster

Abstract

The bottom-up proteomic analysis of cell line and tissue samples to a depth > 10,000 proteins still represents a considerable challenge because of the sheer number of peptides generated by proteolytic digestions and the high dynamic range of protein expression. As a result, comprehensive protein coverage requires multidimensional peptide separation. Recently, off-line hydrophilic strong cation exchange (hSAX) chromatography has proven its merits for high resolution separation of peptides due to its high degree of orthogonality to reversed-phase liquid chromatography. Here we describe the use of hSAX for the deep analysis of tissue proteomes. The protocol includes optimized sample preparation steps (lysis with the aid of mechanical disruption, one-step disulfide bridge reduction and alkylation), setup and operation of hSAX columns and gradients, desalting of hSAX fractions prior to LC-MS/MS analysis, and suggestions for the choice of data acquisition parameters and data analysis using MaxQuant. Application of the protocol to the fractionation of 300 μg human brain tissue digest led to the identification of more than 100,000 unique peptide sequences representing over 10,195 proteins and 9,500 genes in 3 days of measurement time on a Q Exactive Plus mass spectrometer.

Key words Proteomics, Deep fractionation, Chromatography, Strong anion exchange, Tissue proteomics

Abbreviations

ACN	Acetonitrile
AGC	Acquisition gain control
CAA	Chloroacetamide
DTT	Dithiothreitol
FA	Formic acid
FDR	False discovery rate
HCl	Hydrochloric acid
HPLC	High-performance liquid chromatography
hSAX	Hydrophilic strong anion exchange

Lucio Comai et al. (eds.), *Proteomics: Methods and Protocols*, Methods in Molecular Biology, vol. 1550, DOI 10.1007/978-1-4939-6747-6_7, © Springer Science+Business Media LLC 2017

IMAC	Immobilized metal ion affinity chromatography
IT	Injection time
MeOH	Methanol
MS	Mass spectrometer
MS/MS	Tandem mass spectrometry
PBS	Phosphate buffered saline
PSM	Peptide spectrum match
RP	Reversed-phase
SAX	Strong anion exchange
SCX	Strong cation exchange StageTip stop and go extraction tip
TCEP	Tris-(2-carboxyethyl)-phosphin
TFA	Trifluoroacetic acid
Tris	Tris(hydroxymethyl)aminomethane v/v volume/volume
w/w	Weight/weight
ZIC-HILIC	Zwitterionic hydrophilic interaction liquid chromatography

1 Introduction

Despite the fact that tremendous progress has been made recently in mapping out the human proteome [1, 2] and breathtaking advances at all levels of proteomic sample preparation and mass spectrometric instrumentation have been realized [3, 4], identification and quantification of a single proteome to a depth > 10,000 proteins is still a considerable challenge. This has initially been accomplished by two independent research groups in 2011 using the cell lines HeLa [5] and U2OS [6], respectively. At the time, this effort comprised the use of multiple enzymes and the analysis of 72 fractions in 288 h of LC-MS/MS measurement time [5]. Such in-depth proteomic analysis was subsequently extended to a greater number of cell lines [7, 8] but also tissue proteomes [9, 10], the latter of which represents an even greater challenge because protein expression in tissues tends to span a broader dynamic range than cell lines and the analysis is often complicated by the presence of blood, fat or connective tissue. High-resolution two-dimensional peptide separation/fractionation is an efficient means to boost proteome coverage, sequence coverage and quantification performance in bottom-up proteomics experiments. Given that the stationary phase used as the second dimension separation in nano-LC-MS/MS setups is almost exclusively comprised of reversed-phase (RP) material, the first peptide separation dimension should ideally be orthogonal to RP and offer high chromatographic resolution. Many different techniques have been employed for this purpose such as HILIC [11], ZIC-HILIC [12], ERLIC [13], WAX [14], high-pH reversed-phase [15], SAX [16], or SCX [17]. Recently Ritorto et al. have demonstrated the merits of hSAX chromatography [18], which separates peptides primarily based on the number of acidic residues and the stationary phase is characterized by ultralow hydrophobicity. This combination enables orthogonal and robust peptide fractionation with high resolution. We have subsequently adapted the approach for the analysis of phosphoproteomes, taking advantage of the very high retention of negatively charged (phosphate) groups [19].

Here we describe the use of hSAX for the deep characterisation of tissue proteomes. Importantly, with some minor modifications, the procedure is equally applicable to the analysis of cell line digests and phosphoproteome samples. The protocol includes sample preparation steps, hSAX column setup and operation, StageTip desalting of hSAX fractions, data acquisition parameters and instructions on data analysis using MaxQuant [20]. We have optimized lysis conditions (using a bead-beater), disulfide bond reduction (using Tris-(2-carboxyethyl)-phosphin (TCEP), [21]) and alkylation (using chloroacetamide (CAA)) for improved protein extraction and peptide coverage. Applied to the analysis of human brain tissue, the described procedure enabled the identification and quantification of more than 10,000 proteins in 3 days of measurement time on a Q Exactive Plus mass spectrometer starting from only a few mg of tissue.

2 Materials

Unless stated otherwise, all solvents and buffers should be prepared fresh, using ultrapure water and analytical grade reagents. Devices such as centrifuges, vacuum centrifuges/lyophilizer, thermoshaker, or refrigerators (−20/−80 °C) are not explicitly listed. To avoid hSAX column clogging, solvents should be degassed and vacuum-filtered prior to use.

2.1 Preparation of Tissue and Proteome Digest

1. Tissue preparation: Precellys 24 Homogenizer (Bertin Technologies, France), Precellys ceramic kit (1.4 mm "small", 0.5 ml tubes).
2. 550 mM CAA stock solution in water. Dissolve 514 mg CAA in a falcon tube and fill up to 10 ml with water. Prepare 1 ml aliquots and store at −20 °C.
3. 1 M TCEP-HCl stock solution in water: Dissolve 2.87 g of TCEP-HCl in 10 ml of water. Prepare 25 μl aliquots and store at −20 °C.
4. Tissue lysis solution (*see* **Notes 1** and **2**): 50 mM Tris–HCl, pH = 7.6, containing 8 M urea, 10 mM TCEP-HCl, 40 mM CAA, protease and phosphatase inhibitors. A 100 fold stock solution of phosphatase inhibitor cocktail 1, 2, and 3 is commercially available (Sigma Aldrich, Munich, Germany). Prepare a stock solution of 2 M Tris–HCl by dissolving 2.42 g Tris in 5 ml water. Adjust the pH to 7.6 using HCl and fill up to 10 ml with water. Transfer 4.8 g of urea to a 15 ml falcon tube. Add 250 μl Tris–HCl stock solutions, one protease inhibitor

tablet complete mini EDTA-free (Roche, Mannheim, Germany), 100 μl of each phosphatase inhibitor stock solution, 726 μl of the 550 mM CAA stock solution and 10 μl of the 1 M TCEP-HCl stock solution. Fill up to 10 ml with water. Store the tissue lysis solution on ice.

5. Trypsin stock solution: Prepare a stock solution of 1 μg/μl trypsin (sequencing grade modified trypsin, Promega) in 50 mM acetic acid. The trypsin stock can be reused several times and is stored at −80 °C.
6. 50 mM Tris–HCl solution, pH 7.6: Prepare a stock solution of 2 M Tris–HCl by dissolving 2.42 g Tris in 5 ml water. Adjust the pH to 7.6 using HCl and fill up to 10 ml with water. To obtain a 50 mM Tris–HCl solution, transfer 200 μl of the 2 M stock solution to a new 15 ml falcon tube and fill up to 10 ml with water.
7. Sep-Pak C18 peptide purification: 50 mg Sep-Pak cartridges (Waters Corp., Eschborn, Germany). Solvent A: 0.1 % (v/v) FA in water. Solvent B: 60 % (v/v) ACN, 0.1 % (v/v) TFA in water. Store at 4 °C.

2.2 hSAX Chromatography

1. hSAX solvent A: 5 mM Tris, pH 8.5. Fill 900 ml water in a graduated 1 l cylinder. Use a magnetic stirrer to dissolve 0.606 g of Tris(hydroxymethyl)aminomethane (Tris) and adjust the pH to 8.5 with 1 M HCl. Fill up to 1 l with water.
2. hSAX solvent B: 5 mM Tris, 1 M NaCl, pH 8.5. Fill 400 ml water in a graduated 1 l cylinder. Use a magnetic stirrer to dissolve 0.303 g of Tris and 29.221 of sodium chloride and adjust the pH to 8.5 with 1 M HCl. Fill up to 500 ml with water.
3. hSAX analytical column: Dionex IonPac AS24, hydroxide-selective anion-exchange analytical column (2 × 250 mm, Thermo Fisher Scientific, Waltham, USA, Product No. 064153).
4. hSAX guard column: Dionex IonPac AG24, hydroxide-selective anion-exchange guard column (2 × 50 mm, Thermo Fisher Scientific, Waltham, USA, Product No. 064151) (*see* **Note 3**).
5. HPLC system with the following requirements: flow rates ranging from 0.1 ml/min to 1 ml/min; 0.1–1 ml sample loop; UV detector set to read fixed wavelengths of 214 nm and 280 nm (here we used a Dionex Ultimate 3000 system with a flow rate of 0.25 ml/min and a 100 μl sample loop).

2.3 StageTip Desalting of hSAX Fractions

1. StageTip construction: Small, round punch to cut out C-18 disks. 200 μl plastic pipette tip, 1.5 ml reaction vessel, 5 ml Eppendorf CombiTip.
2. Empore Octadecyl C18 47 mm Solid Phase Extraction Disks #2215 (3 M Purification, Eagan, MN, USA).

3. Desalting solvents: Solvent A: 0.1 % (v/v) FA in water. Solvent B: 0.1 % (v/v) FA and 60 % (v/v) ACN in water.

2.4 LC-MS/MS and Data Analysis

1. 0.1 % (v/v) FA in water.
2. LC-MS/MS: nano-HPLC setup coupled to a high resolution mass spectrometer. Here, we used an Eksigent NanoLC-Ultra 1D+ (Eksigent, Dublin, CA) coupled to a Q Exactive Plus mass spectrometer (Thermo Scientific, Bremen, Germany). LC-trap column: 75 μm × 2 cm, packed with 5 μm Reprosil-Pur ODS-3 C-18 material (Dr. Maisch, Ammerbuch, Germany). Analytical column: 75 μm × 42 cm, packed with 3 μm Reprosil-Gold C-18 material (Dr. Maisch, Ammerbuch, Germany).
3. Nano-HPLC solvents: Loading solvent: 0.1 % (v/v) FA in water. Solvent A: 0.1 % (v/v) FA and 5 % (v/v) DMSO [22] in water. Solvent B: 0.1 % (v/v) FA and 5 % (v/v) DMSO in ACN.
4. Data analysis: Freely available MaxQuant [20] software package (e.g., version 1.5.1.0) with the integrated search engine Andromeda [23]. Protein sequence database in FASTA format (e.g., UniprotKB).
5. Spreadsheet editor or the freely available Perseus software package.

3 Methods

3.1 Preparation of Tissue Proteome Digest

1. Add 250 μl of precooled tissue lysis solution to 5–20 mg of wet tissue and transfer into Precellys tubes containing ceramic beads. Mount Precellys tubes in the Precellys 24 bead-milling device and perform tissue lysis and homogenization (5500 rpm, 1 × 25 s, 5 s pause).
2. Use a Bradford assay (or a similar photometric assay) to determine the protein concentration. Store lysates at −80 °C or continue directly. Continue with a lysate volume corresponding to 200–300 μg of total protein (*see* **Notes 4** and **5**).
   - Dilute sample with four volumes of 50 mM Tris–HCl, pH 7.6 to decrease urea concentration to 1.6 M. Add trypsin in a protease-to-protein ratio of 1:100 (w/w) and predigest 4 h in a thermoshaker at 37 °C and 700 rpm. Add another 1:100 (w/w) trypsin and incubate the digestion mixture over night in a thermo-shaker at 37 °C and 700 rpm.
   - Cool samples down to room temperature and acidify the sample to a pH of ~2 by addition of 1 % (v/v) FA (check the pH afterwards). Centrifuge acidified peptides at 14,000 × *g* to precipitate insoluble matter. Use 50 mg Sep-Pak columns and place them into a vacuum manifold (*see* **Note 6**). Prime Sep-Pak columns by adding 1 ml of solvent B. Equilibrate column by adding 2 × 1 ml of solvent A. Transfer the acidi-

fied supernatant to the column and slowly load the sample (*see* **Note 7**). Reapply the flow-through a second time and discard it afterwards. Wash the column with 1 ml solvent A and repeat this step two more times. Elute the peptides with 1 ml solvent B into a 1.5 ml reaction vessel. Use a vacuum centrifuge to dry the desalted digest down. At this point the sample can be stored at −80 °C.

3.2 hSAX Chromatography

Connect the analytical hSAX column and the hSAX guard column to you HPLC system. Upon first time use the column has to be properly flushed with hSAX solvent A until the pressure is stable (*see* **Note 8**). Setup a gradient following the specifications in Table 1. Monitor the UV absorption at fixed wavelengths of 280 and 214 nm and use a flow rate of 0.25 ml/min throughout the gradient.

1. Run a standard digest: A Standard ensures column integrity and proper column equilibration (*see* **Note 9**). Inject ~100 μg in 100 μl hSAX solvent A.
2. Run a blank: Inject 100 μl hSAX solvent A to clean and equilibrate the column and to avoid carry-over from the previous sample (*see* **Note 10**).
3. Sample fractionation (*see* **Note 11**): Dissolve the desalted digest in your 1.5 ml reaction vessel in 105 μl solvent A (*see* **Note 12,** for sonication). Centrifuge the sample at 20,000 ×*g* for 10 min to pellet insoluble debris which might lead to column clogging. Inject 100 μl of the dissolved sample. Use a 96-well plate to collect the eluting fractions in 1 min intervals (0.25 ml/fraction; *see* **Note 13**) starting 2 min into the gradient. Collect a total of 38 fractions (*see* Fig. 1). Freeze the fractions at −80 °C and dry them down using a vacuum centrifuge.

Table 1
Settings for a 50 min hSAX column gradient, including the programmed time, the solvent flow (in ml/min) and percentage of hSAX solvents used

Retention time [min]	hSAX solvent A [%]	hSAX solvent B [%]	Flow [ml/min]
0	100	0	0.25
3	100	0	0.25
27	75	25	0.25
40	0	100	0.25
44	0	100	0.25
45	100	0	0.25
50	100	0	0.25

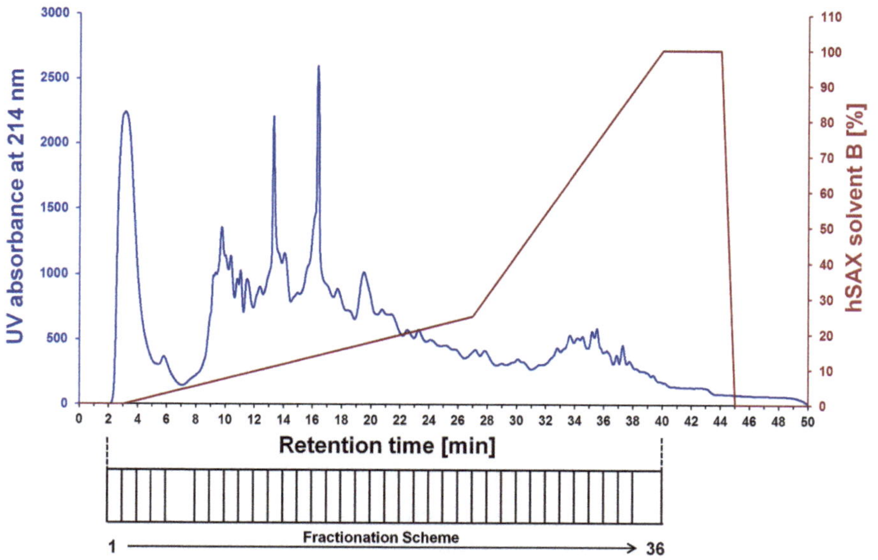

Fig. 1 Typical UV (216 nm) chromatogram of a 50 min hSAX separation using 300 μg of tissue digest and the specified gradient composition (in % of hSAX solvent B). The inset below the chromatogram illustrates the suggested fractionation and pooling scheme applied for this example

4. Run a blank: Inject 100 μl hSAX solvent A to clean the column and to avoid carry-over from the previous sample (*see* **Note 10**).
5. Detach the column which now contains hSAX solvent A and store at 4 °C.

3.3 StageTip Desalting of hSAX Fractions

Given the high salt concentration, it is necessary to desalt the hSAX fractions using C-18 StageTips [24]. Pass all liquids through the tips by centrifugation (~800 × *g*, room temperature, *see* **Note 14**).

1. Resuspend all dried fractions, except for fraction 6 and 38 in 250 μl of solvent A (*see* **Note 15**). Pool fractions according to the scheme depicted in Fig. 1 by transferring the dissolved fraction 5 into the well containing fraction 6 and the dissolved fraction 37 into the well containing fraction 38. This results in a total of 36 fractions.
2. Preparation of C-18 StageTips [24]: Use the small, round punch to cut out five C18 extraction disks from Empore material. Construct a micro-column by packing the disks into a 200 μl pipette tip. Use a sharp scalpel to cut the lid of a 1.5 ml reaction vessel. The reaction vessel will serve as container for the flow-through. Push the micro-column through the cut lid into the 1.5 ml reaction tube. Prepare one tip containing five C-18 disks (Empore Octadecyl C-18 47 mm Solid Phase Extraction Disks #2215, 3 M Purification, Eagan, MN, USA) for each of the 36 hSAX fractions (*see* **Note 16**).

3. Sequentially activate the tips using 250 μl of MeOH, 250 μl of solvent B and 250 μl of solvent A (*see* **Note 17**). Empty the reaction vessels in between.
4. Load one hSAX fraction onto each equilibrated StageTip and reapply the flow-through. Discard the flow-throughs afterwards and wash the columns with 250 μl of solvent A. Empty the reaction vessel.
5. Use 100 μl of solvent B to elute the peptides of the C18 material and into the 1.5 ml reaction vessel. Transfer the eluates into a 96-well plate and dry the sample down using a vacuum centrifuge/lyophilizer. At this point, the plate can be stored at −20 °C.

3.4 LC-MS/MS and Data Analysis

1. Reconstitute the desalted hSAX fractions in 50 μl of 0.1 % FA.
2. Inject 5 μl per fraction (*see* **Note 18**) and wash peptides bound to the trap column for 10 min using loading solvent (0.1 % FA in water) at a flow rate of 5 μl/min. Then, transfer peptides to the analytical column and separate them at a flow rate of 300 nl/min using the following gradient: elute peptides using a linear gradient from 4 to 32 % solvent B for the first 100 min followed by a 10 min wash out and re-equilibration phase (increase to 80 % solvent B within 1 min, hold at 80 % solvent B for 4 min, decrease to 2 % solvent B within 1 min, hold at 2 % solvent B for 4 min).
3. During peptide elution, directly inject peptides into the mass spectrometer via electrospray ionization in positive ionization mode. Suggested parameters for data dependent acquisition on a Q Exactive Plus are specified in Table 2 (*see* **Note 19**).
4. Analyze the data using a proteomics software capable of label-free quantification. All results shown in this chapter are based on peptide identifications by search of raw data against the UniProtKB human database, version July 2013 (88,354 sequences) using the freely available MaxQuant version 1.5.1.0 and its built-in Andromeda search engine. Parameters applied are specified in Table 3.
5. To obtain the number of protein groups and unique genes open the proteinGroups.txt and exclude reverse and contaminant hits. Count the unique entries in the ”Protein IDs” column and the “Gene names” column. Average and report the sequence coverage in percent for the unique proteins. Use the peptides.txt, remove reverse and contaminant hits and subsequently remove the duplicates from the sequence column. Count and report the number of unique sequences. The number of acquired PSMs can be extracted from the “MS/MS identified” column in the summary.txt (*see* Table 4).

Table 2
Suggested parameters for the measurement of hSAX fractions on a Q Exactive Plus

Full MS	
Resolution	70,000
AGC target	3e6
Maximum IT	100 ms
Scan range	360–1300 m/z
MS2	
Resolution	17,500
AGC target	1e5
Maximum IT	50 ms
TopN	20
Isolation window	1.7 m/z
Fixed first mass	NCE 25
Additional settings	
Underfill ratio	1.0%
Charge exclusion	unassigned, 1, 7, 8, >8
Peptide match	Preferred
Exclude isotopes	On
Dynamic exclusion	35.0 s

6. To create an orthogonality plot for hSAX separation (*see* Fig. 2a), filter out modified sequences from the evidence.txt and split the data according to the fraction they are reported in. Separately filter out modified sequence duplicates and plot the number of nonredundant peptides identified per retention time bin for each fraction.
7. To determine the hSAX separation power (*see* Fig. 2b), extract the unique modified sequences per fraction from the evidence.txt. Count the number of fractions each peptide was identified in and plot the percentage against the number of fractions.
8. To obtain the unique peptides per hSAX fraction (*see* Fig. 2c), extract the unique modified sequences per fraction from the evidence.txt. Plot the number of unique modified sequences for each hSAX fraction.
9. Expected results: Table 4 displays the expected results for the analysis of 300 μg brain tissue digest, which was separated into 36 fractions using hSAX chromatography.

Table 3
Search parameters used for data analysis with MaxQuant version 1.5.1.0. In case nothing is specified, default parameters were used

Group-specific parameters	
Type	Standard
Label	No
Variable modifications	Acetyl (Protein N-term), Oxidation (M)
Digestion mode	Specific (Trypsin/P)
Max. missed cleavages	2
Main search peptide tolerance	4.5 ppm
Max. number of modifications per peptide	5
Global parameters	
Database	UniProtKB
Fixed modifications	Carbamidomethyl
PSM FDR	0.01
Protein FDR	0.05
Site decoy fraction	0.01
Min. peptide length	7
Min. score for unmodified peptides	0
Min. score for modified peptides	40
Min. delta score for unmodified peptides	0
Min. delta score for modified peptides	6
MS/MS match tolerance	20 ppm
Second peptide search	Enabled

Table 4
Overview of results typically expected from a 36 fraction hSAX separation of human brain tissue digests, where each fraction was measured using a 2 h LC-MS/MS gradient on a Q Exactive Plus

PSMs, proteins and peptides (summary.txt, proteinGroups.txt and peptides.txt)	
PSMs	473,882
Unique peptides	111,840
Proteins	10,195
Genes	9,516
Average sequence coverage [%]	30.6

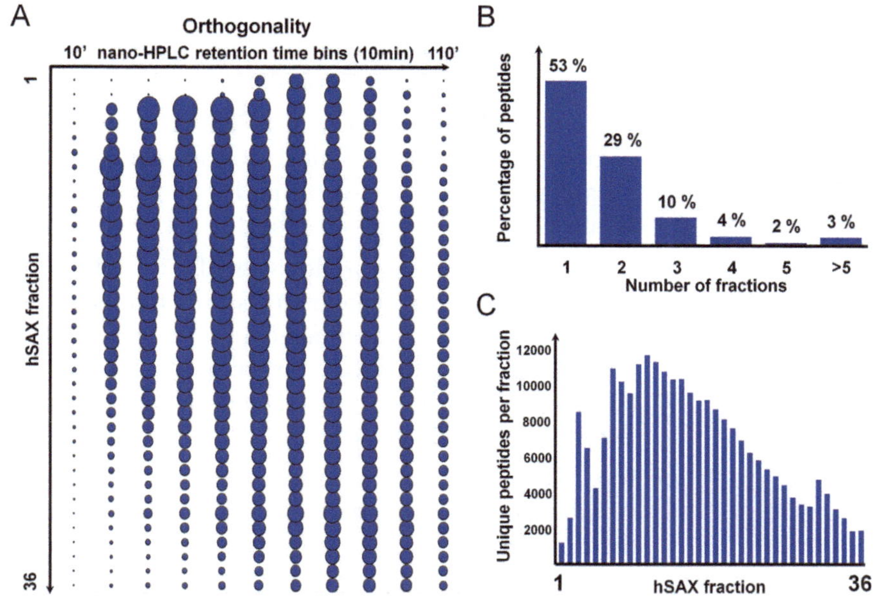

Fig. 2 Two-dimensional peptide separation characteristics. (**a**) Unique peptide sequences per hSAX fraction across the 110 min LC-MS gradient (10 min LC-MS retention time bins). The size of the dots scales with the number of identified peptides. This clearly shows that hSAX separation is highly orthogonal to RP chromatography. (**b**) Separation efficiency of the hSAX fractionation shown as the percentage of peptides found in one or more fractions (the numbers above the bars indicate percentages). (**c**) Number of peptide sequences identified per hSAX fraction

4 Notes

1. The described lysis procedure is not applicable to the analysis of phosphoproteomes because TCEP-HCl interferes with phosphopeptide enrichment. If phosphoproteome enrichment is intended, please refer to Chapter 5 for suitable lysis and digestion conditions. The dried down Fe-IMAC column flow-through and the desalted Fe-IMAC eluate can be reconstituted in hSAX solvent A and is subsequently ready for hSAX separation (*see* **step 3**, Subheading 3.2).
2. The outlined protocol can be easily adapted for the analysis of cell line proteomes. Simply skip the bead beater step and put the lysis solution directly on top of the cells after the culture vessel has been thoroughly rinsed with PBS. Use a cell scraper to remove the cells from the culture vessel and transfer the suspension into a 1.5 ml reaction vessel. Determine the protein concentration and continue with 300 μg. Follow the described procedure starting with **step 3** of Subheading 3.1.
3. The hSAX analytical column can be operated without a guard column. However, a guard column holds back particles, dirt or

insoluble debris and thus protects the analytical column. The exchange of a guard column is considerably cheaper than the exchange of an analytical column.

4. The chemical compatibility of TCEP-HCl and CAA allows the combined reduction and alkylation in one step, which makes for a time saving alternative and does not negatively affect the results [21].
5. Standard proteomic workflows usually include clarification of the lysate prior to protein digestion. Omission of this step actually results in a higher protein/peptide recovery and therefore the identification of a larger number of membrane and nuclear proteins. This makes the use of detergents such as SDS during lysis largely dispensable [21].
6. Sep-Pak sorbent weight has to be chosen according to the amount of digest intended to be desalted. As a rule of thumb, the capacity of Sep-Pack cartridges equals 5 % of the sorbent weight (e.g., 2.5 mg peptide for the 50 mg sorbent weight cartridges and 10 mg for the 200 mg sorbent weight cartridges).
7. Avoid letting the columns run dry. Load the sample slowly onto the Sep-Pak column. Lower the flow rate by adjusting the vacuum at the vacuum manifold. Loading should take at least 10 min to ensure proper binding of the peptides. Reapplying the flow-through increases recovery.
8. Always keep track of the hSAX column pressure at your given flow rate. Pressure increases are early indicators of column clogging or deteriorating column performance. If the increase is too severe consider changing the guard column and/or the analytical column.
9. Prepare a standard digest stock solution in hSAX solvent A according to the procedure described in this protocol. This can be a tissue or a cell line digest. Run a standard prior to each sample batch and compare the chromatograms of the standards to each other in order to spot column deterioration early on.
10. In our experience, it is normal to observe three distinct peaks in every blank run. As the hSAX column was originally designed for the separation of small organic molecules and halo acetic acids [18], we assume that these peaks originate from such components which were not completely removed (e.g., FA).
11. Keep the sample amount below 500 μg in order to be within the limits of the column capacity and avoid overloading. This ensures high resolution peptide separation.
12. In case a visual inspection suggests insufficient dissolution of the sample in hSAX solvent A, a sonicator bath might support sample solubilisation. Dip the 1.5 ml vessel containing the sample into the sonicator bath for 3 × 1 min with 30 s incubations on ice in between the sonication steps.

13. The fraction volume and number is roughly adjusted to the separation power of the hSAX column. In our experience collecting narrower fractionations does not considerably improve protein identification. However, depending on the sample complexity, the available MS machine time and the MS performance, it is also possible to pool the 38 collected fractions into 24 or 12 fractions prior to measurement. Although the protein identification is not severely decreased, the achieved sequence coverage and the number of identified peptides might be considerably lower.
14. Make sure that each tip equilibration step, the washing step, each sample loading step (sample and flow-through application) and each elution step takes approximately 5 min. Since parameters such as C18 material packing density might vary in between experiments, we suggest to separately adjust the centrifugation speed to fit the specified time scale for each experiment.
15. Using volumes of 250 μl prevents columns from running dry even upon prolonged centrifugation. This is especially beneficial for parallelized desalting because not all columns run at the same speed.
16. Although only 36 StageTips are necessary for desalting of hSAX fractions, we recommend preparing and equilibrating 40 StageTips. This way you can save yourself the trouble of starting from the beginning should one or more tips be of insufficient quality.
17. To ensure that the C18 material is not running dry, make sure that there is no air trapped between the applied liquid and the packed C18 material.
18. Although we inject 5 μl per fraction (which is a good starting point) we advise to initially test different injection volumes in order to choose the right amount of loading. This will depend on the hSAX input amount, the capacity of the LC-MS trap and analytical columns and ultimately also on the sensitivity of your mass spectrometer.
19. For label-free experiments, a Top N method should be chosen such that an adequate sampling of the chromatographic peak is assured (~10 MS1 scans per peak). Given the high input amount and dynamic range of full proteome digests, some peptides are highly abundant. Thus the cycle time of the instrument has to be adjusted to the chromatographic peak width which should be determined beforehand. Likewise, the MS dynamic exclusion time should be adjusted to the chromatographic peak width. Since every LC-MS system has somewhat different separation and dead volume characteristics, we suggest adjusting this value to the median peak width at base.

References

1. Wilhelm M, Schlegl J, Hahne H et al (2014) Mass-spectrometry-based draft of the human proteome. Nature 509:582–587
2. Kim M-S, Pinto SM, Getnet D et al (2014) A draft map of the human proteome. Nature 509:575–581
3. Richards AL, Merrill AE, Coon JJ (2015) Proteome sequencing goes deep. Curr Opin Chem Biol 24:11–17
4. Mann M, Kulak NA, Nagaraj N et al (2013) The coming age of complete accurate, and ubiquitous proteomes. Mol Cell 49:583–590
5. Nagaraj N, Wisniewski JR, Geiger T et al (2011) Deep proteome and transcriptome mapping of a human cancer cell line. Mol Syst Biol 7:548
6. Beck M, Schmidt A, Malmstroem J et al (2011) The quantitative proteome of a human cell line. Mol Syst Biol 7:549
7. Geiger T, Wehner A, Schaab C et al (2012) Comparative proteomic analysis of eleven common cell lines reveals ubiquitous but varying expression of most proteins. Mol Cell Proteomics 11:M111.014050
8. Azimifar SB, Nagaraj N, Cox J et al (2014) Cell-type-resolved quantitative proteomics of murine liver. Cell Metab 20:1076–1087
9. Deshmukh AS, Murgia M, Nagaraj N et al (2015) Deep proteomics of mouse skeletal muscle enables quantitation of protein isoforms, metabolic pathways, and transcription factors. Mol Cell Proteomics 14:841–853
10. Wiśniewski R, Dus-Szachniewicz K, Ostasiewicz P et al (2015) Absolute proteome analysis of colorectal mucosa, adenoma and cancer reveals drastic changes in fatty acid metabolism and plasma membrane transporters. J Proteome Res 14:4005–4018
11. Alpert AJ (1990) Hydrophilic-interaction chromatography for the separation of peptides, nucleic acids and other polar compounds. J Chromatogr 499:177–196
12. Boersema PJ, Divecha N, Heck AJR et al (2007) Evaluation and optimization of ZIC-HILIC-RP as an alternative MudPIT strategy. J Proteome Res 6:937–946
13. Hao P, Guo T, Li X et al (2010) Novel application of electrostatic repulsion-hydrophilic interaction chromatography (ERLIC) in shotgun proteomics: comprehensive profiling of rat kidney proteome. J Proteome Res 9:3520–3526
14. Hennrich ML, Groenewold V, Kops GJPL et al (2011) Improving depth in phosphoproteomics by using a strong cation exchange-weak anion exchange-reversed phase multidimensional separation approach. Anal Chem 83:7137–7143
15. Gilar M, Olivova P, Daly AE et al (2005) Two-dimensional separation of peptides using RP-RP-HPLC system with different pH in first and second separation dimensions. J Sep Sci 28:1694–1703
16. Zhou F, Sikorski TW, Ficarro SB et al (2011) Online nanoflow reversed phase-strong anion exchange-reversed phase liquid chromatography-tandem mass spectrometry platform for efficient and in-depth proteome sequence analysis of complex organisms. Anal Chem 83:6996–7005
17. Wolters DA, Washburn MP, Yates JR (2001) An automated multidimensional protein identification technology for shotgun proteomics. Anal Chem 73:5683–5690
18. Ritorto MS, Cook K, Tyagi K et al (2013) Hydrophilic strong anion exchange (hSAX) chromatography for highly orthogonal peptide separation of complex proteomes. J Proteome Res 12:2449–2457
19. Ruprecht B, Koch H, Medard G et al (2015) Comprehensive and reproducible phosphopeptide enrichment using iron immobilized metal ion affinity chromatography (Fe-IMAC) columns. Mol Cell Proteomics 14:205–215
20. Cox J, Mann M (2008) MaxQuant enables high peptide identification rates, individualized p.p.b.-range mass accuracies and proteome-wide protein quantification. Nat Biotechnol 26:1367–1372
21. Kulak NA, Pichler G, Paron I et al (2014) Minimal, encapsulated proteomic-sample processing applied to copy-number estimation in eukaryotic cells. Nat Methods 11:319–324
22. Hahne H, Pachl F, Ruprecht B et al (2013) DMSO enhances electrospray response, boosting sensitivity of proteomic experiments. Nat Methods 10:989–991
23. Cox J, Neuhauser N, Michalski A et al (2011) Andromeda: a peptide search engine integrated into the MaxQuant environment. J Proteome Res 10:1794–1805
24. Rappsilber J, Mann M, Ishihama Y (2007) Protocol for micro-purification, enrichment, pre-fractionation and storage of peptides for proteomics using StageTips. Nat Protoc 2:1896–1906

Chapter 8

High pH Reversed-Phase Micro-Columns for Simple, Sensitive, and Efficient Fractionation of Proteome and (TMT labeled) Phosphoproteome Digests

Benjamin Ruprecht, Jana Zecha, Daniel P. Zolg, and Bernhard Kuster

Abstract

Despite recent advances in mass spectrometric sequencing speed and improved sensitivity, the in-depth analysis of proteomes still widely relies on off-line peptide separation and fractionation to deal with the enormous molecular complexity of shotgun digested proteomes. While a multitude of methods has been established for off-line peptide separation using HPLC columns, their use can be limited particularly when sample quantities are scarce. In this protocol, we describe an approach which combines high pH reversed-phase peptide separation into few fractions in StageTip micro-columns. This miniaturized sample preparation method enhances peptide recovery and hence improves sensitivity. This is particularly useful when working with limited sample amounts obtained from e.g., phosphopeptide enrichments or tissue biopsies. Essentially the same approach can also be applied for multiplexed analysis using tandem mass tags (TMT) and can be parallelized in order to deliver the required throughput. Here, we provide a step-by-step protocol for TMT6plex labeling of peptides, the construction of StageTips, sample fractionation and pooling schemes adjusted to different types of analytes, mass spectrometric sample measurement, and downstream data processing using MaxQuant. To illustrate the expected results using this protocol, we provide results from an unlabeled and a TMT6plex labeled phosphopeptide sample leading to the identification of >17,000 phosphopeptides in 8 h (Q Exactive HF) and >23,000 TMT6plex labeled phosphopeptides (Q Exactive Plus) in 12 h of measurement time. Importantly, this protocol is equally applicable to the fractionation of full proteome digests.

Key words Fractionation, Proteomics, Sample preparation, Mass spectrometry, Phosphorylation, Isotope labeling

Abbreviations

CAN	Acetonitrile
AGC	Acquisition gain control
FA	Formic acid
HCD	Higher energy collision induced dissociation
HPLC	High-performance liquid chromatography
hSAX	Hydrophilic strong anion exchange
IMAC	Immobilized metal ion affinity chromatography

Lucio Comai et al. (eds.), *Proteomics: Methods and Protocols*, Methods in Molecular Biology, vol. 1550,
DOI 10.1007/978-1-4939-6747-6_8, © Springer Science+Business Media LLC 2017

IT	Injection time
MS	Mass spectrometer
MS/MS	Tandem mass spectrometry
NH_4FA	Ammonium formate
RP	Reversed-phase
SAX	Strong anion exchange
SCX	Strong cation exchange
StageTip	Stop and go extraction tip
TEAB	Triethylammonium bicarbonate
TFA	Trifluoroacetic acid
TiO_2	Titanium dioxide
TMT	Tandem mass tag
Ppm	Parts per million
PSM	Peptide spectrum match
pY/pS/pT	Phosphotyrosine, -serine, -threonine
ZIC-HILIC	Zwitterionic hydrophilic interaction liquid chromatography

1 Introduction

Recent advances in mass spectrometric instrumentation enable the identification and quantification of peptides and phosphopeptides at an unprecedented depth [1]. Despite the ever increasing sequencing speed of modern mass spectrometers, single measurements are often not sufficient to fully resolve the (phospho)proteome complexity and thus coverage must generally be improved by some form of two-dimensional peptide fractionation. In shotgun proteomics, the most widely applied second dimension of peptide separation is reversed-phase C18 chromatography directly coupled to tandem mass spectrometry. Many different choices for orthogonal first dimension separations exist: especially strong cation exchange (SCX) [2, 3], which separates peptides by charge, has been widely applied for phosphopeptide (pre)fractionation. But also hydrophilic interaction liquid chromatography (HILIC) [4], hydrophilic strong anion exchange (hSAX) [5] and high pH reversed-phase columns [6] have proven their merit for phosphoproteomic studies. Although off-line 2D approaches using standard inner diameter columns (1–10 mm) are highly robust, efficient and deliver very good analytical depth, fractionation can be time-consuming and parallelization capabilities are limited. In addition, we [5] and others [7] have shown that phosphopeptide fractionation after enrichment requires an increased amount of starting material, due to sample losses that occur during the fractionation process. Here, we outline a simple and robust alternative which combines high pH reversed-phase separation and self-packed StageTip micro-columns [8, 9]. Such high pH reversed-phase columns as well as SAX and SCX materials have recently been used to separate digests of full proteomes [8, 9, 10] and membrane protein preparations [11, 12].

Importantly, the enhanced sensitivity and recovery provided by the high pH reversed-phase micro-column separations is particularly attractive when working with limited sample amounts including but not limited to phosphopeptide enrichments. The described fractionation procedure, which does not require specialized equipment, can be accomplished in a time efficient and highly parallelizable fashion and can easily be combined with TMT labeling [13]. Since TMT labeled (phospho)peptides are generally more hydrophobic than the respective unlabeled peptides, we provide detailed fractionation and pooling schemes for both sample types. This protocol further describes recommended parameters for the mass spectrometric measurement on a Q-Exactive mass spectrometer and outlines how to analyze the obtained data using MaxQuant [14, 15]. Finally, we illustrate the expected results using a phosphoproteome and a TMT6plex labeled phosphoproteome obtained by IMAC enrichment of a cancer cell line digest (*see* Chapter 5). This protocol is equally applicable to the fractionation of full proteome digests, following the exact same fractionation and pooling scheme which is described for TMT labeled phosphopeptides.

2 Materials

Unless stated otherwise, all solvents and buffers are prepared freshly using ultrapure water and analytical grade reagents. Devices such as centrifuges, vacuum centrifuges/lyophilizer, thermoshaker, or refrigerators (−20 °C/−80 °C) are not explicitly listed.

2.1 Sample Preparation

For materials related to lysis, digestion, and phosphopeptide enrichment we refer to the Chapter 5.

2.2 TMT6plex Labeling of Whole Proteome Digests Prior to Phosphopeptide Enrichment

1. Sep-Pak C18 peptide purification before TMT6plex labeling: 50 mg Sep-Pak cartridges (Waters Corp., Eschborn, Germany). Solvent A: 0.1% (v/v) FA in water. Solvent B: 60% (v/v) ACN, 0.1% (v/v) FA in water.
2. TMT6plex labeling reagent: TMT6plex isobaric label reagent set (Thermo Fisher Scientific Inc., Bremen, Germany). This protocol uses 1 mg of each TMT6plex labeling reagent to label 250 μg protein digest per channel (*see* **Note 1**). Allow the vials containing the six different labeling reagents to warm up to room temperature (*see* **Note 2**). Prepare TMT6plex stock solutions by adding 41 μl of the anhydrous ACN (*see* **Note 3**) to each reagent vial (0.8 mg TMT). Vortex vials for 30–60 s and briefly centrifuge. The TMT6plex stock solution should be used immediately or can alternatively be stored at −80 °C for up to 1 week. To increase storage time, the remaining TMT6plex reagents should be dried down using a vacuum centrifuge.

3. 50 mM triethylammonium bicarbonate (TEAB), pH 8.5: Mix 50 μl of 1 M TEAB (Sigma-Aldrich, St. Louis, MO), pH 8.5 with 950 μl of water. Make sure that the pH is not below 8.
4. 5 % hydroxylamine in water.
5. 10 % (v/v) FA and 10 % (v/v) ACN in water.
6. 10 % (v/v) FA in water.
7. Sep-Pak C18 peptide purification after TMT6plex labeling: 50 mg Sep-Pak cartridges (Waters Corp., Eschborn, Germany). Solvent A: 0.07 % (v/v) TFA in water. Solvent B: 50 % (v/v) ACN, 0.07 % (v/v) TFA in water.
8. Vacuum manifold for Sep-Pak desalting.

2.3 High pH Reversed-Phase Micro-Column Fractionation

1. Empore Octadecyl C18 47 mm Solid Phase Extraction Disks #2215 (3 M Purification, Eagan, MN, USA).
2. Small, round punch to cut out C-18 disks (Fig. 1b).
3. 200 μl plastic pipette tip.
4. 1.5 ml reaction vessel.
5. High pH reversed-phase stock solution: 50 mM ammonium formate (NH_4FA, Sigma-Aldrich, St. Louis, MO, Product number 156264, reagent grade 97%), pH 10. Weigh in 315.3 mg of NH_4FA and transfer it into a glass beaker or cylinder. Add water to a volume of 90 ml. Mix with a magnetic stirrer and adjust the pH to 10 using ammonium hydroxide. Fill up to 100 ml with water.

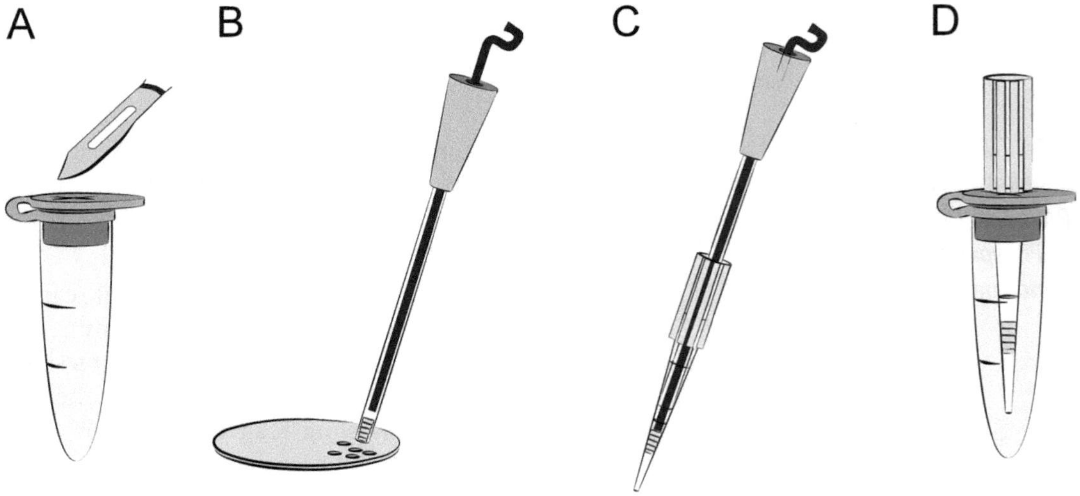

Fig. 1 Construction of a micro-column for high pH reversed-phase fractionation. (**a**) The lid of a 1.5 ml reaction tube is cut crosswise with a small scalpel. (**b**) A punching device, assembled from a syringe and a piece of wire is used to punch out 5 disks of Empore C18 material. (**c**) The punching device is used to push the disks into the 200 μl pipette tip. (**d**) The constructed column is placed in the reaction tube by pushing the tip through the cut lid

6. High pH reversed-phase solvents: Solvent A: Dilute 50 mM high pH reversed-phase stock solution 1:1 with water to obtain 25 mM NH_4FA, pH 10. Solvent B: Dilute high pH reversed-phase stock 1:1 with ACN to obtain 25 mM NH_4FA, 50% ACN, pH 10.
7. Mix high pH reversed-phase solvents A and B to obtain elution solvents (*see* **Note 4**). In case you want to separate an unlabeled phosphoproteome, prepare elution solvents containing 2.5%, 7.5%, 12.5% ACN in 25 mM NH_4FA. For the TMT6plex labeled phospho sample, prepare elution buffers with 5%, 7.5%, 10%, 12.5%, 15%, and 17.5% ACN in 25 mM NH_4FA (*see* **Note 5**). Table 1 displays the mixing scheme for the different elution buffers, sufficient for 10 micro-column elution steps each. The solvents are prepared sequentially; as an example, prepare the 7.5% solvent by diluting 0.6 ml of the 10% ACN elution solvent with 0.2 ml solvent A. For the fractionation of unlabeled phosphopeptides, start by diluting the 50% ACN elution solvent 1:4 with solvent A and continue according to the scheme shown in Table 1.
8. 5 ml Eppendorf CombiTip.
9. Desalting of the high pH reversed-phase micro-column flow-through: Desalting solvent A: 0.07% (v/v) TFA in water. Desalting solvent B: 0.07% (v/v) TFA, 60% (v/v) ACN in water.

2.4 LC-MS/MS and Data Analysis

1. 50 mM citric acid and 1% (v/v) FA in water. Dissolve 105 mg of citric acid monohydrate (VWR, Product No. 20278.298) in 9.90 ml of water and add 100 μl of 100% FA.
2. LC-MS/MS: nano-HPLC setup coupled to a high resolution mass spectrometer. Here, we used an Eksigent NanoLC-Ultra 1D+ (Eksigent, Dublin, USA; to measure fractions of TMT6plex labeled phosphopeptides) or an Thermo Ultimate

Table 1
Mixing scheme for high pH reversed-phase elution solvents

Elution solvent	Take	Add solvent A (ml)	Total [ml]
17.5% ACN	0.7 ml Solvent B (50 % ACN)	1.3	2.0
15.0% ACN	1.2 ml Solvent 17.5% ACN	0.2	1.4
12.5% ACN	1.0 ml Solvent 15.0% ACN	0.2	1.2
10.0% ACN	0.8 ml Solvent 12.5% ACN	0.2	1.0
7.5% ACN	0.6 ml Solvent 10.0% ACN	0.2	0.8
5.0% ACN	0.4 ml Solvent 7.5% ACN	0.2	0.6
2.5% ACN	0.2 ml Solvent 5.0% ACN	0.2	0.4

3000 (Thermo Scientific, Bremen, Germany; to measure fractions of unlabeled phosphopeptides) coupled to an Orbitrap Q Exactive type mass spectrometer (Thermo Scientific, Bremen, Germany). LC-trap column: 75 μm × 2 cm, packed with 5 μm Reprosil-Pur ODS-3 C18 material (Dr. Maisch, Ammerbuch, Germany). Analytical column: 75 μm × 42 cm, packed with 3 μm Reprosil-Gold C18 material (Dr. Maisch, Ammerbuch, Germany).

3. Nano-HPLC solvents: Loading solvent: 0.1% (v/v) FA in water. Solvent A: 0.1% (v/v) FA and 5% (v/v) DMSO in water. Solvent B: 0.1% (v/v) FA and 5% (v/v) DMSO in ACN.
4. Data analysis: Freely available MaxQuant software package (e.g., version 1.5.2.8) with the integrated search engine Andromeda. Protein sequence database in FASTA format (e.g., UniprotKB).
5. Spreadsheet editor or the freely available Perseus software package.

3 Methods

3.1 Sample Preparation

For methods related to lysis, digestion, and peptide desalting using Sep-Pak columns, we refer to Chapter 5. The Fe-IMAC protocol also describes the purification of phosphopeptides using Fe-IMAC columns. The procedure can be readily applied to combined, TMT6plex labeled digests. We recommend using at least 1.5 mg of protein digest for phosphopeptide enrichment (Fe-IMAC input amounts ranging between 1.5 and 3 mg are required to obtain optimal results). For a TMT6plex experiment this translates into 250 μg of protein per channel (*see* **Note 6**).

3.2 TMT6plex Labeling of Whole Proteome Digests Prior to Phosphopeptide Enrichment

1. Desalting of samples prior to TMT6plex labeling (*see* **Note 7**): Acidify digested samples to ~ pH 2 by addition of 100% FA to a final concentration of 1% (v/v) FA (check the pH). Centrifuge peptides at 5000 × *g* to precipitate insoluble matter. Place six 50 mg Sep-Pak columns into a vacuum manifold (*see* **Note 8**). Prime Sep-Pak columns by adding 1 ml of solvent B. Equilibrate column by adding 2 × 1 ml of solvent A. Load the supernatant of the acidified sample slowly onto the column, reapply the flow-through and discard the second flow-through afterwards. Wash the column with 2 × 1 ml of solvent A. Elute peptides of each column with 2 × 150 μl of solvent B into a 1.5 ml reaction vessel. Freeze the samples at −80 °C and make sure that they are still frozen when they are placed in the vacuum centrifuge to dry them down.
2. Labeling reaction (*see* **Note 9**): Reconstitute the six desalted samples in 200 μl of 50 mM TEAB. To start the labeling reaction, add

50 μl of the respective TMT6plex stock solution to each sample, mix by repeatedly pipetting up and down, briefly centrifuge the samples and incubate them for 1 h at 20 °C and 400 rpm. Stop the labeling reaction by adding 20 μl of 5% hydroxylamine to each sample (final concentration of ~0.4% hydroxylamine). Incubate the samples for 15 min at room temperature, shaking at 400 rpm and briefly centrifuge them afterwards.

3. Sample combination (*see* **Note 10**): Combine the six samples by transferring half of each one into two separate 1.5 ml reaction vessels. To acidify the sample to a pH of 2, add 50 μl of 10% FA to each vessel (check the pH). Add 100 μl of 10% FA in 10% ACN to each of the six original vessels, incubate for 5 min and combine the solvents with the pooled sample. Briefly centrifuge both reaction vessels, freeze them at −80 °C, and vacuum-centrifuge to dryness. The two vials can now be stored at -80 °C.
4. Sep-Pak desalting of the combined, labeled sample: Reconstitute peptides in both vessels in 0.5 ml of 0.07 % TFA and pool the sample. Except for the applied solvents, desalting is essentially performed as described in **step 1** of this section. Here, the sample is loaded in 0.07% (v/v) TFA in water and eluted off the Sep-Pak column (50 mg sorbent weight) using 2 × 150 μl of 0.07% TFA, 50% ACN in water. The eluate is subsequently filled up to 0.5 ml with 0.07% TFA in water and can thus directly be applied to Fe-IMAC column based phosphopeptide enrichment.
5. Fe-IMAC enrichment (*see* **Note 11**): *see* Chapter 5. Dry the Fe-IMAC eluate down using a vacuum centrifuge. The StageTip desalting of the phosphopeptide containing Fe-IMAC eluate can be omitted. Store the sample at −80 °C or continue directly with high pH reversed-phase micro-column fractionation.

3.3 High pH Reversed-Phase Micro-Column Fractionation

The following steps for fractionation and desalting are performed by centrifugation of the micro-column at ~800 × *g* (*see* **Note 12**). Pre-cool and store all solvents on ice. Unless stated otherwise, avoid letting the columns run dry.

1. High pH reversed-phase micro-column construction: Use the small, round punch to cut out five C18 extraction disks from Empore material (*see* **Note 13** and Fig. 1). Construct a micro-column by packing the disks into a 200 μl pipette tip. Use a sharp scalpel to cut the lid of a 1.5 ml reaction vessel. The reaction vessel will serve as container for the flow-through. Push the micro-column through the cut lid into the 1.5 ml reaction tube.
2. Dissolve your dried sample in 250 μl solvent A (*see* **Note 14**). Vortex, spin down and store on ice while the column is equilibrated. Check if the pH of the dissolved sample is ~10 using pH indicator strips.

3. Column equilibration (*see* **Note 15**): Add 250 μl of ACN to the top of the micro-column to soak the extraction material and remove air bubbles. Centrifuge the tip and rinse residual ACN using a 5 ml Eppendorf CombiTip. Wash column with 250 μl solvent B followed by 250 μl solvent A and discard the flow-through.
4. Sample loading: Slowly load the sample onto the column. Reapply the flow-through a second time. Transfer the flow-through in a new 1.5 ml vessel and dry it down using a vacuum centrifuge (for flow-through desalting, continue to **step 6**, *see* **Note 16**). Wash the micro-column with 250 μl solvent A and discard the washing fraction.
5. Sample fractionation: The peptides bound to the extraction material are sequentially eluted with increasing concentration of ACN. Use 40 μl of each elution solvent. Let the column run dry in between each step. After each elution step, the eluate is transferred into a 96-well plate. For separation of unlabeled phosphopeptides, sequentially elute peptides from the C18 material with four solvents containing 2.5 %, 7.5 %, 12.5 % and 50 % ACN in 25 mM NH_4FA. The desalted sample flow-through fraction is pooled with the 50 % ACN fraction (*see* **step 6**). For TMT6plex labeled phosphopeptide separation use seven solvents containing 5 %, 7.5 %, 10 %, 12.5 %, 15 %, 17.5 % and 50 % ACN in 25 mM NH_4FA. Combine the desalted sample flow-through fraction (*see* **step 6**) with the 17.5 % ACN fraction and the 5 % ACN fraction with the 50 % ACN fraction.
6. Desalting of the high pH reversed-phase micro-column flow-through (*see* **Note 16**): Dissolve the dried sample in 250 μl of desalting solvent A and keep it on ice while the StageTip is prepared. Check the pH of the dissolved peptide solution and, if required, adjust it to pH 2 using 100 % TFA. Sequentially activate the tip using 250 μl of ACN, 250 μl of desalting solvent B, and 250 μl of desalting solvent A. Empty the 1.5 ml vessel in between each step. Load the dissolved sample and reapply the flow-through. Discard the second flow-through and wash the column with 250 μl desalting solvent A. Use 40 μl of desalting solvent B to elute the peptides from the C18 material. Transfer the eluate into the 96-well plate containing the other fractions (*see* **step 6**) and dry the fractions down using a vacuum centrifuge/lyophilizer. At this point, the sealed plate can be stored at −20 °C.

3.4 LC-MS/MS and Data Analysis

1. Reconstitute the micro-column fractions in 10 μl of 1 % FA in 50 mM citrate (*see* **Note 17**).
2. Inject 10 μl per TMT6plex fraction and 5 μl for fractions containing unlabeled phosphopeptides (*see* **Note 18**). Wash peptides bound to the trap column for 10 min using loading

solvent (0.1 % FA in water) at a flow rate of 5 μl/min. Then, transfer peptides to the analytical column and separate them at a flow rate of 300 nl/min using gradients as follows (*see* **Note 19**): elute unlabeled phosphopeptides using a linear gradient from 2 to 15% solvent B for the first 70 min followed by a linear increase to 27% solvent B within the next 30 min. In contrast, separation of TMT6plex labeled phosphopeptides is performed by an initial increase from 2 to 4% solvent B within the first 2 min followed by a 98 min linear gradient from 4 to 32% solvent B. In both cases the gradient ends with a 10 min wash out phase (increase to 80% solvent B within 2 min, hold at 80% solvent B for 2 min, decrease to 2% solvent B within 2 min, hold at 2% solvent B for 4 min) (Fig. 2).

3. During peptide elution, directly inject peptides into the mass spectrometer via electrospray ionization in positive ionization mode. Suggested parameters for data dependent acquisition and HCD fragmentation on a Q Exactive Plus/HF are specified in Table 2 (*see* **Note 20**).

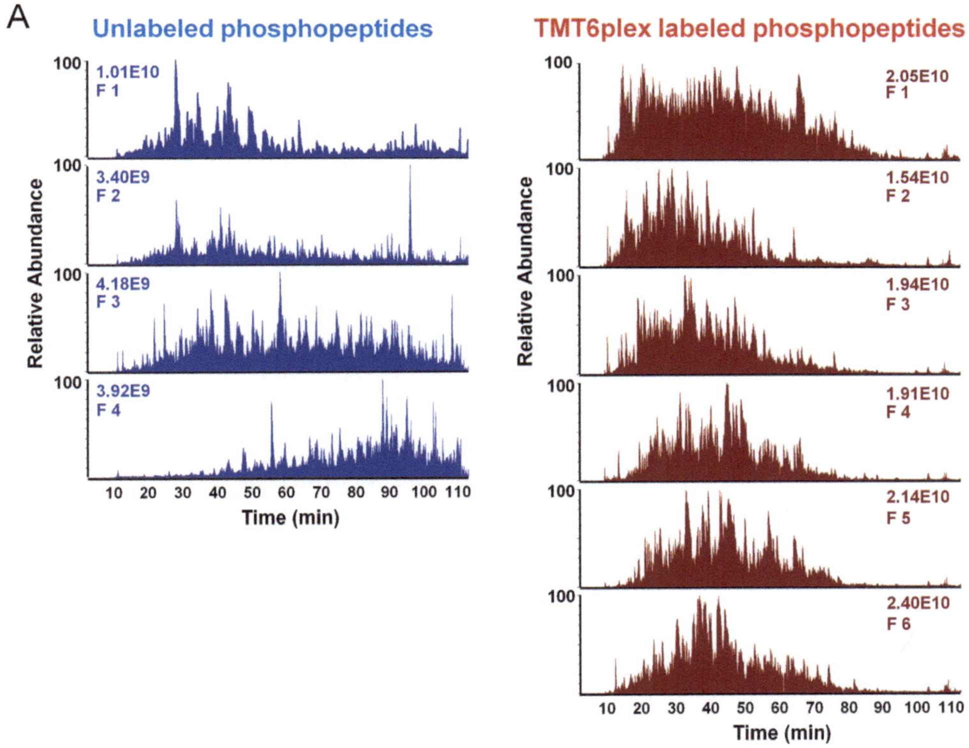

Fig. 2 Total ion current chromatograms displaying peptide elution patterns across the different high pH reversed-phase fractions for unlabeled phosphopeptides (*blue*, Fig. 2a) and TMT6plex labeled phosphopeptides (*red*, Fig. 2b). The number in the corner of each plot indicates the absolute ion current intensity and the respective fraction number

4. Analyze data using a proteomics software capable of label-free and TMT6plex quantification, respectively. All results shown in this chapter are based on peptide identifications by search of raw data against the UniProtKB human database, version July 2013 (88,354 sequences) using the freely available MaxQuant version 1.5.2.8 and its built-in Andromeda search engine. Parameters applied are specified in Table 3.
5. Filter the MaxQuant evidence.txt or the phospho(STY) sites. txt output file for reverse sequences and potential contaminants. To determine the selectivity of the phospho enrichment, the reported number of peptides annotated with a phosphorylation site is divided by the total number of identified sequences. The intensity based selectivity is acquired similarly by dividing the summed intensity of phosphorylated peptides by the total intensity. Filter the "Modified Sequence" column for duplicates to remove redundancies and obtain the number of unique phosphopeptides. Similarly, the phospho (STY)sites.txt is used to determine the number of unique and quantifiable sites. Filter for "Localization probability" ≥0.75 to obtain the number of class I sites [16].
6. To ensure proper TMT6plex labeling, count the number of TMT6plex labeled peptides and divide them by the sum of TMT6plex labeled and non TMT6plex labeled peptides (the labeling efficiency is usually >99 %).
7. To obtain the orthogonality plot for phosphopeptides (Fig. 3), filter out non-phospho sequences from the evidence.txt and split the data according to the fraction they are reported in. Filter out modified sequence duplicates in every fraction separately. Plot the number of nonredundant phosphopeptides identified per retention time bin for each fraction.
8. To determine the separation power of the high pH reversed-phase micro-column, extract the unique phosphorylated sequences per fraction from the evidence.txt (use the modified sequence column). Count the number of fractions each phosphorylated sequence was identified in and plot the percentage against the number of fractions.
9. Expected Results: Table 4 displays the identified phosphopeptides and phosphorylation sites from both experimental approaches.

4 Notes

1. Although the amount of labeling reagent used in this protocol is below that specified by the manufacturer, we found it to be sufficient for complete labeling of peptides.

Table 2
Suggested parameters for the measurement of high pH reversed-phase phosphopeptide fractions on a Q Exactive HF (unlabeled phosphopeptides) and a Q Exactive Plus (TMT6plex labeled phosphopeptides) mass spectrometer

	Unlabeled (QE HF)	TMT6plex labeled (QE Plus)
Full MS		
Resolution	60,000	70,000
AGC target	3e6	3e6
Maximum IT	20 ms	20 ms
Scan range	360–1300 m/z	360–1300 m/z
MS2		
Resolution	30,000	17,500
AGC target	2e5	2e5
Maximum IT	100 ms	100 ms
TopN	15	20
Isolation window	1.7 m/z	1.3 m/z
Isolation offset	0.0 m/z	0.0 m/z
Fixed first mass	100 m/z	100 m/z
NCE	25	33
Additional settings		
Underfill ratio	1.0%	1.0%
Charge exclusion	unassigned, 1, 7, 8, >8	unassigned, 1, 7, 8, >8
Peptide match	Preferred	Preferred
Exclude Isotopes	On	On
Dynamic exclusion	25.0 s	35.0 s

2. Equilibration of TMT reagents to room temperature prior to opening avoids water condensation and hydrolysis of the moisture-sensitive labeling reagents.
3. Make sure the anhydrous ACN bottle is kept water-free at all times. Use a dry syringe pre-filled with Argon to transfer aliquots of ACN to a dry Eppendorf tube pre-filled with argon.
4. Due to the attached TMT6 label, peptides become more hydrophobic. Therefore, it is necessary to adjust the ACN concentration of the high pH reversed-phase elution solvents accordingly.
5. High pH reversed-phase micro-column fractionation can also be applied to the fractionation of full proteome digests using

Table 3
Search parameters used for data analysis with MaxQuant version 1.5.2.8. In case default parameters were used, nothing is specified

	Unlabeled	TMT6plex labeled
Group-specific parameters		
Type	Standard	Reporter ion MS2
Label	No	6plex TMT
Variable modifications	Acetyl (Protein N-term), Oxidation (M), Phospho (STY)	
Digestion mode	Specific (Trypsin/P)	
Max. missed cleavages	2	
Main search peptide tolerance	5 ppm	
Max. number of modifications per peptide	5	
Global parameters		
Database	UniProtKB	
Fixed modifications	Carbamidomethyl	
PSM FDR	0.01	
Min. peptide length	7	
Min. score for (un)modified peptides	0	
Min- delta score for (un)modified peptides	0	
Match between runs	Enabled	
Match time window	1 min	
Alignment time window	20 min	
MS/MS match tolerance	20 ppm	

the same fractionation/elution scheme as described for TMT6plex labeled phosphopeptides. We refer to Chapter 5 for the preparation of proteome digests and advise to use input amounts between 20 and 50 µg as well as five C18 disks per high pH reversed-phase tip. You should expect the identification of ~6000–6500 proteins in 12 h of measurement time on a Q Exactive type mass spectrometer.

6. To increase the multiplexing capacity, TMT10plex labeling reagents can be readily used and following the same procedure. As ten instead of six samples are combined, less input material per channel is required (≥150 µg digest).
7. Avoid letting the columns run dry. Load the sample slowly onto the Sep-Pak column. Lower the flow rate by adjusting the

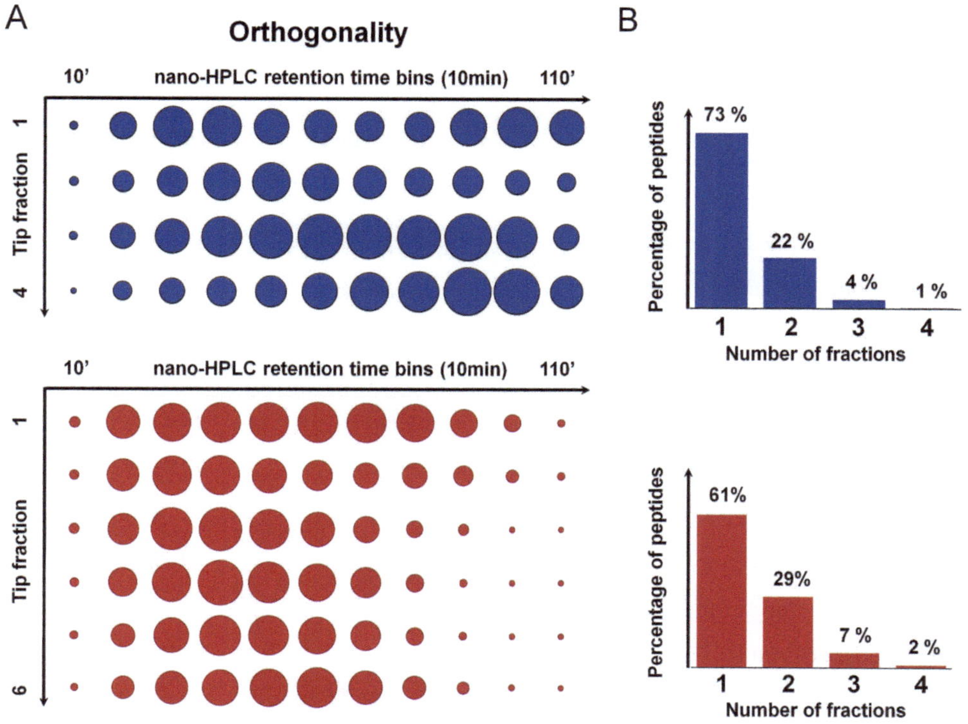

Fig. 3 Orthogonality and peptide separation characteristics of high pH reversed-phase micro-column fractionation. (**a**) Unique phosphopeptide sequences per micro-column fraction across the 110 min LC-MS gradient (*blue*: unlabeled phosphopeptides; *red*: TMT6plex labeled phosphopeptides; grouped into 10 min LC-MS retention time bins). The size of the dots scales with the number of identified phosphopeptides. (**b**) Separation efficiency of the high pH reversed-phase micro-column fractionation shown as the percentage of peptides found in one or more fractions (*blue*: unlabeled phosphopeptides; *red*: TMT6plex labeled phosphopeptides; the numbers above the bars indicate percentages)

vacuum at the vacuum manifold. Loading should take at least 10 min to ensure proper binding of the peptides. Re applying the flow-through increases recovery. Drying down frozen samples improves solubility afterwards.

8. Sep-Pak sorbent weight has to be chosen according to the amount of digest you intend to load. As a rule of thumb, the capacity of Sep-Pack cartridges equals 5 % of the sorbent weight (e.g., 2.5 mg for the 50 mg sorbent weight cartridges and 10 mg for the 200 mg sorbent weight cartridges).
9. TMT reagents are amine-reactive, thus all amine-containing buffers and additives must be removed before labeling. TEAB concentrations >100 mM reduce labeling efficiency. Keep the final ACN concentration during the labeling reaction below 40%.
10. In case of using higher amounts of input digest, the combined sample might have to be split into more than two 1.5 ml reaction vessels.
11. High pH reversed-phase micro-column fractionation is also compatible with other phosphopeptide enrichment methods

Table 4
Summary of expected results for fractionation of TMT6plex and unlabeled phosphopeptides prepared from 1.5 mg and 2 mg of digest, respectively

	Unlabeled	TMT6plex labeled
Phosphopeptides (MaxQuant—evidence.txt)		
Identified unique phosphopeptides	17,140	23,939
Quantified unique phosphopeptides	16,391	22,418
Mono phosphorylated	14,320 (87%)	18,820 (84%)
Multiply phosphorylated	2,071 (13%)	3,598 (16%)
ID based phosphopeptide selectivity	96%	94%
Intensity based phosphopeptide selectivity	93%	99%
Phosphorylation sites (MaxQuant—Phospho(STY)sites.txt)		
Identified phosphorylation sites	12,971	19,359
Quantified phosphorylation sites	11,221	17,681
Class I sites (Loc prob > 0.75)	9,634	14,534
pS sites (class I)	8,208 (85%)	12,352 (85%)
pT sites (class I)	1,368 (14%)	1,463 (10%)
pY sites (class I)	67 (<1%)	728 (5%)

such as TiO_2, IMAC and Ti-IMAC beads in batch or tip format.

12. Make sure that each tip equilibration step, the washing step, each sample loading step (sample and flow-through application) and each elution step takes approximately 5 min. Since parameters such as C18 material packing density might vary in between experiments, we suggest to separately adjust the centrifugation speed to fit the specified time scale for each experiment.
13. In our experience, every C18 disk has a capacity for 5–10 μg of protein digest. We suggest using 5–6 disks per micro-column.
14. Using volumes of 250 μl prevents columns from running dry even upon prolonged centrifugation. This is especially beneficial when parallelized fractionation is intended as not all columns run at the same speed.
15. If only one sample is intended to be fractionated, the procedure can be accelerated by manually pushing the liquids through the tips using a 5 ml Eppendorf CombiTip. The volumes can then be scaled down accordingly (~40 μl per step).
16. The high pH reversed-phase micro-column flow-through can be dried down while fractionation is carried out. Once the frac-

tionation is finished, the 96-well plate containing the eluate fractions is sealed with adhesive foil and stored at −20 °C until the desalting step is finished. Afterwards the plate is thawed and the desalted peptides can be directly eluted into the respective well of the 96-well plate.

17. Citric acid acts as a chelating agent for residual Fe^{3+} ions that might bleed from the Fe-IMAC column. Remaining Fe^{3+} ions can stick to the trap/analytical nano-HPLC columns and deplete phosphopeptides. Ever since we used citrate we have not detected any iron contamination [17]. If in doubt, specify iron as a variable modification during data processing and check if any iron bound peptides are identified.
18. Since TMT6plex labeled phosphopeptide fractionation result in six instead of four fractions, we opted to inject the complete fraction. But it should be sufficient to inject only half of the 10 μl also for the TMT6plex labeled fractions.
19. Unlabeled and TMT6plex labeled phosphopeptides require different nano-HPLC gradients owing to differences in hydrophobicity imparted by the TMT label.
20. For label-free experiments, a Top N method should be chosen such that an adequate sampling of the chromatographic peak is assured (~ten MS1 scans per peak). Thus, the cycle time of the instrument has to be adjusted to the chromatographic peak width. For TMT based experiments, the quantification data is collected in the tandem mass spectrum which generally allows using a higher number of MS2 scans per cycle (Top N). For TMT labeled samples, we further recommend to narrow the isolation window of the quadrupole to reduce ratio compression due to co-eluting and co-isolated features [18]. The collision energy has to be optimized for unlabeled and TMT labeled phosphopeptides separately and is usually higher for TMT labeled samples. Likewise, dynamic exclusion should be optimized for every LC system separately. Since every LC-MS system has somewhat different separation and dead volume characteristics, we suggest adjusting this value to the median peak width at base. For TMT10plex labeled phosphopeptides, the MS2 resolution has to be adjusted to 60,000 [19] in order to resolve the TMT reporter ion isotopologues.

References

1. Wilhelm M, Schlegl J, Hahne H et al (2014) Mass-spectrometry-based draft of the human proteome. Nature 509:582–587
2. Villén J, Gygi SP (2008) The SCX/IMAC enrichment approach for global phosphorylation analysis by mass spectrometry. Nat Protoc 3:1630–1638
3. Gauci S, Helbig AO, Slijper M et al (2009) Lys-N and trypsin cover complementary parts of the phosphoproteome in a refined SCX-based approach. Anal Chem 81:4493–4501
4. McNulty DE, Annan RS (2008) Hydrophilic interaction chromatography reduces the complexity of the phosphoproteome and improves

global phosphopeptide isolation and detection. Mol Cell Proteomics 7:971–980

5. Ruprecht B, Koch H, Medard G et al (2015) Comprehensive and reproducible phosphopeptide enrichment using iron immobilized metal ion affinity chromatography (Fe-IMAC) columns. Mol Cell Proteomics 14:205–215
6. Batth TS, Francavilla C, Olsen JV (2014) Off-line high-pH reversed-phase fractionation for in-depth phosphoproteomics. J Proteome Res 13:6176–6186
7. Kettenbach AN, Gerber SA (2011) Rapid and reproducible single-stage phosphopeptide enrichment of complex peptide mixtures: application to general and phosphotyrosine-specific phosphoproteomics experiments. Anal Chem 83:7635–7644
8. Ishihama Y, Rappsilber J, Mann M (2006) Modular stop and go extraction tips with stacked disks for parallel and multidimensional peptide fractionation in proteomics. J Proteome Res 5:988–994
9. Rappsilber J, Mann M, Ishihama Y (2007) Protocol for micro-purification, enrichment, pre-fractionation and storage of peptides for proteomics using StageTips. Nat Protoc 2:1896–1906
10. Lawrence RT, Perez EM, Hernández D et al (2015) The proteomic landscape of triple-negative breast cancer. Cell Rep 11:630–644
11. Kitata RB, Dimayacyac-Esleta BRT, Choong W-K et al (2015) Mining missing membrane proteins by high-pH reverse-phase StageTip fractionation and multiple reaction monitoring mass spectrometry. J Proteome Res 14:3658–3669
12. Wiśniewski JR, Zougman A, Mann M (2009) Combination of FASP and StageTip-based fractionation allows in-depth analysis of the hippocampal membrane proteome. J Proteome Res 8:5674–5678
13. Thompson A, Schäfer J, Kuhn K et al (2003) Tandem mass tags: a novel quantification strategy for comparative analysis of complex protein mixtures by MS/MS. Anal Chem 75:1895–1904
14. Cox J, Mann M (2008) MaxQuant enables high peptide identification rates, individualized p.p.b.-range mass accuracies and proteome-wide protein quantification. Nat Biotechnol 26:1367–1372
15. Cox J, Neuhauser N, Michalski A et al (2011) Andromeda: a peptide search engine integrated into the MaxQuant environment. J Proteome Res 10:1794–1805
16. Olsen JV, Blagoev B, Gnad F et al (2006) Global, in vivo, and site-specific phosphorylation dynamics in signaling networks. Cell 127:635–648
17. Winter D, Seidler J, Ziv Y et al (2009) Citrate boosts the performance of phosphopeptide analysis by UPLC-ESI-MS/MS. J Proteome Res 8:418–424
18. Ow SY, Salim M, Noirel J et al (2009) iTRAQ underestimation in simple and complex mixtures: "the good, the bad and the ugly". J Proteome Res 8:5347–5355
19. Werner T, Sweetman G, Savitski MF et al (2014) Ion coalescence of neutron encoded TMT 10-plex reporter ions. Anal Chem 86:3594–3601

Chapter 9

Multi-Lectin Affinity Chromatography for Separation, Identification, and Quantitation of Intact Protein Glycoforms in Complex Biological Mixtures

Sarah M. Totten, Majlinda Kullolli, and Sharon J. Pitteri

Abstract

Protein glycosylation is considered to be one of the most abundant post-translational modifications and is recognized for playing key roles in cellular functions. Aberrant N-linked glycosylation has been associated with several human diseases and has prompted the development and constant improvement of analytical tools to separate, characterize, and quantify glycoproteins in complex mixtures extracted from various biological samples (such as blood and tissue). Lectins, or carbohydrate-binding proteins, have been used as valuable tools for enriching for glycoproteins and selecting for specific types of glycosylation. Herein a method using multidimensional intact protein fractionation and LC-MS/MS analysis is described. Immunodepletion is used to remove highly abundant proteins from human plasma, followed by glycoform separation using multi-lectin affinity chromatography, in which specific lectins are chosen to capture and elute specific types of glycosylation. Reversed-phase chromatography prior to digestion is used for further fractionation, allowing for an increased number of protein identifications of moderate- to low-abundant proteins detectable in plasma . This method also incorporates isotopic labeling during alkylation for relative quantitation between two samples (such as a case and control). A bottom-up, tandem mass spectrometry-based proteomics approach is used for protein identification and quantitation, and allows for screening glycoform-specific changes across hundreds of plasma proteins.

Key words Multi-lectin affinity chromatography, Glycoproteomics, Protein glycosylation, Plasma proteomics

1 Introduction

Glycosylation is generally thought to be the most widespread post-translational modification and it is predicted that more than half of mammalian proteins are glycosylated [1, 2]. In eukaryotes, most glycans are attached to proteins at asparagine (N-glycans) or serine/threonine (O-glycans). The method presented herein is focused on the study of N-linked glycoproteins. N-linked glycosylation is a prevalent form of glycans covalently attached to an asparagine given an N-X-S/T motif, where X is any amino acid excluding proline.

Lucio Comai et al. (eds.), *Proteomics: Methods and Protocols*, Methods in Molecular Biology, vol. 1550,
DOI 10.1007/978-1-4939-6747-6_9,

Protein glycosylation is a biologically relevant post-translational modification that plays an important role in disease. For example, glycosylation is aberrant in cancer and glycosylation changes are a widespread feature of tumor progression and are related to cancer grade, invasion, and metastasis [3]. In cancer, key cellular pathways and interactions are disrupted by altered protein glycosylation, resulting in opportunities for metastasis. In addition to a variety of cancers, aberrant glycosylation has been associated with several other human diseases, including Alzheimer's disease, multiple sclerosis, inflammatory bowel disease, and a number of autoimmune diseases [4–9]. New tools for systematic studies of protein glycosylation would substantially improve our understanding of its involvement in cellular functions and its relationship to disease. Mass spectrometry is an emerging tool of choice for studying glycosylation. Among the most well developed approaches include glycomics, in which glycans are released from the proteins and characterized [10–16]. The disadvantage to this method is the loss of protein- and site-specific information. Strategies for the in-depth characterization of protein-specific glycosylation are being developed and are more widely available, however are often limited to studying less complex mixtures, single proteins of interest, or highly abundant serum proteins, making it difficult to apply such strategies to systematically studying complex biological mixtures [17–19]. In recent years, new technologies have been developed to analyze intact glycopeptides, and mass spectrometry-based technologies and data analysis tools are constantly improving [20–22].

Lectins are sugar-binding proteins with specificities for particular glycan structures. Lectins have been used in a variety of formats to study glycoproteins. For example, serial lectin chromatography columns have been used to separate protein glycoforms [23, 24]. Multi-lectin affinity columns have also been used to capture and elute glycoproteins in separate fractions [25–27]. Lectin-based glycoproteomic analysis techniques have been applied for biomarker identification in plasma [27–31], including, for example, lectin arrays for the analysis of fucosylated proteins in ovarian cancer serum [32] and lectins for the analysis of pancreatic cancer serum [33].

The method described here is a protocol for using multidimensional intact protein fractionation and liquid chromatography coupled with tandem mass spectrometry (LC-MS/MS) to detect changes in the relative quantitation of specific glycoforms of moderate- to low-abundant plasma proteins. This protocol incorporates immunodepletion, multi-lectin affinity chromatography (M-LAC), and reversed-phase (RP) chromatography (*see* Fig. 1) to address the issue of the complexity and wide dynamic range of biological mixtures while simultaneously separating lower abundant proteins into separate fractions containing specific glycoforms.

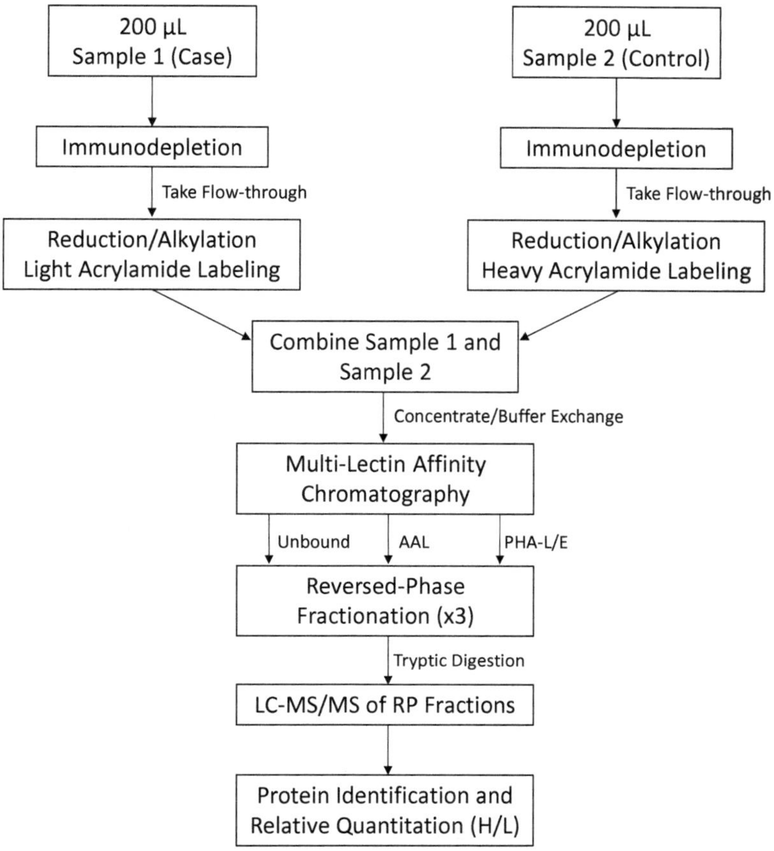

Fig. 1 Multidimensional intact protein fractionation workflow. Plasma is first immunodepleted of 14 abundant proteins, followed by reduction and alkylation. Shown here is an example of how two depleted and isotopically labeled samples can be mixed together to achieve relative quantitation of glycoproteins. Isotopic labeling is performed during the alkylation step using heavy (^{13}C) and light (^{12}C) acrylamide. After samples 1 and 2 are labeled and combined, the protein mixture is separated by multi-lectin affinity chromatography (M-LAC). Glycoproteins are eluted in series from the M-LAC column and further separated by reversed-phase (RP) chromatography. RP fractions are then digested with trypsin and analyzed by LC-MS/MS

By using an M-LAC approach, in which several lectins are combined onto a single column and packed in-house, we can select for specific types of glycosylation, as shown in Fig. 2. Furthermore, by eluting the bound glycoproteins from each individual lectin in multiple elution steps (versus eluting all captured glycoproteins together), specific glycoforms of a given protein are kept in distinct fractions. This provides a way to screen glycosylation changes across hundreds (possibly thousands) of proteins and retain protein-specific information. With a wide variety of agarose bound

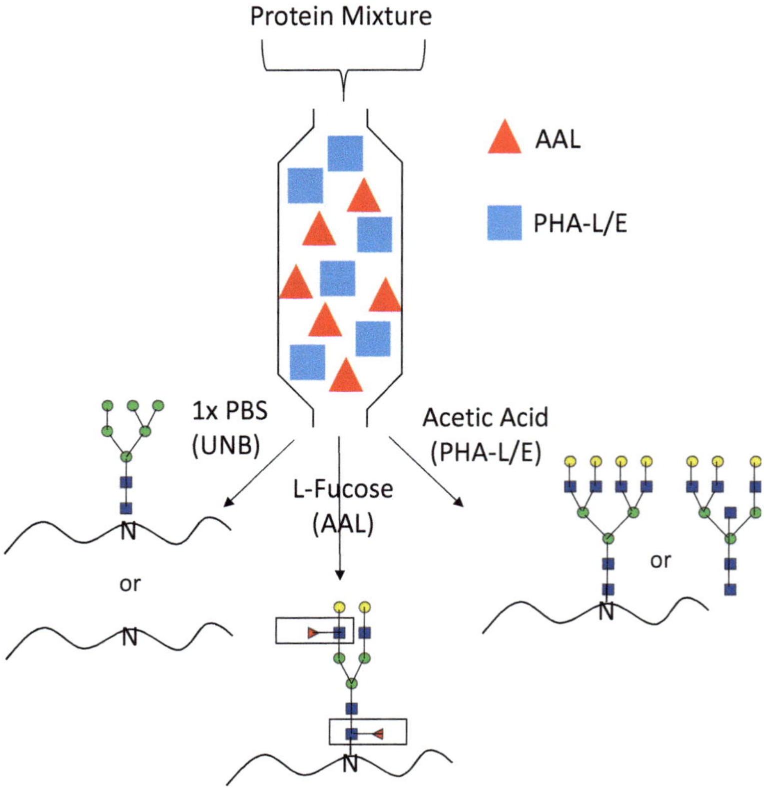

Fig. 2 Multi-lectin affinity chromatography schematic. A mixture of depleted and isotopically labeled proteins is separated on an M-LAC column containing agarose-bound lectins *Aleuria aurantia* lectin (AAL) and *Phaseolus vulgaris* leucoagglutinin (PHA-L) and *Phaseolus vulgaris* erythroagglutinin (PHA-E). Captured glycoproteins are released in a series of elution steps. The flow-through (mobile phase 1× PBS) containing the non- or otherwise-glycosylated proteins is collected first. Next, an L-fucose buffer is used to release fucosylated glycoforms from AAL. AAL preferentially binds core-fucosylated structures, and has a weak affinity for α1-4-linked fucose on the antennae. In the third elution step, low-pH mobile phase (100 mM acetic acid, pH 3.8) is used to release the remaining glycoproteins bound to the PHA-L/E lectins

lectins now commercially available, it is possible to personalize a multi-lectin column by choosing lectins that will select for specific types of protein glycoforms (fucosylation, sialylation, oligomannose, etc.), tailoring the analysis for the specific needs of certain applications. Furthermore, the collected M-LAC fractions are subsequently separated by reversed-phase chromatography on the intact protein level, allowing for deeper analysis and increased number of protein identifications, as shown in Fig. 3. Although this protocol has been optimized for glycoprotein analysis in plasma, it has also been successfully applied to protein extractions from tissue and cell culture extracts.

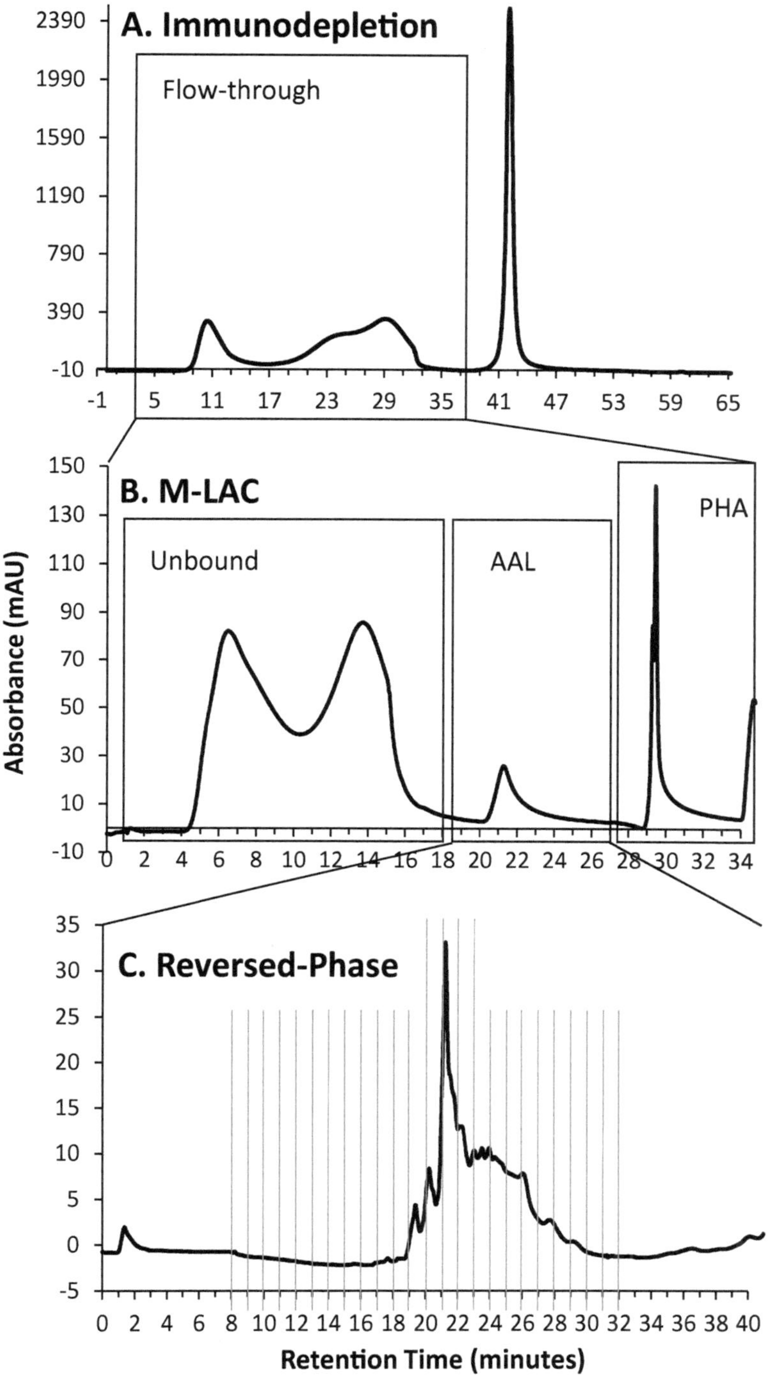

Fig. 3 A series of chromatographic separations on intact plasma proteins. In our multidimensional separation approach, immunodepletion (**a**) is used to deplete plasma of highly abundant proteins. The flow-through from that separation contains moderate- to low-abundant plasma proteins, which are then separated by glycoforms using M-LAC (**b**). Boxes are inset to show suggested fraction collection. Three separate fractions are collected from M-LAC and are further fractionated at the intact protein level by reversed-phase chromatography (**c**). *Inset vertical lines* represent the fractions collected in 1 min intervals

2 Materials

2.1 Depletion of Human Plasma

1. CaptureSelect™ HumanPlasma14 affinity resin (ThermoFisher) (*see* **Note 1** for alternatives).
2. Omnifit® glass columns, 5.6 mL bed volume, 100 mm × 10 mm L × internal diameter (ID) One column has a capacity for 125 μL of plasma. In this protocol, two of these columns are connected and used in tandem for a capacity of 250 μL of plasma.
3. Binding Buffer A: Phosphate-buffered saline (PBS) 1× solution.
4. Elution Buffer B: 100 mM glycine, pH 2.5 with hydrochloric acid (HCl)
5. 200 μL of human plasma of each sample.
6. 3000 nominal molecular weight limit (NMWL) centrifugal filters, 15 mL (Amicon® Ultra—15, Millipore).

2.2 Reduction and Alkylation of Depleted Plasma

1. Bradford protein assay kit (including protein standards and Coomassie dye) and spectrophotometer for measuring protein concentration.
2. Protein denaturation buffer: 8 M urea, 50 mM Tris–HCl, 0.05% octyl β-D-glucopyranoside, pH 7.5, prepared in 100 mM ammonium bicarbonate.
3. Dithiothreitol (DTT) for protein reduction.
4. Acrylamide for protein alkylation. For isotopic labeling, both light (^{12}C) and heavy (^{13}C) acrylamide is required. 1,2,3-$^{13}C3$ (heavy) acrylamide can be purchased from Cambridge Isotope Laboratories. If not labeling, alternative alkylating agents can be used, such as iodoacetamide or iodoacetic acid (not covered in this protocol, *see* **Note 2**).
5. 3000 NMWL centrifugal filters, 4 mL (Amicon® Ultra—4, Millipore) for buffer exchange.
6. Phosphate-buffered saline, 1× solution

2.3 Multi-Lectin Affinity Chromatography

1. Agarose-bound Lectins *Aleuria aurantia* lectin (AAL) and *Phaseolus vulgaris* leucoagglutinin (PHA-L) and *Phaseolus vulgaris* erythroagglutinin (PHA-E) (*see* **Note 3** for alternatives).
2. Omnifit® glass columns, 5.6 mL bed volume, 100 mm × 10 mm L × ID.
3. M-LAC loading buffer: 1× PBS (*see* **Note 4**).
4. 200 mM L-fucose (*see* **Note 5**).
5. 100 mM acetic acid.
6. 3K NMWL centrifugal filters, 4 mL (Amicon® Ultra—4, Millipore).

2.4 Reversed-Phase Chromatography of Intact Proteins from M-LAC Fractions

1. Reversed-phase HPLC analytical column (C8 or similar, 100 mm × 2.1 mm ID).
2. Buffer A: 0.1 % trifluoroacetic acid (TFA) in water.
3. Buffer B: 0.1 % TFA in acetonitrile.

2.5 Tryptic Digestion of Lyophilized Protein from Reversed-Phase (RP) Fractions

1. 50 mM ammonium bicarbonate in 4 % acetonitrile.
2. 200 μL of 0.1 μg/μL trypsin (Promega, approximately 1:20 enzyme to protein).
3. Trypsin resuspension buffer (provided by manufacturer).

2.6 LC-MS/MS Analysis of Tryptic Peptides and Glycopeptides

1. Conical mass spectrometry vials (250 μL, with snap septum lids, or other autosampler vials).
2. Buffer A: 0.1 % formic acid (FA) in water.
3. Buffer B: 0.1 % FA in acetonitrile.
4. C18 analytical column (Picofrit 75 μm ID, New Objective, packed in-house with MagicC18 AQ 5 μm, 100 A resin, packed to 25 cm, or similar).
5. C18 trap column (Thermo Scientific Acclaim PepMap 100, 5 μm particle size, 5 mm length, 300 μm ID).

2.7 Equipment and Instrumentation

1. Depletion, M-LAC separation, and reversed-phase chromatography require a HPLC. The HPLC separations described in this protocol were optimized on an Agilent 1260 Infinity Bio-Inert Quaternary HPLC, equipped with an automatic fraction collector, multiple wavelength UV detector (MWD), and a manual bio-inert injector. Agilent's ChemStation Software (version C.01.05) was used for data collection and analysis.
2. LC-MS/MS experiments require an online nano-HPLC system coupled to a high resolution mass spectrometer with MS/MS capabilities. Protocols described herein were performed on a Dionex Ultimate 3000 RSLC nanoLC system coupled to a LTQ-Orbitrap Elite mass spectrometer (Thermo Scientific).
3. Additional equipment include a microplate absorbance reader or spectrophotometer, a benchtop centrifuge, and a lyophilizer or freeze drying system.

3 Procedures

3.1 Depletion of Human Plasma

CaptureSelect® HumanPlasma14 depletion material is comprised of affinity ligands derived from *Camelidae* antibodies used for the removal of 14 abundant human plasma proteins (albumin, IgG, IgM, IgA, IgE, IgD, and free light chains, transferrin, fibrinogen, α-1 antitrypsin, α-2 macroglobulin, α-1 acid glycoprotein, apolipoprotein A1, and haptoglobulin). CaptureSelect® HP14 immobilized on agarose support resin is commercially available (ThermoFisher Scientific). The efficiency, reproducibility, and specificity of a 5.6 mL CaptureSelect® HP14 column have been previously assessed in our laboratory [34]. The loading capacity of 5.6 mL glass column was determined to be 125 μL of plasma (or 250 μL on two 5.6 mL columns in tandem) [34]. This protocol was optimized on an Agilent 1260 Infinity HPLC, however any HPLC system equipped with two pumps, and a UV detector (monitoring at 280 nm) that can pump up to 5 mL/min can be used. *See* **Note 6** for comments regarding chromatographic gradients and fraction collection times. An example chromatogram of an immunodepletion is shown in Fig. 3a.

1. Gravity pack agarose bound CaptureSelect® HumanPlasma14 resin onto a 5.6 mL Omnifit glass column. Column should be stored at 4 °C and operated at 10 °C. Maximum pressure should not exceed 600 psi. Proteins are detected with UV at 280 nm.
2. For plasma depletion, dilute 200 μL of human plasma 1:5 with 1× PBS. The depletion is performed on two 5.6 mL columns set up in tandem.
3. Load 1 mL of diluted plasma at 0.5 mL/min with Binding Buffer A (1× PBS).
4. Once loaded, maintain the flow rate at 0.5 mL/min for 32 min and then increase to 2.5 mL/min (*see* **Note 6**). Collect flow-through fraction containing the moderate- to low-abundance proteins of interest of the depleted plasma.
5. Elute bound (high-abundance) proteins with 100% Elution Buffer B (100 mM glycine, pH 2.5) at 2.5 mL for 15 min. Collect eluted fraction and freeze for later use if desired.
6. Re-equilibrate column at 2.5 mL/min for 15 min with Binding Buffer A (1× PBS), diverting the flow to waste.
7. Using a 4 mL, 3K NMWL centrifugal filter, concentrate the flow-through fraction (already in 1× PBS) to 100 μL and transfer to a 1.5 mL Eppendorf tube.
8. In isotopic labeling experiments, repeat on a 200 μL aliquot of a second sample to be compared to Sample 1 (i.e., the case sample, as shown in Fig. 1). The two depleted and isotopically labeled samples will be combined, as described in detail in the following sections (*see* **Note 2** for alternative labeling study design).

3.2 Reduction and Alkylation of Depleted Plasma

The amount of depleted protein from plasma is determined using a standard Bradford protein assay. Typically between 0.75 and 1.0 mg of protein is recovered from the depletion of 200 μL of human plasma. Before separation by multi-lectin affinity chromatography, proteins must be reduced and alkylated.

1. Perform a Bradford protein assay to determine protein concentration in depleted and concentrated plasma samples.
2. To the depleted plasma mixture, add the Denaturation Buffer 10:1 (buffer to sample, by volume).
3. Reduce the protein disulfide bonds using DTT by adding 0.616 mg of DTT per mg of protein. Let the reaction run for 2 h at room temperature.
4. Alkylate the reduced protein by adding light acrylamide to Sample 1 at 7.1 mg per mg of reduced protein, and heavy acrylamide at 7.4 mg per mg of reduced protein to Sample 2, as shown in Fig. 1. Leave reaction in the dark at room temperature for one hour (*see* **Note 2**).
5. Once reduced and alkylated, mix the heavy-labeled sample and its paired light-labeled sample together. Concentrate the mixture and buffer exchanged using a 3K Millipore filter into 1× PBS to 1 mL in preparation for multi-lectin affinity chromatography

3.3 Multi-Lectin Affinity Chromatography

Multi-lectin affinity chromatography is used to separate moderate- to low-abundance plasma proteins by specific glycoforms. The method described here combines agarose-bound *Aleuria aurantia* lectin (AAL), *Phaseolus vulgaris* erythroagglutinin and *Phaseolus vulgaris* leucoagglutinin (PHA-E and PHA-L) to capture core fucosylated glycoproteins, and proteins carrying highly branched complex type glycans, respectively, as shown in Fig. 2 (*see* **Note 3**). By eluting each individual fraction in series (versus releasing all glycoproteins bound to all lectins in a single elution step), specific types of glycoforms are separated into distinct fractions depending on their affinity for each lectin (*see* **Note 7**). M-LAC columns are gravity packed with agarose-bound lectins purchased from Vector Labs (Burlingame, CA). Chromatography is performed on an HPLC equipped with a quaternary pump (three pumps are required for M-LAC experiments that use a three buffer system). *See* Fig. 3b for an example M-LAC chromatographic separation and fraction collection.

1. To pack a 5.5 mL Omnifit glass column, combine 2 mL of a 1:1 slurry of AAL in suspension buffer (2 mg of protein per mL of gel) and 2 mL of a 1:1 slurry of PHA-E and PHA-L (for a combined 4 mL of PHA lectin, each 3 mg per mL of gel) in a 15 mL conical tube with gentle swirling (vortexing the slurry could destabilize the media). Rinse each bottle with 1× PBS and add the rinse to the lectin mixture to ensure all of the slurry is used. Slowly load the lectin mixture onto the column and gravity pack a few milliliters at a time, as per the manufacturers' instructions, being sure to never let the column go completely dry.

2. Run a 1× PBS blank across the newly packed column to equilibrate and to check for leaks and pressure issues.
3. When the column is ready, load 1 mL of sample (approximately 1.0–1.5 mg of protein) into a 2 mL loop and inject at 0.5 mL/min with M-LAC loading buffer 1× PBS for 15 min. Collect the flow-through (*see* **Note 6**).
4. Elute the unbound material containing non- or otherwise-glycosylated proteins with 1× PBS at 2.5 mL/min. Collect the eluent. Maximum pressure should not exceed 100 bar.
5. Elute glycoproteins containing core fucosylated glycans (those bound to AAL) from the M-LAC column with 200 mM L-fucose in 1× PBS at 2.5 mL/min. Collect the eluent (*see* **Notes 7** and **8**).
6. Elute glycoproteins containing highly branched, complex-type glycans (those bound to PHA-L/E) by lowering the pH of the mobile phase with 100 mM acetic acid, pH 3.8 at 2.5 mL/min. Collect the eluent.
7. Divert the flow to waste and re-equilibrate the column with 1× PBS for 15 min.
8. Concentrate the three collected fractions (unbound, AAL, and PHA-L/E) using 3K NMWL Amicon centrifugal filters to 200 μL in preparation for reversed-phase fractionation.

3.4 Reversed-Phase Chromatography of Intact Proteins from M-LAC Fractions

Each of the three serial-eluted lectin fractions collected from M-LAC are then further separated by reversed-phase fractionation at the intact protein level. By separating at the protein level, the mixture of proteins is further de-complexed, allowing for deeper analysis and more protein identifications. Furthermore protein glycoforms are confined to more distinct reversed-phase fractions prior to digestion, allowing separation of protein isoforms. Reversed-phase fractionation is performed on a 100 mm × 2.1 mm ID stainless steel column pre-packed with POROS®R2 (Applied Biosystems), 2000 Å particle size poly styrene-divinylbenzene immobile phase. Maximum pressure should not exceed 170 bar.

1. Load the concentrated sample onto a 500 μL sample loop at 0.8 mL/min for 5 min with reversed-phase Solvent A (0.1 % trifluoroacetic acid in water).
2. Separate the proteins using a gradient of increasing organic content in the mobile phase. Between 5 and 38 min, bring reversed-phase Buffer B (0.1 % trifluoroacetic acid in acetonitrile) to 90 %. Hold the gradient at 90 % Buffer B for 2 min, then re-equilibrate the column with 95 % Buffer A for 10 min.
3. Collect reversed-phase fractions in an automatic analytical scale fraction collector (Agilent 1260) into 1 mL racked microtubes.

In total 24, 0.8 mL fractions are collected (one per minute from 8 to 31 min, as shown in Fig. 3c). Fractions can be collected manually if automated fraction collection is unavailable.

4. Repeat **steps 1–3** for remaining M-LAC fractions. Overall, three RP separations are performed (one per M-LAC fraction—unbound, AAL, and PHA-L/E). Per sample, there are 3 sets of 24 RP fractions.
5. Freeze RP fractions at −80 °C and lyophilize.

3.5 Tryptic Digestion of Lyophilized Protein from RP Fractions

1. Reconstitute each of the 24 RP fractions in 50 μL of 50 mM ammonium bicarbonate in 4% acetonitrile. RP fractions containing small amounts of protein can be combined as needed.
2. Digest each RP fraction with 0.5 μg of trypsin at 37 °C for 18 h (or overnight).

3.6 LC-MS/MS Analysis of Tryptic Peptides and Glycopeptides

RP fractions are then analyzed by nano LC-MS/MS. In total, each sample will have 72 RP fractions in total (24 unbound, 24 AAL, and 24 PHA-L/E—or less if some were combined, which is recommended to improve signal in the mass spectrometer).

1. Transfer the 50 μL of each RP fraction to a mass spectrometry vial.
2. Load 15 μL (approximately between 5 and 10 μg of peptides) onto a C18 trap column at 5 μL/min, briefly de-salt and concentrate, then subsequently separate on a 25 cm C18 analytical column (Picofrit 75 μm ID, New Objective, packed in-house with MagicC18 AQ resin). Tryptic peptides are separated using a multistep gradient at a flow rate of 0.6 μL/min in which Buffer B (0.1% FA in acetonitrile) is increased from 0% (100% Buffer A, 0.1% FA in water) to 85% over 120 min. Re-equilibrate the analytical column for 20 min at 98% Buffer A.
3. Ionize nano-LC eluent by electrospray ionization at 2.25 kV with the capillary temperature set to 200 °C.
4. In each MS/MS experiment, perform an initial MS1 scan over an m/z range of 400–1800, followed by 10 data-dependent collision-induced dissociation fragmentation events on the 10 most intense +2 or +3 ions from the MS1 spectrum over an acquisition time of 140 min.
5. Acquired data can be processed by the Computational Proteomics Analysis System (CPAS) [35] pipeline using X! Tandem search algorithm [36, 37] for peptide identifications and the Q3 quantitation algorithm to calculate heavy-to-light ratios for cysteine-containing acrylamide labeled peptides [38]. PeptideProphet [39, 40] and ProteinProphet can be used to validate peptide and protein identifications respectively.

4 Notes

1. Alternative materials for immunodepletion of plasma are available and can be used in this protocol, for example multiple affinity removal column human 14 (Agilent Technologies) and Seppro IgY14 (Sigma-Aldrich). Both of these chemistries deplete plasma of 14 highly abundant proteins, although with slightly different specificities. Serum samples can also be used in lieu of plasma.
2. As described above, isotopic labeling for relative quantitation is incorporated into this protocol during protein alkylation. Heavy (^{13}C) and light (^{12}C) acrylamide are used to label cysteine residues yielding relative ratios of peptides, from which protein levels can be inferred, between two samples, as described in Faca et al. [38]. This method can be used to compare case versus control samples, changes between two time points, or case and controls can be compared against a reference pool. With three ^{13}C, the heavy acrylamide will change the mass of the peptide by 3 Da. The workflow described herein can also be applied to experiments not incorporating isotopic labeling, in which case other alkylating agents, such as iodoacetamide or iodoacetic acid, can be used.
3. One of the great advantages of packing your own multi-lectin affinity column in-house is the ability to incorporate different combinations of lectins to suit the needs of a particular experiment. In this protocol, AAL and PHA-L/E were used to specifically enrich for core-fucosylated glycoforms and glycoforms containing highly branched complex glycan structures, respectively, as shown in Fig. 2. However, there are many additional commercially available agarose-bound lectins (Vector Laboratories) with affinity for a variety of glycan substrates, ranging from those that broadly capture all N-glycan types to those with affinity for very specific structures differing only by glycosidic linkages. Some commonly used lectins include Concanavalin A (ConA), which binds mannose and has a broad affinity for oligomannose, hybrid, and biantennary glycan structures, and wheat germ agglutinin (WGA) that bind N-acetylglucosamine and sialic acid as substrates, recognizing chitobiose at the core of many glycans. These lectins are used to widely capture and enrich for all N-glycans. Alternatively, glycoproteins containing specific glycan structures can be enriched for using lectins such as *Ulex europaeus* agglutinin I (UEA I) or *Lotus tetragonolobus* lectin (LTL) that bind Fuc(α-1,2)Gal (H antigen of ABO blood groups), or *Sambucus nigra* lectin that has preferential affinity for N-acetylneuraminic acids with an α-2,6 linkage to a terminal galactose (see vectorlabs.com for useful information regarding the characteristics of their lectin products).

4. If lectins other than those described here are used in a multi-lectin column, it is important to note they may require different loading buffers. Some lectins do not perform as well in PBS and often require metal ions to activate carbohydrate binding. For example, the widely used ConA lectin performs optimally in loading buffers containing $CaCl_2$, $MnCl_2$, and NaCl. Additionally, the eluting or inhibiting sugar is also specific to the affinity of each lectin. For example, competitive saccharide binding elution buffer for ConA requires α-methylmannoside and α-methylglucoside (whereas L-fucose was the sugar used for release of proteins bound to AAL in this protocol).
5. Fucose buffer should be prepared fresh, ideally the day of use, to prevent bacterial growth. Bacteria will consume the fucose, thereby reducing the buffer concentration.
6. General elution times for each step of the chromatography performed in this protocol is described here, however it is advisable that exact timing for chromatographic gradients and collection time points are optimized to individual columns, sample types, and HPLC systems.
7. Serial elution of glycoproteins from specific lectins during multi-lectin chromatography has advantages and disadvantages. One advantage is being able to collect multiple fractions containing specific glycoforms as opposed to having one "bound" fraction in which all glycoproteins captured by the multiple lectins present in the column are eluted together. Separating proteins by specific glycoforms provides insight into changes in glycosylation between case and control samples. Additionally, collecting multiple fractions during M-LAC separations further fractionate and de-complex biological mixtures containing thousands of proteins and their glycoforms. It should be noted, however, that some of the disadvantages of this method include the nonspecific binding of lectins and ambiguity surrounding the types of glycoforms present in each M-LAC fraction. Elution by competitive saccharide binding may not always be entirely complete, resulting in any remaining glycoproteins bound to that lectin being eluted in the subsequent fraction eluted by low pH. Furthermore, glycoproteins containing multiple types of glycans (for example both fucosylated and highly branched structures) may end up in either or multiple M-LAC fractions depending on how strong the affinity is for either lectin.
8. When bound glycoproteins are serial eluted from a multi-lectin column, the order in which they are eluted is important. Glycoproteins being released from lectins via competitive saccharide binding should always be eluted before the glycoproteins being released from another lectin via low-pH elution (acetic acid mobile phase). The low-pH mobile phase will release all glycoproteins bound to all lectins.

References

1. Wang H, Hanash S (2011) Intact-protein analysis system for discovery of serum-based disease biomarkers. Methods Mol Biol 728:69–85
2. Apweiler R, Hermjakob H, Sharon N (1999) On the frequency of protein glycosylation, as deduced from analysis of the SWISS-PROT database. Biochim Biophys Acta 1473(1):4–8
3. Varki A, Cummings RD, Esko JD, et al., editors. Cold Spring Harbor (NY): Cold Spring Harbor Laboratory Press; 2009. Bookshelf ID: NBK1963 PMID: 20301279
4. Mkhikian H, Grigorian A, Li CF, Chen H-L, Newton B, Zhou RW, Beeton C, Torossian S, Tatarian GG, Lee S-U, Lau K, Walker E, Siminovitch KA, Chandy KG, Yu Z, Dennis JW, Demetriou M (2011) Genetics and the environment converge to dysregulate N-glycosylation in multiple sclerosis. Nat Commun 2:334
5. Palmigiano A, Barone R, Sturiale L, Sanfilippo C, Bua RO, Romeo DA, Messina A, Capuana ML, Maci T, Le Pira F, Zappia M, Garozzo D (2016) CSF N-glycoproteomics for early diagnosis in Alzheimer's disease. J Proteomics 131:29–37
6. Theodoratou E, Campbell H, Ventham NT, Kolarich D, Pucic-Bakovic M, Zoldos V, Fernandes D, Pemberton IK, Rudan I, Kennedy NA, Wuhrer M, Nimmo E, Annese V, McGovern DPB, Satsangi J, Lauc G (2014) The role of glycosylation in IBD. Nat Rev Gastroenterol Hepatol 10:588–600, advance online publication
7. Liang H-C, Russell C, Mitra V, Chung R, Hye A, Bazenet C, Lovestone S, Pike I, Ward M (2015) Glycosylation of human plasma clusterin yields a novel candidate biomarker of Alzheimer's disease. J Proteome Res 14(12):5063–5076
8. Goulabchand R, Vincent T, Batteux F, J-f E, Guilpain P (2014) Impact of autoantibody glycosylation in autoimmune diseases. Autoimmun Rev 13(7):742–750
9. Lauc G, Huffman JE, Pučić M, Zgaga L, Adamczyk B, Mužinić A, Novokmet M, Polašek O, Gornik O, Krištić J, Keser T, Vitart V, Scheijen B, Uh H-W, Molokhia M, Patrick AL, McKeigue P, Kolčić I, Lukić IK, Swann O, van Leeuwen FN, Ruhaak LR, Houwing-Duistermaat JJ, Slagboom PE, Beekman M, de Craen AJM, Deelder AM, Zeng Q, Wang W, Hastie ND, Gyllensten U, Wilson JF, Wuhrer M, Wright AF, Rudd PM, Hayward C, Aulchenko Y, Campbell H, Rudan I (2013) Loci associated with N-glycosylation of human immunoglobulin G show pleiotropy with autoimmune diseases and haematological cancers. PLoS Genet 9(1):e1003225
10. Zaia J (2010) Mass spectrometry and glycomics. OMICS 14(4):401–418
11. Mechref Y, Hu Y, Desantos-Garcia JL, Hussein A, Tang H (2013) Quantitative glycomics strategies. Mol Cell Proteomics 12(4):874–884
12. An HJ, Kronewitter SR, de Leoz MLA, Lebrilla CB (2009) Glycomics and disease markers. Curr Opin Chem Biol 13(5-6):601–607
13. Fujitani N, J-i F, Araki K, Fujioka T, Takegawa Y, Piao J, Nishioka T, Tamura T, Nikaido T, Ito M, Nakamura Y, Shinohara Y (2013) Total cellular glycomics allows characterizing cells and streamlining the discovery process for cellular biomarkers. Proc Natl Acad Sci U S A 110(6):2105–2110
14. Hu Y, Zhou S, Yu C-Y, Tang H, Mechref Y (2015) Automated annotation and quantitation of Glycan by LC-ESI-MS analysis using MultiGlycan-ESI computational tool. Rapid Commun Mass Spectrom 29(1):135–142
15. Ruhaak LR, Miyamoto S, Lebrilla CB (2013) Developments in the identification of glycan biomarkers for the detection of cancer. Mol Cell Proteomics 12(4):846–855
16. Zhou S, Hu Y, DeSantos-Garcia JL, Mechref Y (2015) Quantitation of permethylated N-glycans through multiple-reaction monitoring (MRM) LC-MS/MS. J Am Soc Mass Spectrom 26(4):596–603
17. Wu S-W, Pu T-H, Viner R, Khoo K-H (2014) Novel LC-MS2 product dependent parallel data acquisition function and data analysis workflow for sequencing and identification of intact glycopeptides. Anal Chem 86(11):5478–5486
18. Saba J, Dutta S, Hemenway E, Viner R (2012) Increasing the productivity of glycopeptides analysis by using higher-energy collision dissociation-accurate mass-product-dependent electron transfer dissociation. Int J Proteomics 2012:560391
19. Hong Q, Ruhaak LR, Stroble C, Parker E, Huang J, Maverakis E, Lebrilla CB (2015) A method for comprehensive glycosite-mapping and direct quantitation of serum glycoproteins. J Proteome Res 14(12):5179–5192
20. Mayampurath A, Yu C-Y, Song E, Balan J, Mechref Y, Tang H (2014) Computational framework for identification of intact glycopeptides in complex samples. Anal Chem 86(1):453–463
21. Hu H, Khatri K, Zaia J (2016) Algorithms and design strategies towards automated glycoproteomics analysis. Mass Spectrom Rev n/a-n/a
22. Mayampurath A, Song E, Mathur A, Yu C-y, Hammoud Z, Mechref Y, Tang H (2014) Label-free glycopeptide quantification for

biomarker discovery in human sera. J Proteome Res 13(11):4821–4832

23. Drake PM, Schilling B, Niles RK, Braten M, Johansen E, Liu H, Lerch M, Sorensen DJ, Li B, Allen S, Hall SC, Witkowska HE, Regnier FE, Gibson BW, Fisher SJ (2011) A lectin affinity workflow targeting glycosite-specific, cancer-related carbohydrate structures in trypsin-digested human plasma(). Anal Biochem 408(1):71–85
24. Jung K, Cho W, Regnier FE (2009) Glycoproteomics of plasma based on narrow selectivity lectin affinity chromatography. J Proteome Res 8(2):643–650
25. Gbormittah FO, Hincapie M, Hancock WS (2014) Development of an improved fractionation of the human plasma proteome by a combination of abundant proteins depletion and multi-lectin affinity chromatography. Bioanalysis 6(19):2537–2548
26. Lee LY, Hincapie M, Packer N, Baker MS, Hancock WS, Fanayan S (2012) An optimized approach for enrichment of glycoproteins from cell culture lysates using native multi-lectin affinity chromatography. J Sep Sci 35(18):2445–2452
27. Kullolli M, Hancock WS, Hincapie M (2008) Preparation of a high-performance multi-lectin affinity chromatography (HP-M-LAC) adsorbent for the analysis of human plasma glycoproteins. J Sep Sci 31(14):2733–2739
28. Fanayan S, Hincapie M, Hancock WS (2012) Using lectins to harvest the plasma/serum glycoproteome. Electrophoresis 33(12): 1746–1754
29. Madera M, Mechref Y, Klouckova I, Novotny MV (2007) High-sensitivity profiling of glycoproteins from human blood serum through multiple-lectin affinity chromatography and liquid chromatography/tandem mass spectrometry. J Chromatogr B 845(1):121–137
30. Song E, Zhu R, Hammoud ZT, Mechref Y (2014) LC–MS/MS quantitation of esophagus disease blood serum glycoproteins by enrichment with hydrazide chemistry and lectin affinity chromatography. J Proteome Res 13(11):4808–4820
31. Mechref Y, Madera M, Novotny MV (2008) Glycoprotein enrichment through lectin affinity techniques. In: Posch A (ed) 2D PAGE: sample preparation and fractionation. Humana Press, Totowa, NJ, pp 373–396. doi:10.1007/978-1-60327-064-9_29
32. Wu J, Xie X, Liu Y, He J, Benitez R, Buckanovich RJ, Lubman DM (2012) Identification and confirmation of differentially expressed fucosylated glycoproteins in the serum of ovarian cancer patients using a lectin array and LC-MS/MS. J Proteome Res 11(9):4541–4552
33. Zhao J, Qiu W, Simeone DM, Lubman DM (2007) N-linked glycosylation profiling of pancreatic cancer serum using capillary liquid phase separation coupled with mass spectrometric analysis. J Proteome Res 6(3):1126–1138
34. Kullolli M, Warren J, Arampatzidou M, Pitteri SJ (2013) Performance evaluation of affinity ligands for depletion of abundant plasma proteins. J Chromatogr B Analyt Technol Biomed Life Sci 939:10–16
35. Rauch A, Bellew M, Eng J, Fitzgibbon M, Holzman T, Hussey P, Igra M, Maclean B, Lin CW, Detter A, Fang R, Faca V, Gafken P, Zhang H, Whiteaker J, States D, Hanash S, Paulovich A, McIntosh MW (2006) Computational proteomics analysis system (CPAS): an extensible, open-source analytic system for evaluating and publishing proteomic data and high throughput biological experiments. J Proteome Res 5(1):112–121
36. Craig R, Beavis RC (2004) TANDEM: matching proteins with tandem mass spectra. Bioinformatics 20(9):1466–1467
37. Craig R, Cortens JP, Beavis RC (2004) Open source system for analyzing, validating, and storing protein identification data. J Proteome Res 3(6):1234–1242
38. Faca V, Coram M, Phanstiel D, Glukhova V, Zhang Q, Fitzgibbon M, McIntosh M, Hanash S (2006) Quantitative analysis of acrylamide labeled serum proteins by LC-MS/MS. J Proteome Res 5(8):2009–2018
39. Keller A, Nesvizhskii AI, Kolker E, Aebersold R (2002) Empirical statistical model to estimate the accuracy of peptide identifications made by MS/MS and database search. Anal Chem 74(20):5383–5392
40. Nesvizhskii AI, Keller A, Kolker E, Aebersold R (2003) A statistical model for identifying proteins by tandem mass spectrometry. Anal Chem 75(17):4646–4658

Chapter 10

Parallel Exploration of Interaction Space by BioID and Affinity Purification Coupled to Mass Spectrometry

Geoffrey G. Hesketh, Ji-Young Youn, Payman Samavarchi-Tehrani, Brian Raught, and Anne-Claude Gingras

Abstract

Complete understanding of cellular function requires knowledge of the composition and dynamics of protein interaction networks, the importance of which spans all molecular cell biology fields. Mass spectrometry-based proteomics approaches are instrumental in this process, with affinity purification coupled to mass spectrometry (AP-MS) now widely used for defining interaction landscapes. Traditional AP-MS methods are well suited to providing information regarding the temporal aspects of soluble protein–protein interactions, but the requirement to maintain protein–protein interactions during cell lysis and AP means that both weak-affinity interactions and spatial information is lost. A more recently developed method called BioID employs the expression of bait proteins fused to a nonspecific biotin ligase, BirA*, that induces in vivo biotinylation of proximal proteins. Coupling this method to biotin affinity enrichment and mass spectrometry negates many of the solubility and interaction strength issues inherent in traditional AP-MS methods, and provides unparalleled spatial context for protein interactions. Here we describe the parallel implementation of both BioID and FLAG AP-MS allowing simultaneous exploration of both spatial and temporal aspects of protein interaction networks.

Key words Mass spectrometry, BioID, Biotin, Streptavidin, FLAG tag, Proximity labeling, Affinity purification, Proteomics, Protein interactions, Protein network, Protein identification

1 Introduction

Understanding how proteins associate with one another is paramount to discovering their molecular function. While multiple experimental strategies have been devised for this purpose [1], the coupling of immunoprecipitation or other affinity purification strategy with mass spectrometry (referred to as AP-MS) is particularly powerful at detecting interactions for proteins of interest under near-endogenous conditions [2]. Building on initial successes in the model organism *S. cerevisiae* [3, 4], the use of AP-MS in human cells has grown from initially small-scale experiments consisting of only one or a few "bait" proteins, to scales that are approaching the genomics realm [5, 6]. Multiple computational tools have been

Lucio Comai et al. (eds.), *Proteomics: Methods and Protocols*, Methods in Molecular Biology, vol. 1550, DOI 10.1007/978-1-4939-6747-6_10, developed that can help identify "true" interactions from contaminants, by comparing the quantitative mass spectrometry results to other purifications performed under the same conditions [7, 8], by comparing against well-defined negative control purifications [2], or a combination of both [9]. As a result, the quality as well as the quantity of protein interactions identified by AP-MS has grown considerably over the past several years.

A key strength of AP-MS is its versatility: it is compatible with a number of tagging strategies and cell types or, provided that proper controls are designed, with immunoprecipitation of endogenous proteins [10]. Providing that the interactions withstand the purification conditions, AP-MS can capture relative changes in interactions imparted by changes in growth conditions, for example stimulation by growth factors, or treatment with pharmacological inhibitors. This enables the analysis of protein interaction kinetics, especially when coupled to robust quantitative approaches such as Multiple/Selected Reaction Monitoring [11, 12], Data Independent Acquisition (DIA, e.g., SWATH, [13, 14]), MS1 intensity measurements [15, 16], or isobaric and isotopic labeling strategies [17–19].

While the temporal aspects of protein–protein interactions can be captured by AP-MS, using this approach to understand the spatial component of interactions has been much more challenging. The central problem lies in the fact that cell lysis must be performed prior to affinity purification, which inevitably results in mixing of the cellular subcompartments and loss of spatial identity. In some small scale studies, this has been to some extent mitigated by combining subcellular fractionation with AP-MS [20], but this approach is neither very robust, nor scalable, and can only provide information on the few subcellular fractions that are generated. We note that emerging approaches such as those employing cell-permeable cross-linkers can in principle generate spatial information for protein interactions (e.g., [21]), but to date, the coverage of crosslinking information for any protein (or protein complex) is sparse. Alternative approaches that do not involve a mass spectrometer at all (e.g., Proximity Ligation Assay, PLA, Fluorescence Resonance Energy Transfer, FRET, or Protein-fragment Complementation Assay, PCA) have been classically used for characterizing the spatial components of protein–protein interactions [1], but these methods require prior knowledge of the two interaction partners or the capability to execute large scale screens.

Besides the loss of spatial identity in AP-MS (and all biochemical strategies employing post-cell lysis purification), another challenge has been the analysis of interactions for proteins that are refractory to solubilization in standard lysis buffers used for AP-MS. These notably include membrane proteins, which require higher concentrations of detergents for extraction from the lipid bilayer than is typically used to maintain protein–protein interactions in AP-MS. While it is indeed possible to systematically alter detergent types and concentrations prior to performing AP-MS [22], or to screen for multiple conditions of detergents, salts and buffers [23], it is not clear what interactions are maintained under these conditions. Furthermore, this systematic condition screening, while sometimes itself informative [23], drastically reduces the overall throughput of the process.

Complementary approaches to AP-MS with the potential to improve spatial resolution have emerged in the past few years. The first of these, which will be reviewed here, is termed BioID and consists of the fusion of a bait protein to an enzyme, namely a mutated bacterial biotin ligase (BirA*) capable of inducing the activation of biotin to biotin-AMP, but does not directly couple this activated biotin to a specific sequence [24]. As such, BirA* creates a cloud of activated biotin-AMP in the vicinity of the bait that can covalently react with lysine ε-amines, tagging the proximity partners for the bait of interest. A related approach, APEX, uses a peroxidase that can mediate the transfer of biotin (from biotin-phenol) to tyrosine residues upon stimulation with peroxide [25]. Since the bait's neighbors become covalently labeled with biotin, there is no need to maintain protein–protein interactions during lysis and purification, and indeed, harsher lysis conditions can be employed. In the original BioID manuscript, proximity partners for the nuclear lamina component LMNA were identified [24], and this work was followed by a number of papers describing the applicability of the approach for the identification of components of the nuclear pore [26], chromatin [27, 28], plasma membrane or junctions [29, 30], centrosome and cilium [30, 31].

Parallel AP-MS and BioID across several studies revealed their complementary nature [27, 29, 32]. Notably, the long labeling times of BioID (6–24 h is typical) make it less than ideal for studying interaction dynamics (e.g., growth factors elicit multiple interactomic changes in less than a minute), but it is superior to AP-MS for probing the interactomes of insoluble structures. Interestingly, BioID can also be used to capture cycling/transient interactions, such as those involving a posttranslational recognition event such as phosphorylation [29], and to identify substrates for certain classes of enzymes such as E3 ligases [32] and phosphatase regulatory subunits [33].

The protocols presented here build upon the complementarity of BioID and AP-MS [27, 29, 32] (*see* Fig. 1 for overall workflow). We describe the cloning and generation of a single pool of cells that harbors both the BirA* tag for BioID and a FLAG tag for detection of bait expression and for AP-MS. Here, we provide a version of the BioID protocol, which we have optimized for membrane-associated proteins. We also briefly describe the FLAG protocol (which only bears minor modification from our previous method article [34]; note that this tag is significantly larger than the single or 3× FLAG used in our previous Methods article, which may affect folding and/or expression of some proteins).

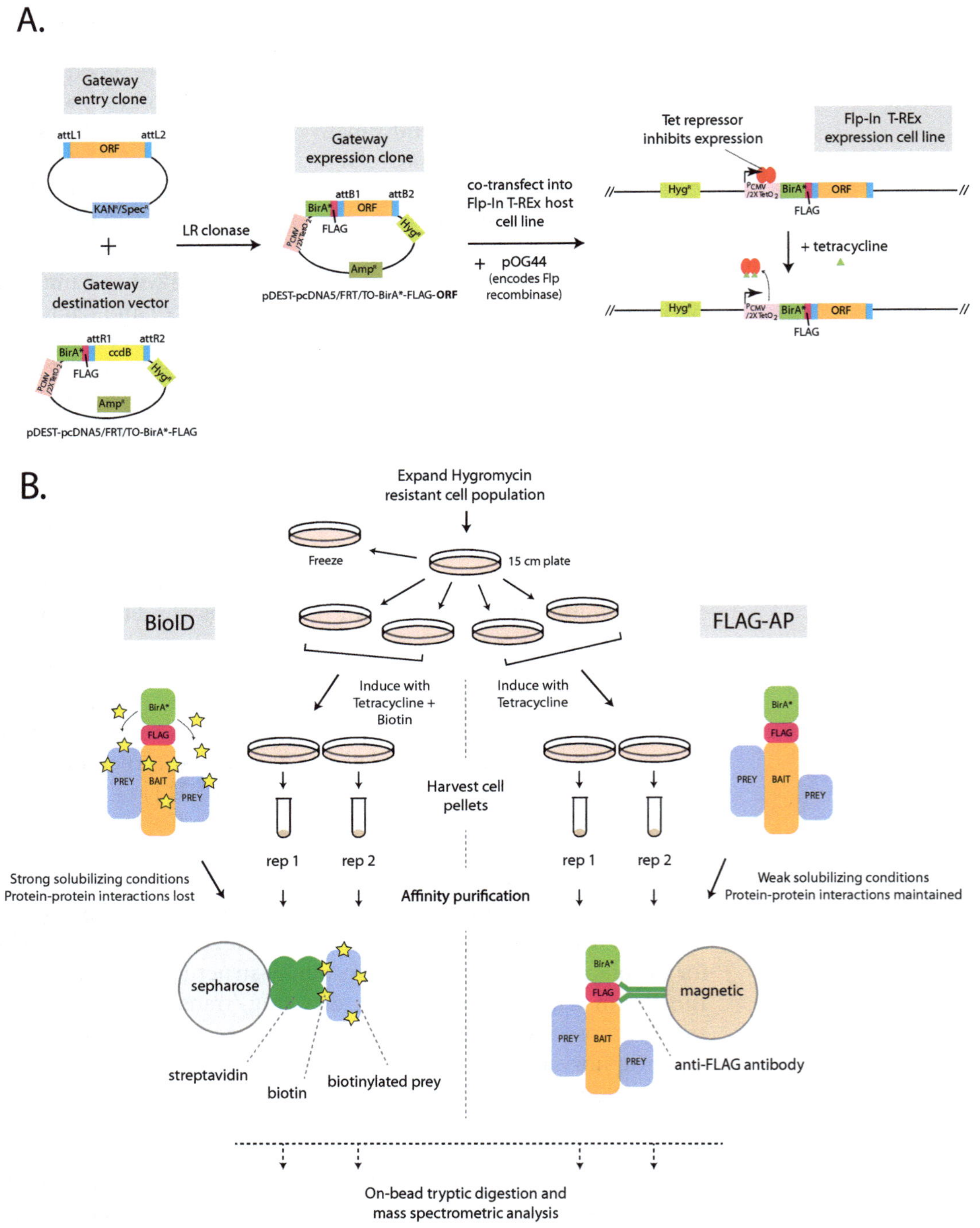

Fig. 1 (**a**) Simplified diagram shows steps to generate pDEST-pcDNA5-FRT/TO-BirA*-FLAG-ORF expression plasmid and to integrate this construct at a specific genomic locus in Flp-In™ T-REx™ cell line. In the first step, the Open Reading Frame (ORF) of your gene of interest in a Gateway entry vector is transferred to the pDEST-pcDNA5/FRT/TO-BirA*-FLAG vector through LR reaction. Confirmed expression plasmid (pDEST-pcDNA5/FRT/TO-BirA*-FLAG-ORF) can then be co-transfected into Flp-In™ T-Rex™ host cell line with pOG44 plasmid, which facilitates integration of the expression plasmid at a predestined FRT locus. Successful integration at the FRT locus provides resistance to hygromycin. Once integrated, BirA*-FLAG-ORF expression is

2 Materials

2.1 Expression Constructs

1. Gateway entry vector for the protein(s) of interest (*see* Subheading 3). Either "open" (no stop codon) or "closed" (with stop codon) ORFs should be used depending on whether C-terminal or N-terminal fusions, respectively, are desired.
2. Gateway destination vector (pDEST-pcDNA5-BirA*-FLAG_Nterm for N-terminal fusions or pDEST-pcDNA5- BirA*FLAG_Cterm for C-terminal fusions). These can be requested through the Gingras lab. Non-Gateway versions of these vectors can be requested through the Raught lab.
3. Gateway LR Clonase II (Invitrogen, Cat #11791-100).
4. Competent bacteria (e.g., DH5α).
5. SOC and LB media.
6. LB-ampicillin (100 μg/mL) agar plates.
7. Ampicillin stock (100 mg/mL).
8. DNA miniprep kit.
9. BsrGI restriction enzyme and standard DNA agarose gel system (for clone validation).

2.2 Generation of Pooled Stable Cell Lines

1. Flp-In T-REx HEK293 cells (Invitrogen, for cell culture).
2. DMEM (high glucose, with pyruvate and L-glutamine) supplemented with 10% FBS, 100 U/mL penicillin/streptomycin (Flp-In and T-REx cassettes can be selected for by the addition of Zeocin (100 μg/mL) and Blasticidin (3 μg/mL), respectively, to the media).
3. pOG44 vector (encodes Flp recombinase) (1 μg/mL), purified expression clone (100 ng/μL), transfection reagent.
4. Standard growth media supplemented with 200 μg/mL Hygromycin B.

Fig. 1 (continued) constitutively repressed by the Tet repressor until tetracycline is supplemented. For complete details, please see the Gateway and Flp-In™ T-Rex™ manual available on the Invitrogen website. (**b**) Schematic representation of the parallel BioID and FLAG AP-MS workflow. Hygromycin resistant clones are pooled and expanded to five 15 cm plates; one for generating frozen stock, and two biological replicates for BioID, the two biological replicates for FLAG-AP. For BioID, pooled clones expressing BirA*-FLAG-ORF construct and parallel control cell lines are grown to approximately 80 % confluence, and treated with tetracycline and biotin (4–24 h) to induce bait expression and biotinylation of proximal proteins. Cells are lysed under strong solubilizing conditions (protein–protein interactions are lost) and AP is performed with streptavidin-sepharose beads (buffer conditions are outlined in Fig. 2). For FLAG-AP, cells are grown to approximately 80 % confluence, and treated with tetracycline (24 h) to induce bait expression. Cells are lysed under gentle solubilizing conditions (protein–protein interactions are maintained) and AP is performed with anti-FLAG magnetic beads. In both cases, proteins are digested on-bead for subsequent mass-spectrometric analysis

2.3 Biotin–Streptavidin Affinity Purification (BioID)

Use "clean" (non-autoclaved) pipette tips and tubes throughout procedure (*see* **Note 1**). Use ultrapure/HPLC grade water for all buffers and reagents.

1. BioID Lysis Buffer Stock, for 50 mL: 50 mM Tris pH 7.5 (2.5 mL of 1 M stock), 150 mM NaCl (1.5 mL of 5 M stock), 0.4% SDS (2 mL of 10% stock), 1% IGEPAL CA-630 (or other Nonidet P-40 substitute) (5 mL of 10% stock), 1.5 mM $MgCl_2$ (150 μL of 0.5 M stock), 1 mM EGTA (200 μL of 0.25 M stock), fill up to 50 mL with water.
2. BioID Lysis Buffer Complete (*see* **Note 2**): Per 1 mL Lysis Buffer add (freshly) 2 μL of 500× stock of Protease Inhibitors (PI) (Cat #P8340, Sigma-Aldrich), and 1 μL Benzonase (250 U/μL) (Cat #71205, EMD-Millipore) (alternatively Turbo Nuclease (Cat #9207, BioVision) may be used).
3. BioID Wash Buffer (for 10 mL): 2% SDS (2 mL of 10% stock), 50 mM Tris pH 7.5 (500 μL of 1 M stock), fill up to 10 mL with water.
4. Biotin stock solution (for 20 mL of 20 mM which is a 400× stock): 100 mg Biotin, add 2 mL of 30% NH_4OH and place on ice. Slowly add 5 mL of 1 N HCl, wait 5 min, and repeat for a total of 18 mL added. Store at 4 °C protected from light (*see* **Note 3**).
5. Tetracycline stock solution (10 mg/mL which is a 10,000× stock): 100 mg in 10 mL 50% EtOH (anhydrous) in water. Store at −20 °C protected from light (*see* **Note 4**).
6. Streptavidin-sepharose beads "high performance" (Cat #17-5113-01, GE Healthcare).
7. Ammonium bicarbonate (ABC) (50 mM, pH 8.0): 200 mg in water (*see* **Note 5**).
8. Trypsin from porcine pancreas (proteomics grade) (*see* **Note 6**): for the solution to digest on the Streptavidin-sepharose beads, resuspend one 20 μg tube (Cat #T6567, Sigma-Aldrich) in a final volume of 2 mL ABC (concentration 10 ng/μL). For spike-in trypsin addition, resuspend a 20 μg tube of trypsin in 100 μL ABC (concentration 0.2 μg/μL). When using trypsin "singles" (Cat #T7575, Sigma-Aldrich) instead, resuspend each in 10 μL ABC and dilute appropriately to achieve desired final concentration.
9. Formic acid (mass spectrometry grade): 5% and 10% stocks freshly prepared in HPLC grade water and stored in clean glass vials (Cat #14-955-319, Fisher Scientific).
10. Autosampler vials (Cat #160134, Dionex) and caps (Cat #03-391-43, Fisher Scientific).

2.4 FLAG Affinity Purification (FLAG-AP)

1. FLAG Lysis Buffer: 50 mM Hepes–KOH pH 8.0, 100 mM KCl, 2 mM EDTA, 0.1% NP40, and 10% glycerol, supplemented with 1 mM PMSF, 1 mM DTT, and 1× protease

inhibitor cocktail (Cat #P8340, Sigma-Aldrich) immediately prior to using.

2. FLAG Rinsing Buffer: 20 mM Tris–HCl (pH 8.0) and 2 mM $CaCl_2$.
3. Trypsin Digest Buffer: 20 mM Tris–HCl (pH 8.0).
4. Anti-FLAG M2 magnetic beads (Cat #M8823, Sigma-Aldrich).
5. Trypsin from porcine pancreas (proteomics grade): resuspend a 20 μg tube (Cat #T6567, Sigma-Aldrich) in 200 μL Trypsin Digest Buffer (concentration 0.1 μg/μL). If using trypsin "singles" (Cat #T7575, Sigma-Aldrich), resuspend each tube in 10 μL Trypsin Digest Buffer (concentration 0.1 μg/μL).
6. Formic acid: (mass spectrometry grade) 50% stock prepared in HPLC grade water and stored in clean glass vial (Cat #14-955-319, Fisher Scientific).

3 Methods

3.1 Preparation of Expression Constructs Using Gateway Cloning

A Gateway "entry" clone is defined as the open reading frame (ORF) for the protein of interest cloned into a Gateway donor vector such as pDONR233 (Invitrogen). For N-terminal fusions, we recommend using an entry clone harboring a stop codon to prevent cloning scars [35]; C-terminal fusions should be generated with "open" entry vectors. The protocol assumes that an entry clone for Gateway cloning is available for the protein of interest (and that it has been sequence-verified); a miniprep provides a sufficient amount. New "Entry" clones can be generated through PCR amplification of inserts with flanking attB sites and recombination into a donor vector with attP sites by Gateway "BP" reaction. Entry clones may also be generated via a BP reaction with an expression clone and a donor vector. *Alternatively, BirA*-FLAG vectors compatible with standard restriction enzyme/ligation cloning can be used.*

1. For the Gateway LR reaction, combine the following: (a) Entry clone (for LR reaction) (ideally 150 ng, although less can be used); (b) Destination vector (1 μL of a 150 ng/μL stock); (c) Water or TE pH 8.0 up to 4 μL.
2. Thaw LR Clonase on ice and briefly mix/vortex, then add 1 μL of LR Clonase to reaction (total volume of 5 μL), mix and spin down. Incubate reaction at RT for 1 hour up to overnight (*see* **Note** 7).
3. Add 1 μL of Proteinase K solution to reaction and incubate for 10 min at 37 ° C.
4. To transform, add 1–2 μL of LR reaction to 20–50 μL chemically competent *E. coli* (e.g., DH5α) and incubate on ice for

approximately 15 min. Heat shock at 42 °C for 30 s and return to ice for 2 min. Add a 10× volume of SOC medium at RT and place in a 37 °C shaking incubator for 60 min. Plate appropriate volumes (typically 40–100 μL will yield good colony density) onto LB-agar plates containing 100 μg/mL ampicillin. Incubate plates at 37 °C overnight.

5. Expand single colonies in 2 mL LB + ampicillin media overnight in 37 °C shaking incubator. Prepare DNA mini-preps of the clones, following the manufacturer's instructions. A diagnostic digest can be performed with BsrGI restriction enzyme to check insert size by analysis with DNA-gel electrophoresis (Gateway recombination sites flanking the ORF contain BsrGI cut sites. Be sure to check if insert contains an internal BsrGI site in order to appropriately interpret the digestion results). If the original entry clone was sequenced, we do not sequence the destination vector. Label and store the constructs (Laboratory Information Management Systems such as OpenFreezer [36] are useful for this process).

3.2 Generation of Pools of Stable Cell Lines

Our groups have been primarily using the Flp-In T-REx HEK293 cell system from Invitrogen for interaction proteomics (other Flp-In T-Rex cells can alternatively be used with the same constructs). The system enables tetracycline-inducible expression of the transgenes expressed at a single copy (from the same Flp Recombination Target (FRT) containing locus) through a Flp-mediated recombination event. This system enables robust establishment of cell lines and pools of cell lines, even for proteins whose expression is toxic for cell growth (*see* **Note 8**).

1. On DAY 0, plate approximately 6×10^5 cells in wells of 6-well plates such that they are approximately 60–80% confluent at the time of transfection (use media that does not contain antibiotics for optimal cell viability). In order to transfect with pcDNA5-FRT-TO based vectors, ensure that media does not contain Zeocin, to which resistance is lost upon successful recombination.
2. On DAY 1, transfect cells with 1 μg pOG44 plasmid (expresses the Flp recombinase) and 100 ng pDEST-pcDNA5-BirA*-FLAG-ORF (aka, the expression clone) construct using mammalian cell transfection reagent of choice according to manufacturer's directions (*see* **Note 9**). Remember to transfect your selected controls in parallel (*see* **Note 10**).
3. On DAY 2, split each well into a 10 cm dish in complete media (*see* **Note 11**).
4. On DAY 3, aspirate media and replace with complete media supplemented with 200 μg/mL Hygromycin B to begin selection of integrated cells. Be careful to avoid cross-contamination during the selection process.

5. Allow selection to occur for approximately 1–2 weeks, changing selection media as required to remove dead cells and to allow stable colonies to grow.
6. Once colonies are clearly visible by eye (a few mm in diameter) pool them into either 6 cm (<20 colonies), 10 cm (approximately 50–100 colonies), or 15 cm (high colony density) plates and continue to expand in selection media (*see* **Note 12**).
7. Scale cells up until a near confluent 15 cm dish is achieved, and then split this plate into additional 15 cm plates. If parallel BioID and FLAG AP-MS experiments are to be performed in biological duplicates, re-plate into 5 × 15 cm plates (one plate to be frozen, and two to be processed by BioID and two to be processed by FLAG AP-MS; *see* Fig. 1. Distribute the volume into the destination plates such that the two replicates for FLAG-AP or BioID are neither induced at the same time nor harvested at the same time. For example: resuspend trypsinized plate into 12 mL and distribute 3 mL to freeze back, 3 mL for each of the biological replicates 1 (BioID and FLAG), and 1.5 mL for each of the biological replicates 2. If only FLAG or BioID is to be performed, or if a different number of replicates is needed, adjust this last splitting accordingly.

3.3 Preparing Cell Pellets

The protocol for the preparation of the cell pellets and the BioID and FLAG-AP purification assumes that only identification and quantification of the proteins is desired. Note that cell pellets are typically harvested at room temperature for BioID, whereas pellets should be harvested on ice with chilled PBS for FLAG-AP. *For projects that involve the characterization of specific posttranslational modifications, extra care should be taken* (e.g., *to inactivate phosphatases in the case of phosphorylation). This can be done through the addition of inhibitors, handling all steps with ice-cold reagents, and shortening some of the incubation times. Also note that if the analysis of specific peptides is needed, it is recommended to include alkylation and reduction steps prior to mass spectrometric identification.*

1. Grow 1 × 15 cm plate of cells stably expressing Flp-In BirA*-FLAG tagged construct (or suitable control) to approximately 80% confluence in complete media (*see* **Note 13**).
2. Add 1 μg/μL of tetracycline (2 μL of a 10 mg/mL stock per 20 mL media) for both BioID and FLAG-AP analysis and 50 μM biotin (50 μL of a 20 mM stock per 20 mL media) for BioID only. Incubate the cells for 24 h (*see* **Note 14**). For BioID, note that you can pre-build protein expression by incubating with tetracycline for 12–24 h prior to adding biotin for 4–12 h. Keep in mind, however, that you should always time the incubation of your controls and samples carefully to enable comparative studies.

3. Pre-weigh pre-labeled 2 mL tubes and write weight on side (*see* **Note 15**). This is important, as the quantity of lysis buffer used is based on the weight of the "dry" cell pellet. Prepare other equipment to harvest cells: beaker of PBS, beaker of 70% EtOH, beaker for waste, bottle of PBS (at room temperature), clean rubber spatula, paper towels and Kimwipes, 1 mL pipette and tips.
4. Remove media into waste (either by pouring media into beaker, or using vacuum aspirator), then rinse cells gently and briefly with approximately 20 mL PBS and remove to waste. Tilt plate on its side for approximately 30 s and pipette/aspirate off remaining PBS.
5. Add 1 mL fresh PBS and scrape cells with spatula (pre-wetted in PBS beaker) to one edge of the tilted plate. Suspend cells in the PBS by repeated aspirations with a 1 mL pipette and add suspension to labeled 2 mL tube
6. Clean spatula well with EtOH and wipe with Kimwipes. Return the spatula to the beaker of PBS.
7. Repeat with remaining plates, washing spatula as above between each plate.
8. Centrifuge cells at a maximum of 500×*g* for 5 min at room temperature (as low as 200–300×*g* for 2–3 min can also be used). 500×*g* corresponds to 2300 rpm on an Eppendorf 5415D benchtop centrifuge.
9. Carefully remove all visible supernatant with 1 mL pipette (*see* **Note 16**) and freeze pellet on dry ice and store at −80 °C until ready to process.

3.4 Parallel BioID and FLAG-AP Purification

The BioID protocol has been optimized for the capture of proximity partners for membrane-associated proteins and other non-soluble cellular structures (such as RNA granules and bodies), and differs slightly from those we previously reported for signaling molecules, identification of E3 ligase substrates, chromatin associated proteins or the centrosome/cilia, notably in the concentration of detergents used ([27, 29, 31, 32]; *see* Fig. 2, and **Note 17**). The protocol starts from 1×15 cm plate per replicate (normally corresponds to a dry cell pellet weighing approximately 75–150 mg), which is sufficient for up to six injections on AB Sciex TripleTOF mass spectrometers (e.g., enabling acquisition of technical replicates, or parallel Data Dependent and Data Independent Acquisition runs [13]; *see* **Note 18**).

3.4.1 Biotin–Streptavidin Affinity Purification (BioID) and on-Bead Trypsin Digest

Steps 1–8 should be performed at 4 °C and/or on ice (ensure that all solutions are chilled prior to use), with the remaining steps being carried out at RT.

A.

	1. RIPA	2. BioID Lysis Buffer
Tris-HCl (pH 7.5)	50 mM	50 mM
NaCl	150 mM	150 mM
SDS	**0.1%**	**0.4%**
IGEPAL CA-630	-	**1%**
Triton X-100	**1%**	-
sodium deoxycholate	**0.5%**	-
EDTA	**1 mM**	-
EGTA	1 mM	1 mM
$MgCl_2$	-	**1.5 mM**

B.

	1. mild wash	2. harsh wash
1x	RIPA buffer	BioID Wash Buffer
2x	FLAG Lysis Buffer	BioID Lysis Buffer
3x	ABC Buffer	ABC Buffer

C.

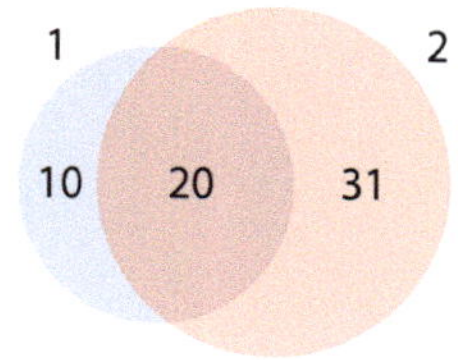

Fig. 2 (**a**) Detailed comparison of the composition of buffers previously reported: (1) RIPA lysis buffer used for BioID in our previous study [29], and (2) the modified BioID Lysis Buffer described in this protocol. (**b**). Comparison of wash steps after streptavidin affinity-purification (AP): (*1*) mild wash steps as previously reported [9], and (*2*) the harsh wash steps described in this protocol. (**c**) Number of high confidence preys identified using BirA-FLAG-DCP1A as bait using the two different methods outlined in A. and B.; (*1*) RIPA lysis + mild wash post-AP and (*2*) BioID Lysis Buffer + harsh wash post-AP

1. Weigh tubes and subtract tube weight to determine cell pellet weight.
2. Thaw cell pellets on ice and add BioID Lysis Buffer (complete with protease inhibitors and benzonase added (*see* **Note 19**) at a 4:1 ratio vol:wt, e.g., 400 μL buffer for a 100 mg pellet).
3. Break up pellet with pipette (by repeated aspirations) and then freeze–thaw once by freezing samples on dry ice, then immediately moving tube to a 37 °C water bath with agitation, *just* until the visible ice is about to thaw.
4. Immediately place the samples on an end-over-end rotator at 4 °C and incubate for 30 min to allow benzonase to fully degrade nucleic acids and for complete solubilisation. Save a 20 μL aliquot to monitor bait expression by Western blot. To determine the percentage of solubilisation, centrifuge this 20 μL as in **step 5** below and analyze the supernatant and the pellet separately (there should be essentially no visible pellet or insoluble proteins with this protocol).
5. Centrifuge for 20 min at 16,100 × *g* at 4 °C (*see* **Note 20**).
6. Collect supernatant in new 1.5 mL tube.
7. Prepare the Streptavidin-sepharose beads. Streptavidin-sepharose beads "high performance" are supplied as a 60% slurry (in a 20% ethanol solution). 20 μL bed volume of beads are used per sample (this corresponds to approximately 35 μL slurry). Wash the desired volume of beads three times with 1 mL BioID Lysis Buffer (in a 1.5 mL tube). Resuspend in BioID Lysis Buffer to the original 60% slurry volume (e.g., by adding

15 μL to the 20 μL packed beads). Beads are best pipetted using a tip cut at an angle with a clean razor blade; when preparing beads for multiple samples, prepare a bit extra to account for losses due to pipetting.

8. Add 20 μL dry bead equivalent as bead slurry (35 μL) to each sample. Ensure that bead slurry is homogeneously mixed prior to pipetting beads for each sample. This is effectively achieved by gently vortexing the tube (at lowest vortex intensity required to effectively suspend beads) for a few seconds immediately prior to pipetting. Rotate samples "end-over-end" overnight at 4 °C (*see* **Note 21**).
9. Centrifuge beads at 500 × *g* for 2 min (*see* **Note 22**).
10. Remove supernatant with a vacuum line equipped with a fine pipette tip (e.g., P20 or "gel loading" tip), being careful to not disturb or lose beads.
11. Transfer beads in 500 μL of BioID Lysis Buffer into a new 1.5 mL tube (*see* **Note 23**). Centrifuge at 500 × *g* for 30 s and remove the supernatant.
12. For all washes, add 500 μL of the indicated wash solution, and suspend beads with brief, gentle vortexing/mixing. Centrifuge beads at 500 × *g* for 30 s and remove supernatants as above. Perform the following washes: (a) Wash once with BioID Wash Buffer; (b) Wash twice with BioID Lysis Buffer; (c) Wash three times with 50 mM ammonium bicarbonate (ABC) solution (*see* **Note 24**).
13. Add 100 μL trypsin solution (i.e., 1 μg trypsin in 100 μL ABC) to beads and incubate 4 h with rotation at 37 °C (*see* **Notes 25** and **26**).
14. Add a fresh 1 μg of trypsin (5 μL if stock is prepared as described in Materials) to each sample and continue incubation overnight with rotation at 37 °C.
15. Centrifuge beads at 500 × *g* for 2 min and retrieve 100 μL of the supernatant (containing the digested peptides) into a new 1.5 mL tube.
16. Add 100 μL of HPLC grade water to the beads and gently vortex/mix. Centrifuge and collect 100 μL of supernatant as above, and pool with peptide solution collected in previous step.
17. Add formic acid to digested peptides to a final concentration of 2% (use 50 μL of 10% stock yielding 250 μL total solution) and vortex briefly.
18. Centrifuge at 16,100 × *g* for 5 min to eliminate debris and any residual beads that may have carried over, then transfer exactly 230 μL (being careful not to disrupt the debris "pellet" in the remaining 20 μL) to new 1.5 mL tubes.

19. Dry by centrifugal evaporation without heat (in a Speed-Vac vacuum centrifuge or similar).
20. Store dried peptides at −80 °C until ready to analyze by mass spectrometry.
21. When ready to analyze, warm tube to RT and centrifuge dry tube at 16,100 × *g* for 5 min to ensure that the peptides are concentrated at the bottom of the tube.
22. Resuspend peptides in 10 μL 5 % formic acid (FA) and vortex aggressively for at least 20–30 s, then centrifuge at 16,100 × *g* for 5 min.
23. Mass spectrometric analysis (also see below): For analysis on AB Sciex Triple-TOF instruments, pipette 4.5 μL of 5 % FA into a new autosampler tube and then add 1.5 μL of resuspended/centrifuged peptide solution for a single injection. Set MS method to pick up 5 μL of this solution (*see* **Note 27** for analysis on other instruments).
24. Any unused sample should be stored (do not re-dry) at −80 °C. To rerun stored samples, thaw tubes to RT, vortex, centrifuge, and take sample for injection (as in **step 23**).

3.4.2 FLAG Affinity Purification (FLAG-AP) and on-Bead Trypsin Digest

This protocol was developed to perform affinity purification from 1 × 15 cm plate prepared in Subheading 3.3. In this procedure cells are lysed by passive lysis assisted by freeze–thaw and affinity purification is performed on a magnetic bead support (*see* ref. [34] for alternative protocol using an agarose bead support and for protocol optimization). Ensure that all solutions are chilled on ice prior to use.

1. Weigh tubes and subtract tube weight to determine cell pellet weight.
2. Thaw cell pellets on ice and add FLAG Lysis Buffer at a 4:1 ratio vol (μL): wt (mg) (e.g., 400 μL buffer to a 100 mg pellet).
3. Break up pellet with pipette (by repeated aspirations) and then freeze–thaw once by freezing samples on dry ice, then immediately moving tube to a 37 °C water bath with agitation, *just* until the visible ice is about to thaw.
4. Centrifuge at 16,100 × *g* for 20 min at 4 °C and transfer the supernatants to fresh tubes (remove or avoid the lipid layer on top of the lysate if present). Save a 20 μL aliquot to monitor bait expression by Western blot.
5. Prepare the anti-FLAG M2 magnetic beads. Use 25 μL per sample (supplied by vendor as a 50 % slurry). Wash the required volume of beads 3–4× with 1 mL of FLAG Lysis Buffer (using a magnetic tube rack). Resuspend the beads in FLAG Lysis Buffer to make a 50 % slurry and distribute 25 μL of this into each sample.

6. Incubate this mixture for 2–3 h at 4 °C with "end-over-end" rotation.
7. Centrifuge samples at 500 × *g* for 1 min to recover beads that may have stuck to the lid. Remove the supernatant with a pipette and transfer the beads to a clean 1.5 mL tube using 1 mL of FLAG Lysis Buffer.
8. Using a magnetic tube rack, wash beads once with 1 mL FLAG Lysis Buffer, followed by one wash with FLAG Rinsing Buffer (resuspending the beads in buffers by pipetting up and down four times). The washing steps should be done as quickly as possible while allowing approximately 30 s for beads to magnetize before aspirating supernatant—a complete wash cycle should take between 1 and 2 min (*see* **Note 28**).
9. After the last wash is aspirated, centrifuge the sample at 500 × *g* for 1 min to pellet the beads and remove any remaining liquid with a fine pipette.
10. Add 7.5 μL of trypsin (100 ng/μL solution = 750 ng trypsin) to the beads and incubate at 37 °C overnight on a rotator.
11. The following day, centrifuge the samples at 500 × *g* for 1 min, magnetize the beads for 30 s and transfer the supernatant to a fresh tube, then add an additional 2.5 μL of trypsin (250 ng) to the collected supernatant and incubate for 4 h at 37 °C (no agitation required).
12. Add 1 μL of 50 % formic acid solution to the samples (approximately 5 % final concentration). The sample can be directly analyzed by mass spectrometry (centrifuge at 16,100 × *g* for 5 min prior to retrieving sample).
13. Mass spectrometric analysis: For analysis on AB Sciex TripleTOF instruments load approximately 5 μL of sample.
14. Any unused sample may be stored at −80 °C.

3.4.3 Mass Spectrometry and Data Analysis

Mass spectrometric analysis uses a standard LC-MS/MS set-up, which we have also described elsewhere [13, 27, 29]. Briefly, using an autosampler (Eksigent), samples are loaded onto fused silica capillary columns (0.75 μm ID, 350 μm OD) pre-loaded with 10 cm of C18 reversed phase material (3.5 μm diameter). Ionized peptides are emitted by nanoelectrospray ion source in-line with a nano-HPLC system and analyzed using either Data Dependent (DDA) or Data Independent (DIA) Acquisition methods. Important considerations include identifying and mitigating potential carryover issues, and being able to identify which of the identified proteins are likely to represent bona fide interactors (or proximity partners in the case of BioID) and which are likely to be contaminants.

We attempt to address these considerations by: (1) the inclusion of negative controls in our experimental design (*see* **Note 10**); (2) the processing of at least two biological replicates per bait analyzed, reporting only those interactions confidently detected across both replicates; (3) the inclusion of long wash cycles of the LC column between samples (*see* **Note 29**); (4) the randomization of acquisition order on the LC-MS/MS across both replicates such that if carryover is present (column retention of the most abundant peptides contaminating subsequent samples), it should not carry through to our final interaction list; 5) the acquisition of quantitative data which, when analyzed through the pipeline briefly described below, enables consistent scoring and reporting of the interactions.

All data processing and analysis is done using software developed and maintained in-house (ProHits [37]), which is open-source and can be run using a "virtual machine" [38], or on a local server. ProHits enables storing, tracking and annotating experiments; simple visualization interfaces permit rapid analysis of each sample side-by-side, and the identification of possible issues (carryover, low recovery of the bait or its known interactors, etc.). High confidence interactions are determined by comparing bait samples against appropriate control samples (*see* **Note 10**) using the statistical tool SAINT [9, 39]. A user-friendly way to run SAINT analysis and other interaction scoring tools without having to install any software is through the Contaminant Repository for Affinity Purification (CRAPome) website [40]. Our laboratories have deposited to the CRAPome multiple negative controls from HEK293 FLAG and BioID experiments that can be used to supplement user-generated controls in order to aid in the removal of nonspecific / background preys. Results from SAINT analysis (through ProHits, the CRAPome, or as a stand-alone version) can all be prepared for publication through a suite of visualization tools [41] developed by our group.

4 Notes

1. Autoclaving may result in a residue being deposited on to plasticware, which may result in contamination during MS analysis. As sterility is not essential for the protocols (other than during cell culture), use tips and tubes directly as received from the manufacturer. Keep all plasticware and reagents protected from dust and other environmental contaminants as much as possible and use gloves for all preparation steps.
2. Protease inhibitors and nucleases are only added to the working volume of lysis buffer that will be directly added to the cell pellets. Sum the weights of all pellets and prepare only the volume of lysis buffer that will be required (plus extra volume for pipetting error).
3. Extreme care must be taken in preparing biotin solutions, as it can easily precipitate from the solution if the pH is raised too rapidly or is not kept adequately chilled during preparation

(this may be reversed by the readdition of a small amount of ammonium hydroxide). In our experience the biotin stock solution is very stable and can be used for approximately 6 months to a year without any change in potency when prepared and stored appropriately.

4. Tetracycline stocks can be stored long term at −20 °C, but a small amount will gradually precipitate out of solution. Warm the solution to RT prior to use, avoiding the settled particulate. This gradual loss of tetracycline from solution has no apparent impact on the induction of expression, since the concentration used is well in excess of what is required to inhibit the Tet repressor. To minimize this, consider aliquoting to avoid multiple freeze–thaw cycles.
5. The pH of ammonium bicarbonate solution rises upon storage (with pH ~7.8 when freshly made and ≥ 8.5 after storage for a few weeks). Make solution freshly and readjust pH as necessary by mixing older solutions to have a working pH around 8–8.2 for reproducible trypsin digestion.
6. We have not noticed major differences in digestions when using the 20 μg trypsin vial or the trypsin Singles from Sigma-Aldrich. The selection of the enzyme is primarily based on cost efficiency for the number of samples to be analyzed.
7. The LR reaction is very efficient and we routinely scale the enzyme quantity in half.
8. Some degree of leaky expression can be expected in the absence of tetracycline induction, and this may be heterogeneous across the cell population (i.e., some sporadic cells will exhibit much higher expression than the general population as assessed by immunofluorescence). If leakiness of a toxic protein is detected, consider using a tetracycline-free source of Fetal Bovine Serum (such as the Clontech "Tet Approved FBS" product). Because of the integration at a single locus, expression across individually isolated clones is relatively uniform. As such, we use pools of clones for all proteomics experiments described here. If isolated clones are desired (e.g., for functional studies), they can easily be obtained by letting colonies on selection grow to > 1 mm diameter and picking them using sterile filter papers dipped in trypsin (we use the Scienceware 3.2 mm disks (Cat #Z374431, Sigma-Aldrich) for this purpose and transfer each clone into a prepared 96- or 48-well plate).
9. We use JetPrime (Cat #11407, Polyplus Transfection) which is a cost effective reagent to use with the HEK293 Flp-In T-REx system. For a 6 well format we use 200 μL JetPrime buffer and 2 μL transfection reagent. Other transfection reagents can be used instead.

10. The inclusion of negative controls in the experimental design is essential to distinguish the true interactors of the protein of interest from the nonspecific (contaminant, or background) proteins in both FLAG-AP and BioID experiments. There are a number of sources of background, including nonspecific protein binding to the resin, the affinity tag, or to any overexpressed protein. For BioID, additional sources of background contaminants include ubiquitous endogenously biotinylated proteins and nonspecific proximal biotinylation by BirA*. A FLAG pull-down experiment would typically include at least one type of negative control (at the minimum, the "empty" BirA*-FLAG construct, expressed at levels at least as great as the FLAG-BirA*-tagged bait). However, to properly model the background of the BioID experiment—and especially the endogenous biotinylation—it is necessary to also include cells not transfected with BirA* (e.g., transfected with another construct such as FLAG-GFP, 3×FLAG or left untransfected). These non-BirA* cells should be treated with biotin to maximize endogenous biotinylation. Depending on the biological question, it may be important to also include other controls for the BioID experiments; for example, to increase the stringency of filtering, subcellular compartment specific controls can be generated through fusion of BirA* to sequence tags for the compartment of interest (e.g., a nuclear localization signal [27]). These "compartment-specific" controls may be treated as true negative controls when scoring interactions, but can also be used as a secondary filter (e.g., by analyzing the quantitative fold enrichment of each potential interactor against the "compartment-specific" controls).
11. HEK293 Flp-In T-REx cells are loosely adherent and therefore care must be taken not to disrupt cells during media changes.
12. Cells in the center of dense colonies will have slower proliferation rates and therefore it is desirable to split plates before allowing colonies to grow too large in order to allow cells to proliferate more efficiently. The selection of the destination vessels should be determined empirically (the estimates provided are only guidelines).
13. Hygromycin may be omitted from the plates used for induction of protein expression.
14. We have reduced biotin labeling times to as low as 4 h. However, if bait expression is low, more material (i.e., 2 or more 15 cm plates) may need to be combined to achieve an adequate signal if short labeling times are used.
15. In our experience, most brands of tubes exhibit variation in their weights (by as much as 50 mg or more) and therefore accurate cell weight can only be achieved by obtaining the tare weight of each tube individually.

16. It is important to remove all residual liquid in order to obtain an accurate cell pellet weight.
17. The power of the BioID protocol rests on the strength of the biotin–streptavidin interaction, which is one of the strongest known biological interactions (K_d ~10^{-14} M) and is highly resistant to disruption by SDS. Due to the strength of this interaction, the lysis buffer employed here contains a higher concentration of SDS (0.4%) and employs a high concentration SDS wash step (2%). This is in contrast to typical buffers used in affinity purification protocols, which depend on native protein–protein interactions to be maintained. The higher concentration of SDS in the lysis buffer, which is highly denaturing, ensures near complete solubilization of membrane proteins (including those with multiple transmembrane domains) and other insoluble cellular structures (including RNA granules and bodies). The high concentration of SDS in the wash helps remove contaminating (i.e., non-biotinylated) proteins, thus improving the signal-to-noise ratio. Note that we have also omitted EDTA (which chelates Mg^{2+}) and included $MgCl_2$ which improves the activity of Benzonase in our revised buffer (*see* Fig. 2).
18. We provided here the amounts loaded on the AB Sciex TripleTOF instruments equipped with Eksigent Ultra systems and 10 cm C18 fused silica capillary columns (0.75 μm ID, 350 μm OD). Other instrument configurations will require optimization of this amount. Notably, we are loading more peptides for analysis on ThermoScientific Velos Orbitrap or Orbitrap Elite mass spectrometers (*see* **Note 27**).
19. Benzonase and protease inhibitors are only added to the volume of BioID Lysis Buffer that will specifically be used to lyse the cells (aka "complete"). For washing steps with BioID Lysis Buffer, Benzonase and protease inhibitors are not added (aka "stock").
20. When establishing the protocol in the laboratory for the first time, or if initial results are unsatisfactory and troubleshooting must be performed, take an aliquot of the unbound fraction and of the washes to analyze by immunoblotting. A white, fibrous looking pellet should result (only 1–2 mm in diameter), and likely corresponds to highly insoluble aggregated protein.
21. Note that incubations as short as 3 h are routinely performed in our group (see notes in the cell pellet preparation regarding the analysis of posttranslational modifications). Ensure however, that you are consistent across any experiments that need to be compared.
22. Changing the tube at this step helps eliminate residues that may have adhered onto the tube walls.

23. The general idea of the washes in ABC is to eliminate the detergents that would interfere with the mass spectrometric identification and to equilibrate beads in a buffer suitable for trypsin digestion that is volatile and will be evaporated upon sample drying. If residual detergents are detected in the mass spectrometer, consider using fresh tubes for the last two washes in ABC.
24. For on-bead trypsin digestion we use a "rotating drum" style rotator (Cel-Gro Tissue Culture Rotator, Thermo Scientific) placed inside a 37 °C incubator.
25. For protein-level analysis, no alkylation/reduction steps are performed. This will result in a loss of cysteine-containing peptides from the analysis. If the identification of cysteine-containing peptides is desired, introduce a reduction and alkylation step prior to mass spectrometric analysis.
26. For the analysis of BioID samples on Thermo OrbiTrap Velos or Orbitrap Elite instruments, pipette 5 μL of undiluted peptide solution into new autosampler tube and set MS method to pick up 1 μL less than volume in tube (e.g., 5 μL of 6 μL total solution).
27. Alternatively, the stringency of the washes can be increased by washing the beads 3x with 1 mL FLAG Lysis Buffer, followed by 2× with 1 mL FLAG Rinsing Buffer.
27. Performing database searches of bovine serum albumin (BSA) or other control samples after washes and between runs enables further identification of carryover issues and optimization of your washing conditions.

Acknowledgment

We thank Wade H Dunham and Zhen-Yuan Lin for optimization of the FLAG protocol, all members of the Gingras and Raught laboratories for help in optimizing the BioID protocol and for helpful discussions, and Boris Dyakov and Cassandra Wong for comments on the manuscript. This work is funded by the Canadian Institutes of Health Research (Foundation grant FDN143301 to A.-C.G.; salary support to PST), the Natural Sciences and Engineering Research Council of Canada (Discovery grant to A.-C.G; salary support to JYY), and a Genome Canada Genome Innovation (GIN) network (through the Ontario Genomics Institute OGI-069 to A.-C.G.). Salary awards are from the Canada Research Chairs Program (ACG and BR) and a Basic Research Fellowship from Parkinson Canada (GGH).

References

1. Snider J, Kotlyar M, Saraon P, Yao Z, Jurisica I, Stagljar I (2015) Fundamentals of protein interaction network mapping. Mol Syst Biol 11(12):848. doi:10.15252/msb.20156351
2. Gingras AC, Gstaiger M, Raught B, Aebersold R (2007) Analysis of protein complexes using mass spectrometry. Nat Rev Mol Cell Biol 8(8):645–654. doi:10.1038/nrm2208
3. Gavin AC, Bosche M, Krause R, Grandi P, Marzioch M, Bauer A, Schultz J, Rick JM, Michon AM, Cruciat CM, Remor M, Hofert C, Schelder M, Brajenovic M, Ruffner H, Merino A, Klein K, Hudak M, Dickson D, Rudi T, Gnau V, Bauch A, Bastuck S, Huhse B, Leutwein C, Heurtier MA, Copley RR, Edelmann A, Querfurth E, Rybin V, Drewes G, Raida M, Bouwmeester T, Bork P, Seraphin B, Kuster B, Neubauer G, Superti-Furga G (2002) Functional organization of the yeast proteome by systematic analysis of protein complexes. Nature 415(6868):141–147. doi:10.1038/415141a
4. Ho Y, Gruhler A, Heilbut A, Bader GD, Moore L, Adams SL, Millar A, Taylor P, Bennett K, Boutilier K, Yang L, Wolting C, Donaldson I, Schandorff S, Shewnarane J, Vo M, Taggart J, Goudreault M, Muskat B, Alfarano C, Dewar D, Lin Z, Michalickova K, Willems AR, Sassi H, Nielsen PA, Rasmussen KJ, Andersen JR, Johansen LE, Hansen LH, Jespersen H, Podtelejnikov A, Nielsen E, Crawford J, Poulsen V, Sorensen BD, Matthiesen J, Hendrickson RC, Gleeson F, Pawson T, Moran MF, Durocher D, Mann M, Hogue CW, Figeys D, Tyers M (2002) Systematic identification of protein complexes in Saccharomyces cerevisiae by mass spectrometry. Nature 415(6868):180–183. doi:10.1038/415180a
5. Hein MY, Hubner NC, Poser I, Cox J, Nagaraj N, Toyoda Y, Gak IA, Weisswange I, Mansfeld J, Buchholz F, Hyman AA, Mann M (2015) A human interactome in three quantitative dimensions organized by stoichiometries and abundances. Cell 163(3):712–723. doi:10.1016/j.cell.2015.09.053
6. Huttlin EL, Ting L, Bruckner RJ, Gebreab F, Gygi MP, Szpyt J, Tam S, Zarraga G, Colby G, Baltier K, Dong R, Guarani V, Vaites LP, Ordureau A, Rad R, Erickson BK, Wuhr M, Chick J, Zhai B, Kolippakkam D, Mintseris J, Obar RA, Harris T, Artavanis-Tsakonas S, Sowa ME, De Camilli P, Paulo JA, Harper JW, Gygi SP (2015) The bioplex network: a systematic exploration of the human interactome. Cell 162(2):425–440. doi:10.1016/j.cell.2015.06.043
7. Breitkreutz A, Choi H, Sharom JR, Boucher L, Neduva V, Larsen B, Lin ZY, Breitkreutz BJ, Stark C, Liu G, Ahn J, Dewar-Darch D, Reguly T, Tang X, Almeida R, Qin ZS, Pawson T, Gingras AC, Nesvizhskii AI, Tyers M (2010) A global protein kinase and phosphatase interaction network in yeast. Science 328(5981):1043–1046. doi:10.1126/science.1176495
8. Sowa ME, Bennett EJ, Gygi SP, Harper JW (2009) Defining the human deubiquitinating enzyme interaction landscape. Cell 138(2):389–403. doi:10.1016/j.cell.2009.04.042
9. Choi H, Larsen B, Lin ZY, Breitkreutz A, Mellacheruvu D, Fermin D, Qin ZS, Tyers M, Gingras AC, Nesvizhskii AI (2011) SAINT: probabilistic scoring of affinity purification-mass spectrometry data. Nat Methods 8(1):70–73. doi:10.1038/nmeth.1541
10. Dunham WH, Mullin M, Gingras AC (2012) Affinity-purification coupled to mass spectrometry: basic principles and strategies. Proteomics 12(10):1576–1590. doi:10.1002/pmic.201100523
11. Bisson N, James DA, Ivosev G, Tate SA, Bonner R, Taylor L, Pawson T (2011) Selected reaction monitoring mass spectrometry reveals the dynamics of signaling through the GRB2 adaptor. Nat Biotechnol 29(7):653–658. doi:10.1038/nbt.1905
12. Zheng Y, Zhang C, Croucher DR, Soliman MA, St-Denis N, Pasculescu A, Taylor L, Tate SA, Hardy WR, Colwill K, Dai AY, Bagshaw R, Dennis JW, Gingras AC, Daly RJ, Pawson T (2013) Temporal regulation of EGF signalling networks by the scaffold protein Shc1. Nature 499(7457):166–171. doi:10.1038/nature12308
13. Lambert JP, Ivosev G, Couzens AL, Larsen B, Taipale M, Lin ZY, Zhong Q, Lindquist S, Vidal M, Aebersold R, Pawson T, Bonner R, Tate S, Gingras AC (2013) Mapping differential interactomes by affinity purification coupled with data-independent mass spectrometry acquisition. Nat Methods 10(12):1239–1245. doi:10.1038/nmeth.2702
14. Collins BC, Gillet LC, Rosenberger G, Rost HL, Vichalkovski A, Gstaiger M, Aebersold R (2013) Quantifying protein interaction dynamics by SWATH mass spectrometry: application to the 14-3-3 system. Nat Methods 10(12):1246–1253. doi:10.1038/nmeth.2703
15. Hubner NC, Bird AW, Cox J, Splettstoesser B, Bandilla P, Poser I, Hyman A, Mann M (2010) Quantitative proteomics combined with BAC TransgeneOmics reveals in vivo

protein interactions. J Cell Biol 189(4):739–754. doi:10.1083/jcb.200911091

16. Roncagalli R, Hauri S, Fiore F, Liang Y, Chen Z, Sansoni A, Kanduri K, Joly R, Malzac A, Lahdesmaki H, Lahesmaa R, Yamasaki S, Saito T, Malissen M, Aebersold R, Gstaiger M, Malissen B (2014) Quantitative proteomics analysis of signalosome dynamics in primary T cells identifies the surface receptor CD6 as a lat adaptor-independent TCR signaling hub. Nat Immunol 15(4):384–392. doi:10.1038/ni.2843
17. Blagoev B, Kratchmarova I, Ong SE, Nielsen M, Foster LJ, Mann M (2003) A proteomics strategy to elucidate functional protein-protein interactions applied to EGF signaling. Nat Biotechnol 21(3):315–318. doi:10.1038/nbt790
18. Hilger M, Mann M (2012) Triple SILAC to determine stimulus specific interactions in the Wnt pathway. J Proteome Res 11(2):982–994. doi:10.1021/pr200740a
19. Pagliuca FW, Collins MO, Lichawska A, Zegerman P, Choudhary JS, Pines J (2011) Quantitative proteomics reveals the basis for the biochemical specificity of the cell-cycle machinery. Mol Cell 43(3):406–417. doi:10.1016/j.molcel.2011.05.031
20. Lavallee-Adam M, Rousseau J, Domecq C, Bouchard A, Forget D, Faubert D, Blanchette M, Coulombe B (2013) Discovery of cell compartment specific protein-protein interactions using affinity purification combined with tandem mass spectrometry. J Proteome Res 12(1):272–281. doi:10.1021/pr300778b
21. Kaake RM, Wang X, Burke A, Yu C, Kandur W, Yang Y, Novtisky EJ, Second T, Duan J, Kao A, Guan S, Vellucci D, Rychnovsky SD, Huang L (2014) A new in vivo cross-linking mass spectrometry platform to define protein-protein interactions in living cells. Mol Cell Proteomics 13(12):3533–3543. doi:10.1074/mcp.M114.042630
22. Babu M, Vlasblom J, Pu S, Guo X, Graham C, Bean BD, Burston HE, Vizeacoumar FJ, Snider J, Phanse S, Fong V, Tam YY, Davey M, Hnatshak O, Bajaj N, Chandran S, Punna T, Christopolous C, Wong V, Yu A, Zhong G, Li J, Stagljar I, Conibear E, Wodak SJ, Emili A, Greenblatt JF (2012) Interaction landscape of membrane-protein complexes in Saccharomyces cerevisiae. Nature 489(7417):585–589. doi:10.1038/nature11354
23. Hakhverdyan Z, Domanski M, Hough LE, Oroskar AA, Oroskar AR, Keegan S, Dilworth DJ, Molloy KR, Sherman V, Aitchison JD, Fenyo D, Chait BT, Jensen TH, Rout MP, LaCava J (2015) Rapid, optimized interactomic screening. Nat Methods 12(6):553–560. doi:10.1038/nmeth.3395
24. Roux KJ, Kim DI, Raida M, Burke B (2012) A promiscuous biotin ligase fusion protein identifies proximal and interacting proteins in mammalian cells. J Cell Biol 196(6):801–810. doi:10.1083/jcb.201112098
25. Rhee HW, Zou P, Udeshi ND, Martell JD, Mootha VK, Carr SA, Ting AY (2013) Proteomic mapping of mitochondria in living cells via spatially restricted enzymatic tagging. Science 339(6125):1328–1331. doi:10.1126/science.1230593
26. Kim DI, Birendra KC, Zhu W, Motamedchaboki K, Doye V, Roux KJ (2014) Probing nuclear pore complex architecture with proximity-dependent biotinylation. Proc Natl Acad Sci U S A 111(24):E2453–E2461. doi:10.1073/pnas.1406459111
27. Lambert JP, Tucholska M, Go C, Knight JD, Gingras AC (2015) Proximity biotinylation and affinity purification are complementary approaches for the interactome mapping of chromatin-associated protein complexes. J Proteomics 118:81–94. doi:10.1016/j.jprot.2014.09.011
28. Dingar D, Kalkat M, Chan PK, Srikumar T, Bailey SD, Tu WB, Coyaud E, Ponzielli R, Kolyar M, Jurisica I, Huang A, Lupien M, Penn LZ, Raught B (2015) BioID identifies novel c-MYC interacting partners in cultured cells and xenograft tumors. J Proteomics 118:95–111. doi:10.1016/j.jprot.2014.09.029
29. Couzens AL, Knight JD, Kean MJ, Teo G, Weiss A, Dunham WH, Lin ZY, Bagshaw RD, Sicheri F, Pawson T, Wrana JL, Choi H, Gingras AC (2013) Protein interaction network of the mammalian Hippo pathway reveals mechanisms of kinase-phosphatase interactions. Sci Signal 6(302):rs15. doi:10.1126/scisignal.2004712
30. Firat-Karalar EN, Rauniyar N, Yates JR 3rd, Stearns T (2014) Proximity interactions among centrosome components identify regulators of centriole duplication. Curr Biol 24(6):664–670. doi:10.1016/j.cub.2014.01.067
31. Gupta GD, Coyaud E, Goncalves J, Mojarad BA, Liu Y, Wu Q, Gheiratmand L, Comartin D, Tkach JM, Cheung SW, Bashkurov M, Hasegan M, Knight JD, Lin ZY, Schueler M, Hildebrandt F, Moffat J, Gingras AC, Raught B, Pelletier L (2015) A dynamic protein interaction landscape of the human centrosome-cilium interface. Cell 163(6):1484–1499. doi:10.1016/j.cell.2015.10.065
32. Coyaud E, Mis M, Laurent EM, Dunham WH, Couzens AL, Robitaille M, Gingras AC, Angers S, Raught B (2015) BioID-based

identification of Skp cullin F-box (SCF)beta-TrCP1/2 E3 ligase substrates. Mol Cell Proteomics 14(7):1781–1795. doi:10.1074/mcp.M114.045658

33. Cheng YS, Seibert O, Kloting N, Dietrich A, Strassburger K, Fernandez-Veledo S, Vendrell JJ, Zorzano A, Bluher M, Herzig S, Berriel Diaz M, Teleman AA (2015) PPP2R5C couples hepatic glucose and lipid homeostasis. PLoS Genet 11(10):e1005561. doi:10.1371/journal.pgen.1005561

34. Kean MJ, Couzens AL, Gingras AC (2012) Mass spectrometry approaches to study mammalian kinase and phosphatase associated proteins. Methods 57(4):400–408. doi:10.1016/j.ymeth.2012.06.002

35. Banks CA, Boanca G, Lee ZT, Florens L, Washburn MP (2015) Proteins interacting with cloning scars: a source of false positive protein-protein interactions. Sci Rep 5:8530. doi:10.1038/srep08530

36. Olhovsky M, Williton K, Dai AY, Pasculescu A, Lee JP, Goudreault M, Wells CD, Park JG, Gingras AC, Linding R, Pawson T, Colwill K (2011) OpenFreezer: a reagent information management software system. Nat Methods 8(8):612–613. doi:10.1038/nmeth.1658

37. Liu G, Zhang J, Larsen B, Stark C, Breitkreutz A, Lin ZY, Breitkreutz BJ, Ding Y, Colwill K, Pasculescu A, Pawson T, Wrana JL, Nesvizhskii AI, Raught B, Tyers M, Gingras AC (2010) ProHits: integrated software for mass spectrometry-based interaction proteomics. Nat Biotechnol 28(10):1015–1017. doi:10.1038/nbt1010-1015

38. Liu G, Zhang J, Choi H, Lambert JP, Srikumar T, Larsen B, Nesvizhskii AI, Raught B, Tyers M, Gingras AC (2012) Using ProHits to store, annotate, and analyze affinity purification-mass spectrometry (AP-MS) data. Curr Protoc Bioinformatics Chapter 8:Unit8.16. doi:10.1002/0471250953.bi0816s39

39. Teo G, Liu G, Zhang J, Nesvizhskii AI, Gingras AC, Choi H (2014) SAINTexpress: improvements and additional features in significance analysis of INTeractome software. J Proteomics 100:37–43. doi:10.1016/j.jprot.2013.10.023

40. Mellacheruvu D, Wright Z, Couzens AL, Lambert JP, St-Denis NA, Li T, Miteva YV, Hauri S, Sardiu ME, Low TY, Halim VA, Bagshaw RD, Hubner NC, Al-Hakim A, Bouchard A, Faubert D, Fermin D, Dunham WH, Goudreault M, Lin ZY, Badillo BG, Pawson T, Durocher D, Coulombe B, Aebersold R, Superti-Furga G, Colinge J, Heck AJ, Choi H, Gstaiger M, Mohammed S, Cristea IM, Bennett KL, Washburn MP, Raught B, Ewing RM, Gingras AC, Nesvizhskii AI (2013) The CRAPome: a contaminant repository for affinity purification-mass spectrometry data. Nat Methods 10(8):730–736. doi:10.1038/nmeth.2557

41. Knight JD, Liu G, Zhang JP, Pasculescu A, Choi H, Gingras AC (2015) A web-tool for visualizing quantitative protein-protein interaction data. Proteomics 15(8):1432–1436. doi:10.1002/pmic.201400429

Chapter 11

LUMIER: A Discovery Tool for Mammalian Protein Interaction Networks

Miriam Barrios-Rodiles, Jonathan D. Ellis, Benjamin J. Blencowe, and Jeffrey L. Wrana

Abstract

Protein–protein interactions (PPIs) play an essential role in all biological processes. In vivo, PPIs occur dynamically and depend on extracellular cues. To discover novel protein–protein interactions in mammalian cells, we developed a high-throughput automated technology called LUMIER (*LU*minescence-based *M*ammalian *I*nt*ER*actome). In this approach, we co-express a Luciferase (LUC)-tagged fusion protein along with a Flag-tagged protein in an efficiently transfectable cell line such as HEK-293T cells. The interaction between the two proteins is determined by co-immunoprecipitation using an anti-Flag antibody, and the presence of the LUC-tagged interactor in the complex is subsequently detected via its luciferase activity. LUMIER can easily detect transmembrane protein partners, interactions that are signaling- or splice isoform-dependent, as well as those that may occur only in the presence of posttranslational modifications. Using various collections of Flag-tagged proteins, we have generated protein interaction networks for several TGF-β family receptors, Wnt pathway members, and have systematically analyzed the effect of neural-specific alternative splicing on protein interaction networks. The results have provided important insights into the physiological and functional relevance of some of the novel interactions found. LUMIER is highly scalable and can be used for both low- and high-throughput strategies. LUMIER is thus a valuable tool for the identification and characterization of dynamically regulated PPIs in mammalian systems. Here, we describe a manual version of LUMIER in a 96-well format that can be easily implemented in any laboratory.

Key words Protein–protein interaction, LUMIER, Mammalian cells, Signaling pathways, Transmembrane proteins, Binary complex, Ternary complex

1 Introduction

All cells respond to extracellular cues from their environment. The cellular reponse to such triggers is orchestrated via diverse signaling pathways that rely on dynamic interactions between intracellular proteins. These intracellular effectors may possess enzymatic activities (i.e., kinases, phosphatases, ubiquitin ligases) or contain specific structural domains that allow their association with other proteins. Over the last few decades, several methods have been

Lucio Comai et al. (eds.), *Proteomics: Methods and Protocols*, Methods in Molecular Biology, vol. 1550,
DOI 10.1007/978-1-4939-6747-6_11, © Springer Science+Business Media LLC 2017

developed to screen for novel binary protein interactions. An important and low cost assay has been the yeast-two-hybrid method (Y2H) that relies on a transcriptional reporter readout that is stimulated when two proteins that are independently fused to a DNA binding domain or a transcription activation domain interact in a yeast cell [1]. Y2H was first applied to yeast protein–protein interactions (PPIs) and later applied to mammalian proteins [2, 3]. Other examples of methods exploiting the Y2H principle of PPIs indirectly detected through the transcriptional activation of a reporter gene include MAPPIT [4], MYTH [5] and MaMTH [6]. Another class of PPI detection methods are based on the reconstruction of an enzyme or a fluorescent protein that has been "split" in two parts with each part fused to the potential interacting partners. When the putative partners come into close proximity, the enzymatic activity or fluorescence is restored. In this category, Split dihydrofolate reductase, split-fluorescent protein (such as GFP) or split-luciferase methods have been reported [7–9]. Finally, examples of a different class of methods to detect PPIs which rely on the physical interaction of partners that have been fused to fluorescent tags or an enzyme for detection include FRET/BRET [10], NAPPA [11], and LUMIER [12]. The evidence for the interaction is provided by detection of the attached fluorescent signal or enzymatic activity fused to one of the partners.

We first reported LUMIER, which stands for *LU*minescence-based *M*ammalian *I*nt*ER*actome as a high-throughput assay with a luminescence readout [12]. For this method, we co-express a Luciferase (LUC)-tagged fusion protein (Bait) along with a Flag-tagged protein (Prey) in HEK-293T cells, or any other efficiently transfectable cell line. The interaction between the two proteins is determined by co-immunoprecipitation using an anti-Flag antibody, and subsequently detecting the presence of the LUC-tagged interactor in the complex by measuring luciferase activity (Fig. 1a). We have used different luciferases to perform LUMIER assays in manual (low throughput) or automated screens (high-throughput). During the development of LUMIER, we used *Renilla* luciferase (RLUC or RL, 36 kDa) fused to TGF-β core pathway components to uncover novel interactions in this pathway, including the associations between the TGF-β receptor with Occludin or Par6 [12, 13]. Furthermore, we observed the dynamic gain and loss of interactions for pathway components Smad2 and Smad4 upon signal activation [12]. We also used Firefly luciferase (64 kDa) on Baits to unveil novel interactions for members of the Wnt signaling pathway such as the interaction between β-catenin and Ube2m [14]. Moreover, the introduction of genetically modified and smaller luciferases such as NanoLuc (19 kDa) [15], with an optimized substrate plus enhanced sensitivity, allows the use of luminometers without ultra-sensitive photomultipliers and facilitates an increased throughput, a key feature for network biology studies and drug screens.

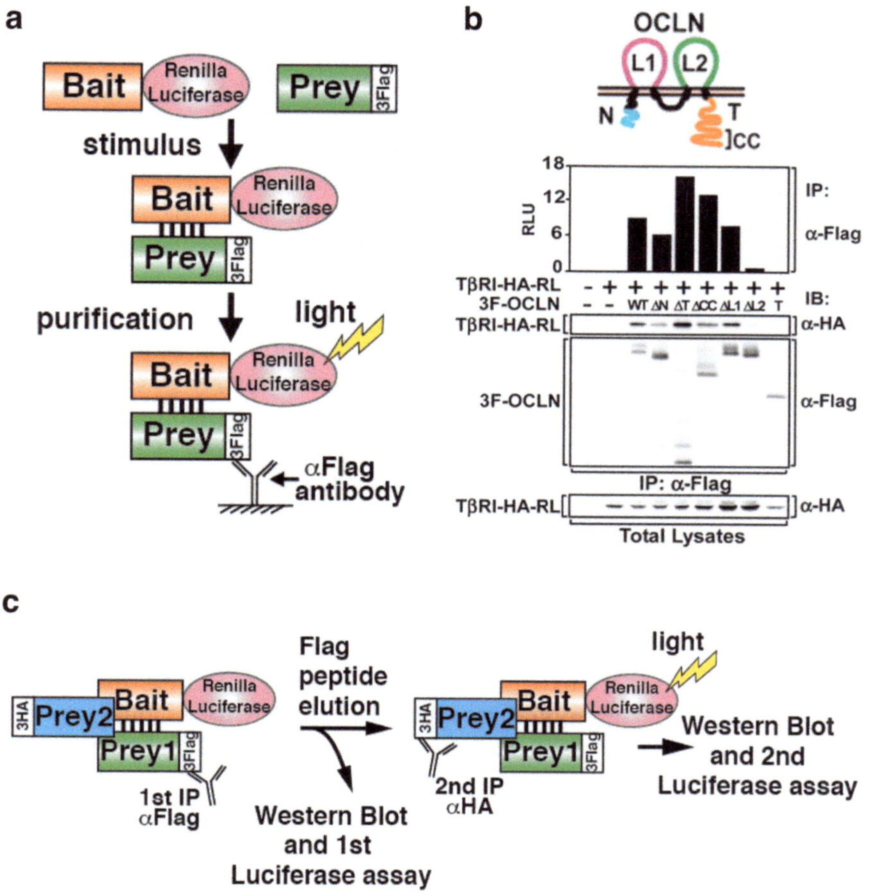

Fig. 1 (**a**) LUMIER strategy for the detection of protein–protein interactions. A protein of interest (Bait) tagged with *Renilla* luciferase (RL or RLUC) is co-expressed in mammalian cells along with a 3Flag-tagged putative partner (Prey). If the two proteins interact (in the presence or absence of a stimulus), the complex can be immunoprecipitated using the anti-Flag antibody. The presence of the Bait is detected by light emission upon *Renilla* substrate addition. (**b**) LUMIER strategy for domain mapping. TBRI-HA-RL wild type was expressed alone or together with wild type (WT) or mutant Occludin (OCLN) harboring the indicated domain deletions (see schematic), and interactions were measured by LUMIER (*top panel*) or by immunoblotting with antibody against HA. Expression of proteins was confirmed by immunoblotting total cell lysates, as indicated. From Barrios-Rodiles M. et al., 2005 *Science* (307):1621–1625. Reprinted with Permission from AAAS. (**c**) LUMIER strategy for the detection of a ternary complex. The RLUC-tagged Bait is co-expressed with a 3Flag-tagged Prey1 and a second 3HA-tagged Prey2. A first immunoprecipitation is performed using the anti-Flag antibody and an aliquot is taken to perform a luciferase assay and western blot to confirm the presence of the Bait and the Flag-tagged Prey1. Following elution using Flag-peptide to release the complex from the beads, a second immunoprecipitation is conducted using the anti-HA antibody. A second luciferase assay and western blot is perfomed to determine the presence of the Bait and the HA-tagged Prey2 (*see* ref. 19)

In addition to its simple set up, LUMIER has several advantages. It can easily detect interactions for transmembrane proteins [16] including seven-transmembrane domain receptors (GPCRs) [17]. LUMIER detects PPIs dependent on posttranscriptional modifications such as phosphorylation that change dynamically during cell signaling [12]. Moreover, mapping the domain responsible for a PPI can be quickly achieved with this assay [12, 18] (Fig. 1b), while a ternary protein complex can be detected using two-step immunoprecipitations [19] (Fig. 1c). We have taken advantage of the semiquantitative nature of LUMIER to detect differential PPIs due to tissue-dependent alternative splicing events [20]. This work revealed that approximately one-third of neural-specific exons either disrupt or promote specific PPIs, many of which are important for neural cell functions. Furthermore, recent protocol modifications combine LUMIER with ELISA measurements to quantify the Flag-tagged protein in the immunoprecipitation to generate a more quantitative evaluation of interaction strength. Quantification of the Flag-tagged partner also allows the immunoprecipitation of the Flag-tagged protein to be verified in each well and thus false negatives can be identified [21]. In summary, since we first developed the assay in 2005 [12], LUMIER has proven to be a useful tool for the systematic mapping and functional characterization of mammalian PPI networks. This method has been employed in several laboratories for manual or high-throughput applications [22–25], and we have previously described the protocol to conduct LUMIER automated screens [3, 12]. In this chapter we focus on the implementation of LUMIER for manual experiments that can be easily performed in any research laboratory.

2 Materials

2.1 Cell Culture Reagents

1. Human Embryonic Kidney 293T (HEK-293T) cells.
2. Plasmids for subcloning Baits and Preys (Fig. 2; *see* **Notes 1** and **2**).
   (a) pCMV5-hRLUC Amp^R, N or C terminal humanized *Renilla* luciferase tag (AF362545).
   (b) pCMV5-3Flag Amp^R, N or C terminal 3Flag tag.
   (c) 3Flag-hRLUC Amp^R, this plasmid encodes *Renilla* luciferase tagged with 3Flag at the N-terminus.
   (d) pCMV5 Amp^R, plasmid without any insert.
3. Dulbeccos's Modified Eagles's Medium—high glucose (DMEM): DMEM (500 ml, SIGMA), 10% fetal calf serum, 5 ml penicillin/streptomycin (100×, 10,000 U/ml, Life Technologies). Ready-to-use media are stored at 4 °C.
4. Trypsin–EDTA (0.5%, 10×, Life Technologies), diluted to 1× in sterile Dulbecco's Phosphate-Buffered Saline (DPBS).
5. Sterile DPBS, no calcium, no magnesium.
6. Serum- and antibiotic-free DMEM.
7. QIAGEN PolyFect transfection reagent.

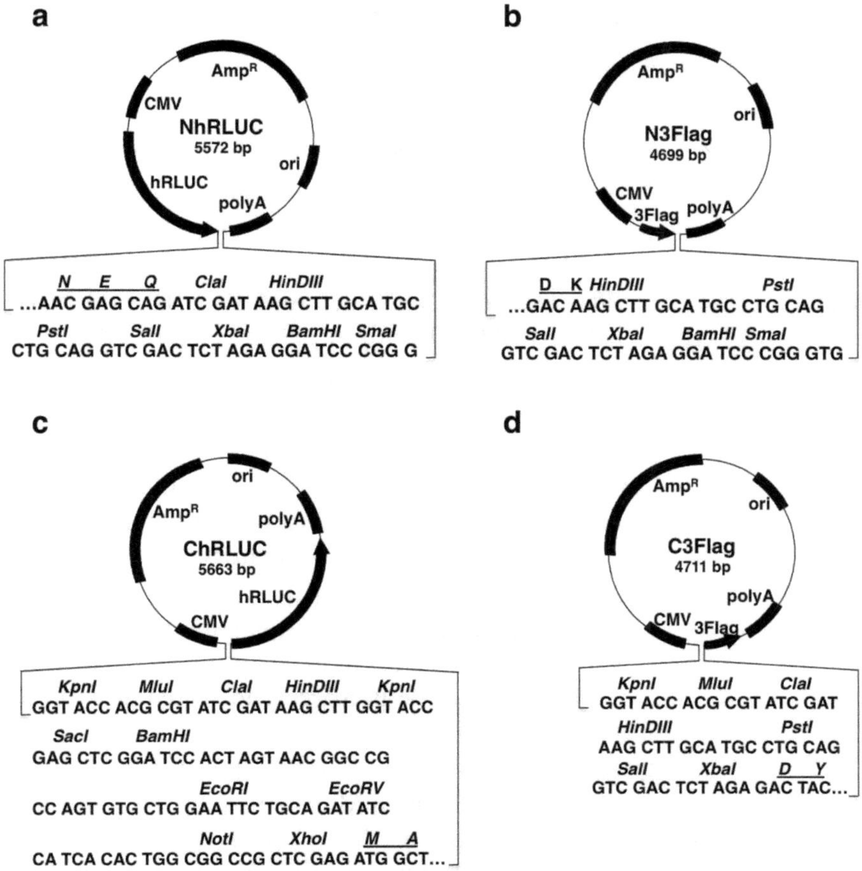

Fig. 2 Plasmids used in LUMIER. The vectors are pCMV5-based for high protein expression when transiently transfected into HEK-239T cells. Only unique restriction enzyme sites at the multiple cloning site (MCS) are shown. (**a** and **b**) NhRLUC and N3Flag are used to tag the Bait and Prey at the N-terminus. There is no stop codon at the 3′ end sequence of *Renilla* and N3Flag. (**c** and **d**) ChRLUC and C3Flag are used to tag the Bait and Prey at the C-terminus. The MCS of ChRLUC is a composite of pCMV5 and pCDNA3 polylinker regions. There is a stop codon at the 3′ end of Renilla and 3Flag. The GenBank accession number for the humanized Renilla luciferase (hRLUC) sequence is AF362545. The amino acid sequence for the 3Flag is: MDYKDHDGDYKDHDIDYKDDDDK. The cytomegalovirus (CMV) promoter drives high levels of gene expression for the Bait or Prey. The AmpR gene encodes beta-lactamase for ampicillin resistance selection in bacteria

2.2 LUMIER and ELISA Reagents

1. Monoclonal M2 anti-Flag-antibody (Sigma).
2. *Renilla*-Glo Assay System (Promega).
3. Monoclonal Anti-Flag M2-Peroxidase (HRP) clone M2 (Sigma).
4. ELISA pico substrate (Pierce).

2.3 Equipment

1. HandEvac aspirator 8-channel with ejector for disposable tips (Argos EV514 and EV520).
2. Multichannel pipets (for 10 μl, 20–300 μl, and 1 ml).

3. Transparent seals (QIAGEN).
4. Orbital shaker for plates.
5. Luminometer.
6. Corning, poly-L-lysine coated 96-well plates for cell culture.
7. Greiner Lumitrac 600 flat-bottom 96-well plates.
8. COSTAR White round-bottom 96-well plates.

2.4 Buffers

1. *PBS*: 137 mM NaCl, 2.7 mM KCl, 10 mM Na_2HPO_4, and 1.8 mM KH_2PO_4. Adjust pH to 7.4 with HCl.
2. Blocking Buffer: 3% bovine serum albumin, 5% sucrose, 0.5% Tween 20 in PBS.
3. Lysis Buffer: 50 mM Tris–HCl pH 7.4, 150 mM NaCl, 1 mM tetrasodium EDTA, 0.5% Triton X-100. A 10× stock can be stored at 4 °C and diluted to 1× just before lysis when adding phosphatase and protease inhibitors.
4. Phosphatase Inhibitors (final concentrations): 10 mM sodium pyrophosphate tetrabasic decahydrate, 1 mM sodium orthovanadate, 25 mM NaF. Stock solutions at 10×, 100× and 20× respectively, can be prepared and stored at 4 °C to be added just before lysis.
5. Protease Inhibitors: 1 mM phenylmethanesulfonylfluoride (PMSF; stock 100 mM in ethanol), 10 μg/ml pepstatin A (stock 1 mg/ml in DMSO), 100 μg/ml soy trypsin inhibitor (stock 10 mg/ml in Tris–EDTA buffer). 10 μg/ml leupeptin, 10 μg/ml antipain, 50 μg/ml aprotinin, 100 μg/ml benzamidine hydrochloride (stocks 1 mg/ml, 1 mg/ml, 5 mg/ml, and 10 mg/ml, respectively in Tris–EDTA buffer). Stocks for protease inhibitors are stored at −20 °C.
6. Tris–EDTA Buffer for protease inhibitors: 10 mM Tris–HCl pH 7.4, 1 mM tetrasodium EDTA.
7. Wash buffer: 50 mM Tris–HCl pH 7.4, 150 mM NaCl, 1 mM tetrasodium EDTA, and 0.1% Triton X-100.
8. Working *Renilla*-Glo substrate: Dilute the 100× substrate to 1× using the provided buffer in the kit.
9. PBS-T: 0.05% Tween 20 in PBS.
10. Buffer to dilute Flag-HRP antibody (Flag-HRP buffer): 5% Tween 20 and 1% fetal bovine serum in PBS.
11. Working Super signal ELISA pico substrate: Reagent A and B are mixed 1:1 ratio and then diluted 1:5 with distilled water.

3 Methods

See Fig. 3a for assay workflow.

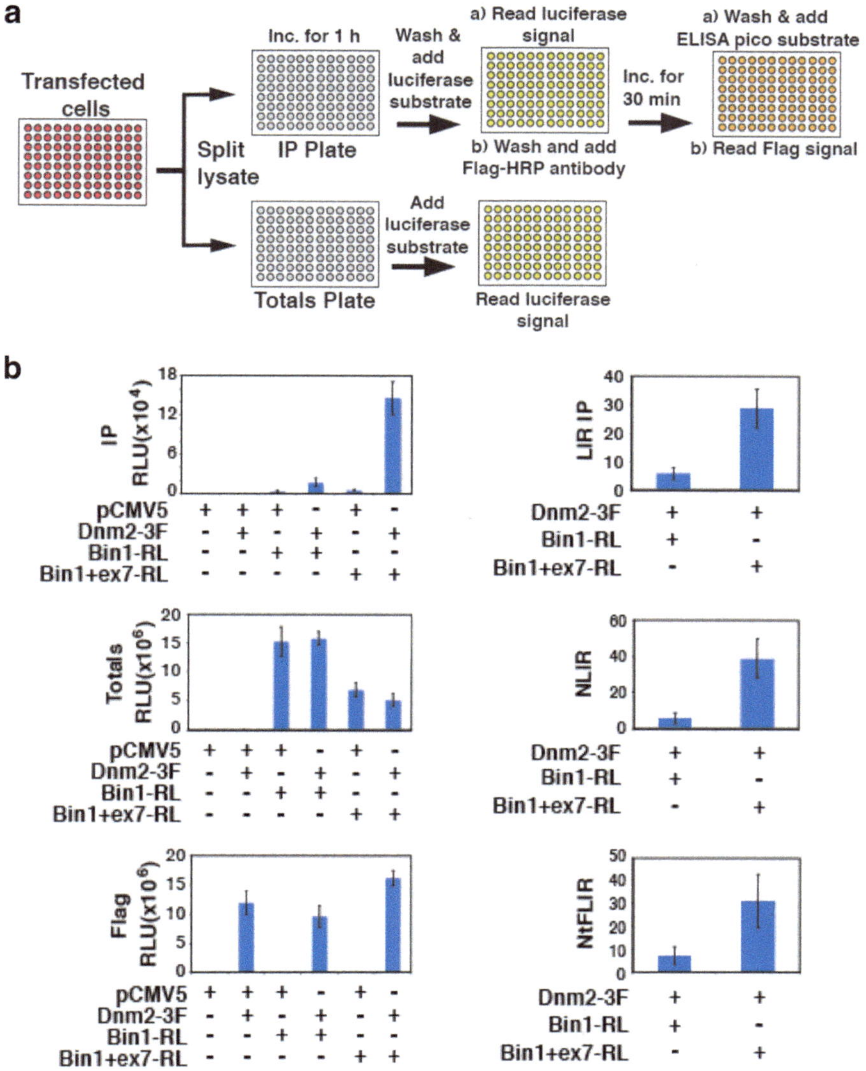

Fig. 3 (**a**) A LUMIER workflow diagram. Forty-eight hours post-transfection in 96-well plates, the cells are lysed. Lysate is split into Immunoprecipitation (IP) plate and Totals plate. After 1 h incubation the IP plate is washed and Renilla substrate added to IP and Totals plates to assess the interaction and to measure the expression levels of the Bait. Subsequently, an anti-Flag-HRP-conjugated antibody is added to the IP plate to measure the expression levels of the 3Flag-tagged Prey. After lysis, LUMIER results can be obtained within 4 h. (**b**) LUMIER data shows that exon 7 of Bridging Integrator 1 (Bin1) promotes an interaction with Dynamin 2 (Dnm2). HEK-293T cells were transfected as indicated. Dnm2-3Flag was used as prey while Bin1-RL (without exon 7) or Bin1 + ex7-RL (with exon 7) were used a baits. Averages for *Renilla* luciferase signals from IP plate, Totals plate and Flag-HRP measurements are shown on the *left-side graphs* (error bars show ± standard deviations from four technical replicates). The LIR IP, NLIR, and NtFLIR values calculated as described in Data Analysis are shown on the *right-side graphs* indicating that exon 7 promotes the interaction between Bin1 and Dnm2

3.1 Day 1: Plate HEK-293T Cells and Coat Plates with Anti-Flag Antibody

1. Manually plate 15,000 HEK293T cells per well in 70 μl/well, in poly-L-lysine coated 96-well plates (*see* **Note 3**) in DMEM with 10% fetal bovine serum and antibiotics (penicillin/streptomycin). Incubate overnight at 37 °C.

2. Coat white-flat-bottom Lumitrac plates with 100 μl/well of anti-Flag antibody in 1× PBS at 10 ng/μl (IP plate). Seal plates and store at 4 °C overnight. No shaking is necessary.

3.2 Day 2: Transfect Cells with LUMIER Constructs and Block Plates

1. Transfect every Bait-Prey combination at least in duplicate (quadruplicates are recommended). Dilute 100 ng of 3Flag-tagged cDNA (Prey) and 100 ng of the *Renilla* luciferase (RL) tagged cDNA (Bait) in DMEM without serum or antibiotics. Transfection is performed using PolyFect (*see* **Note 4**) according to 96-well plate manufacturer's protocol, using a 1:4 ratio of DNA (μg): PolyFect (μl). Maintain cells at 37 °C for 48 h. Include as negative controls the following conditions: empty vector only (pCMV5) and a second condition where the RLUC-tagged Bait is present, while the 3Flag-tagged Prey is substituted by empty vector (pCMV5). As a positive control to ensure the assay is working transfect 3Flag-hRLUC (*see* **Notes 5** and **6** for this and other recommended controls).
2. Aspirate the liquid from Lumitrac plates using the 8-channel aspirator and pat-dry plate on a paper towel to remove all liquid. Dispense 250 μl/well of Blocking Buffer. Shake at low speed in orbital shaker for 1 h at room temperature. Aspirate all liquid and pat-dry on paper towel. Store sealed plate at 4 °C.

3.3 Day 4: Lyse Cells and Perform Immunoprecipitation (IP) and Flag-Tagged Prey Measurement

1. Forty-eight hours post-transfection aspirate media from cells and dispense 150 μl/well of Lysis Buffer that includes phosphatase and protease inhibitors (*see* **Note 7**). Shake plate at low speed for 40 min at 4 °C.
2. Transfer 90 μl of lysate to anti-Flag-coated plate (Lumitrac; IP plate) and 20 μl to a round bottom-white plate (white round-bottom; Totals plate). Incubate both plates at 4 °C for 1 h, without shaking.
3. Aspirate liquid from IP plate using the 8-channel aspirator and dispense 250 μl of Wash buffer for each wash. Repeat wash step 5–6 times. Aspirate liquid from last wash and flick off remaining liquid. Dispense 100 μl/well of Working *Renilla*-Glo substrate to IP plates, incubate for 10 min at room temperature, and read in ENVISION luminometer 100 ms/well to detect interactions (*see* **Note 8**).
4. Measure expression levels of the *Renilla*-tagged proteins. Dispense 20 μl/well of Working *Renilla*-Glo substrate to Totals plate containing 20 μl of lysate, incubate for 10 min at room temperature, and read in ENVISION luminometer 100 ms/well.
5. Aspirate the liquid from IP plate and flick to remove all liquid. Wash IP plate eight times with 250 μl/well of PBS-T each time. Dispense 100 μl/well of anti-Flag-HRP antibody at 1:20,000 dilution (*see* **Note 9**) in Flag-HRP Buffer. Incubate for 30 min at 4 °C without shaking.

6. Aspirate all liquid and wash eight times with 250 μl/well PBS-T. Aspirate liquid from last wash and flick off remaining liquid. Dispense 100 μl/well of Working Super signal ELISA pico substrate. Incubate at room temperature for 7 min and read in ENVISION for 20 ms/well.

3.4 Data Analysis

The Normalized LUMIER Intensity Ratio (NLIR) value for each interaction is calculated as follows: The LUMIER Intensity Ratio for each Bait/Prey combination is first determined from the luciferase activity in the immunoprecipitates (LIR IP). This LIR IP is calculated as the luciferase intensity in IP plate when each Bait (RLUC-fusions) and Prey (3Flag-tagged cDNAs) are co-transfected, over the luciferase intensity from immunoprecipitates when the Bait (RLUC-fusion) is co-transfected with empty vector (pCMV5).

LIR IP = signal IP (Bait + Prey)/signal IP (Bait + pCMV5).

The LIR TOT is caculated in a similar manner from the luciferase activity measured in the total cell lysates (Totals plate) of each Bait/Prey combination over the luciferase activity from total cell lysates of each Bait/empty vector.

LIRTOT = signal Totals (Bait + Prey)/signal Totals (Bait + pCMV5).

The Normalized LIR (NLIR) is the ratio of LIR IP/LIR TOT for each Bait/Prey combination. In general, an NLIR value equal or higher than 3 is considered as a positive interaction.

To give a more quantitative interaction score a Flag Intensity Ratio (FIR) is calculated to account for changes in the immunoprecipitation of Flag-tagged Prey protein with different RLUC-tagged Baits. This FIR is calculated with the Flag-HRP signal value from immunoprecipitates when Bait and Prey are co-transfected, over the Flag-HRP signal value from immunoprecipitates when the Prey is co-transfected with empty vector.

FIR = signal Flag-HRP (Bait + Prey)/signal Flag-HRP (Bait + pCMV5).

A final Normalized to Flag LIR value (NtFLIR) is the ratio of NLIR/FIR for each interaction tested. *See* Fig. 3b for an example on data analysis.

4 Notes

1. The plasmids to tag the Bait with *Renilla* luciferase and the Prey with 3Flag are pCMV5-based. There is also a Gateway® version for each.
2. The Baits and Preys can be tagged at either N- or C-terminus. This is an important consideration when designing a LUMIER experiment. If the Bait of interest has an important domain at

one end (i.e., PDZ-binding domain or a kinase domain), place the luciferase tag at the opposite end. Alternatively, place the tag at both ends and test the performance of the two Baits (N- and C-terminally tagged) in preliminary experiments using a known Flag-tagged partner as positive control.

3. Regular 96-well TC plates can be coated with poly-L-lysine (Sigma cat. P-1399) at 50 μg/ml dissolved in PBS and filter-sterilized. Dispense 100 μl/well of diluted poly-L-lysine and incubate at room temperature for 1 h. Rinse twice with sterile PBS and plate cells immediately after PBS has been removed. Stock of poly-L-Lysine can be made at 2 mg/ml in PBS and stored at −20 °C. Coating the plates is strongly recommended because during transfection the HEK-293T cells can easily dettach and increase variability.
4. Other transfection reagents alternative to PolyFect can be used with HEK-293T cells, including the Calcium Phosphate method which is low cost, or Lipofectamine 2000 (Life Technologies). PolyFect is recommended because of its high transfection efficiency, cells can be transfected in the presence of antibiotics and there is no need to change media the next day.
5. To ensure that the assay is working, a very good positive control to use is a plasmid encoding 3Flag-hRLUC. When this plasmid is transfected in HEK-293T cells, a high signal for Renilla luciferase activity is detected in the IP plate and Totals plate upon addition of *Renilla*-Glo substrate, but also from Flag-HRP after the addition of the ELISA pico substrate.
6. To distinguish whether a 3Flag-tagged partner is interacting with Renilla luciferase within the Bait fusion, include as control a condition where the 3Flag-tagged cDNA is co-transfected with the ChRLUC plasmid (Fig. 2c) which has luciferase activity in total lysates but is not fused to any Bait. In high-throughput screens out of ~560 Preys only few proteins were found to interact with Renilla luciferase (UBA52, UBC, UBB, RPS27A, SQSTM1, and ECSIT).
7. A 10× lysis buffer stock can be prepared and diluted to 1× while adding phosphatase and protease inhibitos just before use. Instead of PMSF, AEBSF (Sigma cat. A8456) can be used at a final concentration of 0.5 mM, because it is a more stable serine protease inhibitor.
8. A Luminometer with injectors can be used if a luminometer with a high sensitivity detector like ENVISION (Perkin Elmer) is not available. For this, the *Renilla* Luciferase Assay system kit (Promega), which contains a 'flash-kinetics' substrate with high sensitivity must be used. This substrate has to be dispensed by a luminometer's injectors because the signal obtained with this kit has a very short half-life (~2 min), in contrast to

the *Renilla*-Glo substrate, whose signal half-life is greater than 60 min at room temperature.

9. Comparison between different lot numbers of a rabbit polyclonal Flag antibody conjugated to HRP from Abcam to a mouse monoclonal Flag antibody conjugated to HRP from Sigma showed that the Sigma antibody performs significantly better in ELISA experiments in terms of signal to noise ratio.

Acknowledgments

This work was supported by funding from the Canadian Institutes of Health Research (CIHR) (J.L.W. and B.J.B.), the Alzheimer's Society of Canada (B.J.B.), and the Krembil Foundation (J.L.W.). We would also like to thank Dr. Saranya Kittanakom for the plasmid schematics framework.

References

1. Fields S, Song O (1989) A novel genetic system to detect protein-protein interactions. Nature 340(6230):245–246
2. Lievens S, Lemmens I, Tavernier J (2009) Mammalian two-hybrids come of age. Trends Biochem Sci 34(11):579–588
3. Braun P, Tasan M, Dreze M, Barrios-Rodiles M, Lemmens I, Yu H, Sahalie JM, Murray RR, Roncari L, de Smet AS, Venkatesan K, Rual JF, Vandenhaute J, Cusick ME, Pawson T, Hill DE, Tavernier J, Wrana JL, Roth FP, Vidal M (2009) An experimentally derived confidence score for binary protein-protein interactions. Nat Methods 6(1):91–97
4. Eyckerman S, Verhee A, der Heyden JV, Lemmens I, Ostade XV, Vandekerckhove J, Tavernier J (2001) Design and application of a cytokine-receptor-based interaction trap. Nat Cell Biol 3(12):1114–1119
5. Stagljar I, Korostensky C, Johnsson N, te Heesen S (1998) A genetic system based on split-ubiquitin for the analysis of interactions between membrane proteins in vivo. Proc Natl Acad Sci U S A 95(9):5187–5192
6. Petschnigg J, Groisman B, Kotlyar M, Taipale M, Zheng Y, Kurat CF, Sayad A, Sierra JR, Mattiazzi Usaj M, Snider J, Nachman A, Krykbaeva I, Tsao MS, Moffat J, Pawson T, Lindquist S, Jurisica I, Stagljar I (2014) The mammalian-membrane two-hybrid assay (MaMTH) for probing membrane-protein interactions in human cells. Nat Methods 11(5):585–592
7. Pelletier JN, Campbell-Valois FX, Michnick SW (1998) Oligomerization domain-directed reassembly of active dihydrofolate reductase from rationally designed fragments. Proc Natl Acad Sci U S A 95(21):12141–12146
8. Wilson CG, Magliery TJ, Regan L (2004) Detecting protein-protein interactions with GFP-fragment reassembly. Nat Methods 1(3):255–262
9. Remy I, Michnick SW (2006) A highly sensitive protein-protein interaction assay based on Gaussia luciferase. Nat Methods 3(12): 977–979
10. Boute N, Jockers R, Issad T (2002) The use of resonance energy transfer in high-throughput screening: BRET versus FRET. Trends Pharmacol Sci 23(8):351–354
11. Ramachandran N, Hainsworth E, Bhullar B, Eisenstein S, Rosen B, Lau AY, Walter JC, LaBaer J (2004) Self-assembling protein microarrays. Science 305(5680):86–90
12. Barrios-Rodiles M, Brown KR, Ozdamar B, Bose R, Liu Z, Donovan RS, Shinjo F, Liu Y, Dembowy J, Taylor IW, Luga V, Przulj N, Robinson M, Suzuki H, Hayashizaki Y, Jurisica I, Wrana JL (2005) High-throughput mapping of a dynamic signaling network in mammalian cells. Science 307(5715):1621–1625
13. Ozdamar B, Bose R, Barrios-Rodiles M, Wang HR, Zhang Y, Wrana JL (2005) Regulation of the polarity protein Par6 by TGFbeta receptors controls epithelial cell plasticity. Science 307(5715):1603–1609

14. Miller BW, Lau G, Grouios C, Mollica E, Barrios-Rodiles M, Liu Y, Datti A, Morris Q, Wrana JL, Attisano L (2009) Application of an integrated physical and functional screening approach to identify inhibitors of the Wnt pathway. Mol Syst Biol 5:315
15. Hall MP, Unch J, Binkowski BF, Valley MP, Butler BL, Wood MG, Otto P, Zimmerman K, Vidugiris G, Machleidt T, Robers MB, Benink HA, Eggers CT, Slater MR, Meisenheimer PL, Klaubert DH, Fan F, Encell LP, Wood KV (2012) Engineered luciferase reporter from a deep sea shrimp utilizing a novel imidazopyrazinone substrate. ACS Chem Biol 7(11): 1848–1857
16. Xu G, Barrios-Rodiles M, Jerkic M, Turinsky AL, Nadon R, Vera S, Voulgaraki D, Wrana JL, Toporsian M, Letarte M (2014) Novel protein interactions with endoglin and activin receptor-like kinase 1: potential role in vascular networks. Mol Cell Proteomics 13(2):489–502
17. Kittanakom S, Barrios-Rodiles M, Petschnigg J, Arnoldo A, Wong V, Kotlyar M, Heisler LE, Jurisica I, Wrana JL, Nislow C, Stagljar I (2014) CHIP-MYTH: a novel interactive proteomics method for the assessment of agonist-dependent interactions of the human beta(2)-adrenergic receptor. Biochem Biophys Res Commun 445(4):746–756
18. Varelas X, Miller BW, Sopko R, Song S, Gregorieff A, Fellouse FA, Sakuma R, Pawson T, Hunziker W, McNeill H, Wrana JL, Attisano L (2010) The Hippo pathway regulates Wnt/beta-catenin signaling. Dev Cell 18(4):579–591
19. Beyer TA, Weiss A, Khomchuk Y, Huang K, Ogunjimi AA, Varelas X, Wrana JL (2013) Switch enhancers interpret TGF-beta and Hippo signaling to control cell fate in human embryonic stem cells. Cell Rep 5(6): 1611–1624
20. Ellis JD, Barrios-Rodiles M, Colak R, Irimia M, Kim T, Calarco JA, Wang X, Pan Q, O'Hanlon D, Kim PM, Wrana JL, Blencowe BJ (2012) Tissue-specific alternative splicing remodels protein-protein interaction networks. Mol Cell 46(6):884–892
21. Taipale M, Krykbaeva I, Koeva M, Kayatekin C, Westover KD, Karras GI, Lindquist S (2012) Quantitative analysis of HSP90-client interactions reveals principles of substrate recognition. Cell 150(5):987–1001
22. Ryzhakov G, Teixeira A, Saliba D, Blazek K, Muta T, Ragoussis J, Udalova IA (2013) Cross-species analysis reveals evolving and conserved features of the nuclear factor kappaB (NF-kappaB) proteins. J Biol Chem 288(16): 11546–11554
23. Blasche S, Mortl M, Steuber H, Siszler G, Nisa S, Schwarz F, Lavrik I, Gronewold TM, Maskos K, Donnenberg MS, Ullmann D, Uetz P, Kogl M (2013) The E. coli effector protein NleF is a caspase inhibitor. PLoS One 8(3):e58937
24. Deng Q, Wang D, Li F (2014) Detection of viral protein-protein interaction by microplate-format luminescence-based mammalian interactome mapping (LUMIER). Virol Sin 29(3): 189–192
25. Tahoun A, Siszler G, Spears K, McAteer S, Tree J, Paxton E, Gillespie TL, Martinez-Argudo I, Jepson MA, Shaw DJ, Koegl M, Haas J, Gally DL, Mahajan A (2011) Comparative analysis of EspF variants in inhibition of Escherichia coli phagocytosis by macrophages and inhibition of E. coli translocation through human- and bovine-derived M cells. Infect Immun 79(11): 4716–4729

Chapter 12

Dual-Color, Multiplex Analysis of Protein Microarrays for Precision Medicine

Solomon Yeon, Florian Bell, Michael Shultz, Grace Lawrence, Michael Harpole, and Virginia Espina

Abstract

Generating molecular information in a clinically relevant time frame is the first hurdle to truly integrating precision medicine in health care. Reverse phase protein microarrays are being utilized in clinical trials for quantifying posttranslationally modified signal transduction proteins and cellular signaling pathways, allowing direct comparison of the activation state of proteins from multiple specimens, or individual patient specimens, within the same array. This technology provides diagnostic and therapeutic information critical to precision medicine. To enhance accessibility of this technology, two hurdles must be overcome: data normalization and data acquisition. Herein we describe an unamplified, dual-color signal detection methodology for reverse phase protein microarrays that allows multiplex, within spot data normalization, reduces data acquisition time, simplifies automated spot detection, and provides a stable signal output. This method utilizes Quantum Nanocrystal fluorophore labels (Qdot) substituted for organic fluorophores coupled with an imager (ArrayCAM) that captures images of the microarray rather than sequentially scanning the array. Streamlining and standardizing the data analysis steps with ArrayCAM high-resolution, dual mode chromogenic/fluorescent array imaging overcomes the data acquisition hurdle. The spot location and analysis algorithm provides certain parameter settings that can be tailored to the particular microarray type (fluorescent vs. colorimetric), resulting in greater than 99% spot location sensitivity. The described method demonstrates equivalent sensitivity for a non-amplified Qdot immunoassay when using automated vs. manual immunostaining procedures.

Key words Fluorescence, Multicolor detection, Precision medicine, Posttranslational modification, Protein microarray, Protein phosphorylation, Qdot nanocrystal, Receptor tyrosine kinase, Reverse phase protein microarray, Signal transduction

1 Introduction

How do we provide functional molecular information about an individual person's tumor or disease state? What is the state of a patient's cellular signal transduction cascades before, during, and after therapy? What are the on and off target effects of molecular targeted inhibitors? These questions represent a small fraction of the complex proteomic information that can be gleaned from reverse phase protein microarrays (RPPA) [1–6].

Lucio Comai et al. (eds.), *Proteomics: Methods and Protocols*, Methods in Molecular Biology, vol. 1550,
DOI 10.1007/978-1-4939-6747-6_12, © Springer Science+Business Media LLC 2017

Precision medicine requires rapid, precise, and reproducible technology for integrating genetic, proteomic, and phenotypic information into actionable treatment strategies [7–10]. Our laboratory has developed and standardized Reverse Phase Protein Microarrays (RPPAs) as a technology for studying posttranslationally modified proteins and their unmodified protein forms in the context of complex signal transduction networks [6, 11–17]. RPPAs provide functional epigenetic protein information that is not discernable from gene expression arrays. Several laboratories in the USA offer RPPA technology for precision medicine (Table 1).

Table 1
Examples of US laboratories performing reverse phase protein microarray technology

Laboratory	Services and tests	Sample requirements/method of lysate prep	Turnaround time
George Mason University[a] Center for Applied Proteomics and Molecular Medicine	Translational Research and clinical trials	*Sample*: 1. Cell/tissue lysates 2. Macro-dissected tissue samples 3. Laser capture microdissected samples 4. Stem cells/cell lines 5. Body fluids (serum, urine, vitreous) *Lysate Prep*: Laser Capture Microdissection	Investigator/project dependent
Baylor College of Medicine[b] The Cancer Prevention and Research Institute of Texas (CPRIT) Cancer proteomics and metabolomic Core Facility	Research for BMC Faculty only; Two-tiered services	*Sample*: 1. Cell/tissue lysates 2. Macro-dissected tumor samples 3. Laser capture microdissected samples 4. Stem cells 5. Body fluids (serum and urine)	Information not provided
MD Anderson RPPA Core Facility[c]	Clinical diagnostics and research Functional Proteomics 1. Human samples—panel of ≥200 antibodies 2. Mouse samples—panel of ≥150 antibodies	*Sample*: 1. Cultured cells, mouse xenografts and frozen human tissue. 2. Not currently accepting FFPE samples. 3. Client must prepare RPPA samples for submission according to prescribed protocol *Lysate Prep*: Lysate sample preparation from tissue by electric homogenizer Lysate sample preparation from frozen tissue by Precellys Homogenizer	8–10 weeks

(continued)

Table 1
(continued)

Laboratory	Services and tests	Sample requirements/method of lysate prep	Turnaround time
Theranostics Health[d]	Clinical Diagnostics; Research and development	*Sample*: 1. Tumor from core needle biopsy or open biopsy which have been fixed in 10% neutral buffered formalin within 20 min of biopsy 2. Paraffin blocks—cut 4–5 μm, mounted on uncharged slides *Lysate Prep*: Laser Capture Microdissection	Performed weekly on Tuesday—final report available electronically in 10 business days

[a]http://capmm.gmu.edu/
[b]http://www.mdanderson.org/education-and-research/resources-for-professionals/scientific-resources/core-facilities-and-services/functional-proteomics-rppa-core/index.html
[c]https://www.bcm.edu/centers/cancer-center/research/shared-resources/cprit-cancer-proteomics-and-metabolomics/reverse-phase-proteinarray
[d]http://www.theranosticshealth.com/protemic-services/

RPPA technology can be applied to any cellular lysate, or protein-containing body fluid, including bacterial lysates. The lysate/protein specimen is immobilized (printed) on a nitrocellulose substratum, typically using a robotic printer [12, 18]. Controls, calibrators, and standards can be printed on the same array, providing a complete analytical test on one array. RPPAs enable comparison of the functional state of multiple or single specimens within the same array. Each spot printed on an array is representative of the proteome in that specimen. Robotic arraying devices and automated slide stainers facilitate reliable technical precision of the RPPA [12, 18–20]. The RPPA detection methods depend on the analyte concentration, the type of microarray imaging system available in the laboratory, and the type of sample being investigated. Typically, identification of low abundance cell signaling proteins and their posttranslationally modifications, such as phosphorylation, have required signal amplification and optimized detection methods (fluorescent or colorimetric).

To enhance accessibility of RPPA technology, two hurdles must be overcome: data normalization and data acquisition [2, 21–25]. The most labor intensive process in RPPA methodology is the spot finding and quantification. Generating health-related information in a clinically relevant time frame is the first hurdle to providing precision medicine. Molecular information regarding the specific functional defects in a tumor is worthless unless that information can be delivered and utilized promptly. Herein we describe an unamplified, dual-color signal detection methodology for reverse phase protein microarrays that allows multiplex, within spot data normalization, reduces data acquisition time, simplifies automated spot detection, and provides a stable signal output. This method utilizes Quantum Nanocrystal fluorophore labels (Qdot) substituted for organic fluorophores coupled with an imager (ArrayCAM) that captures images of the microarray rather than sequentially scanning the array.

Quantum nanocrystals (Qdots) are spherical compound semiconductor particles that exhibit luminescence from electronic states related to the particle's material composition and size [26, 27]. Due to the quantum confinement of electronic states in the crystals, Qdots absorb light within a broad band from the ultraviolet to visible range and emit fluorescent light within narrow bands from the visible to near infrared. Each Qdot species is tuned to emit light in a specific band for the purpose of fluorescence detection [28]. Qdot emission bands are narrow compared to emission bands from organic fluorophores. When used as labels for immunoassay detection, Qdots have a number of advantages relative to organic fluorophores, including greater brightness, resistance to photobleaching and quenching, and long-term stability. Additionally, all Qdots can be excited with a common laser wavelength and this can contribute to simplified and lower-cost instrumentation for fluorescence detection.

Because of their narrow fluorescence emission bands, Qdots may be applied to the specific problem of multiplexed labeling within the same microarray [29] or the same well of a microtiter plate [30]. Individual interrogation molecules may be conjugated to particular Qdot emitters and all emitters can be activated with a common excitation source. Through the use of separate emission filters, each label can be individually detected. This method is particularly effective when a digital camera is used to image the microarray. In the current work, we have established a methodology for multiplexing the detection of several analytes simultaneously and within the same spot in the microarray, using Qdot labels and ArrayCAM, a high-resolution, high-sensitivity imaging system (Fig. 1).

This chapter describes two new methods for processes relevant to protein microarrays: (1) manual and automated non-amplified immunostaining procedures using Qdot nanocrystal conjugated antibodies, and (2) spot finding/quantification with an ArrayCAM chromogenic/fluorescent array imager. Manual immunostaining methods describe RPPA detection strategies without the use of an automated slide stainer, whereas the automated immunostaining method describes adaptation of immunostaining to an automated slide stainer (Autostainer, Dako). This chapter also describes use of the ArrayCAM microarray imager to sequentially identify multiple fluorescent endpoints within the same microarray spot. The method provides equivalent results to separate assays with a single fluorescent/colorimetric endpoint (Fig. 2).

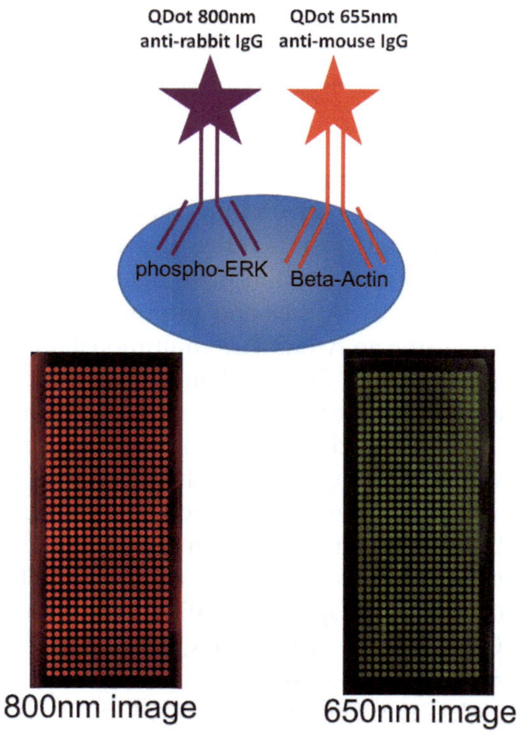

Fig. 1 Dual-color quantum dot (Qdot) detection scheme. Qdot nanocrystals with different emission wavelengths are conjugated to either anti-mouse immunoglobulin (IgG) (Qdot 655 nm) or to anti-rabbit IgG (Qdot 800 nm). Both Qdot conjugated antibodies are mixed in a single detection cocktail, permitting within spot detection of two different analytes

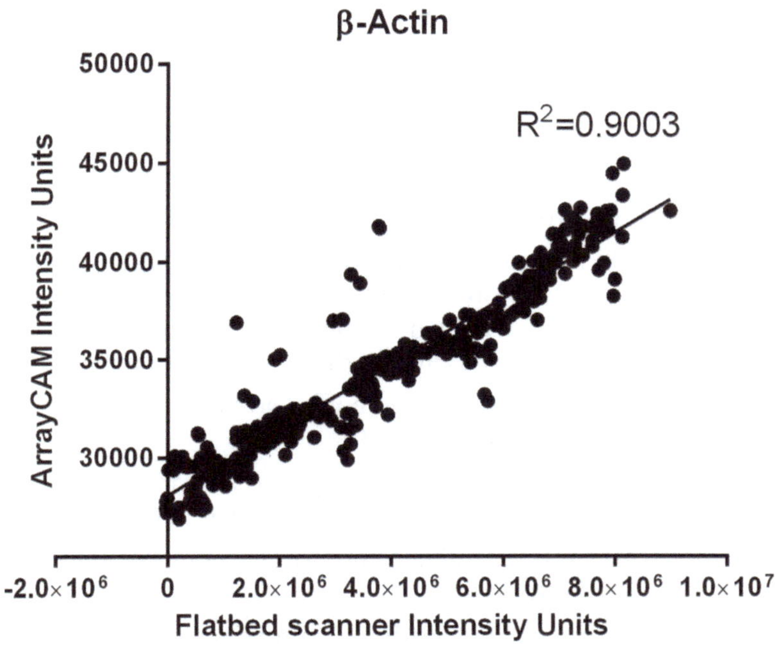

Fig. 2 Equivalent resolution in the detection of relevant biologically pathways previously observed using colorimetric detection can be achieved on the ArrayCAM platform. Strong concordance in spot signal intensities between a β-Actin stained array imaged on a flatbed scanner (*x*-axis, Image Quant ver 5.1 software spot analysis) and the ArrayCAM imager (*y*-axis, ArrayCAM software spot analysis). β-Actin was detected with a catalyzed signal amplification method using colorimeteric detection [18]

2 Materials

Reverse phase protein microarray technology consists of four main steps: sample preparation/cell lysis, array printing, array immunostaining, and spot finding and quantification. The focus of this chapter is on immunostaining with dual-color, intra-spot signal detection by manual and automated methods, and spot quantification using an ArrayCAM multimodal imager. The staining reagents are designed for simultaneous detection of two IgG antibody probes of different species at two spectrally distinct wavelengths.

RPPA slides containing immobilized protein specimens are simultaneously probed with a cocktail of mouse and rabbit primary antibodies against the targets of interest. One protein target (antibody) is used for data normalization, while the second target (antibody) binds to an analyte of interest. The primary antibodies are detected with species specific anti-IgG conjugated to either 800 nm Qdot (rabbit) or 655 nm Qdot (mouse). After staining, the RPPA slides are read using an ArrayCAM instrument with excitation by a violet laser and imaging at the Qdot's respective wavelength.

2.1 Manual Dual-Color Qdot Immunostaining

1. Reverse phase protein microarrays printed with whole cell lysates, serum, vitreous, laser capture microdissected cell lysates, peripheral blood mononuclear cells, or other protein containing body fluids.
2. ProPlate slide chambers and spring clips (Grace Bio-Labs): 1-well format.
3. Orbital shaker.
4. Pipettes (1 μL–50 mL delivery volume).
5. Plastic trays/petri dishes with lids.
6. Type 1 reagent grade water (dH_2O).
7. Antigen retrieval solution, 10% sodium hydroxide in water, 10×. Store at 4 °C. Dilute to 1× prior to use: Prepare 100 mL of 1× Antigen Retrieval by adding 10 mL of 10× Antigen Retrieval to 90 mL dH_2O. Mix thoroughly. Store at 4 °C.
8. Q-Block blocking buffer, proprietary composition (Grace Bio-Labs). Store at 4 °C.
9. Primary antibodies against analytes of interest (anti-rabbit and/or anti-mouse, unconjugated) (*see* **Note 1**).
10. Antibody diluent proprietary composition (Grace Bio-Labs). Store at 4 °C.
11. 10× Wash Buffer Q, proprietary composition (Grace Bio-Labs). Store at 4 °C. Prior to use prepare 400 mL of 1× Wash Buffer Q by adding 40 mL of 10× Wash Buffer Q to 360 mL

of type 1 reagent grade water (dH_2O). Mix thoroughly. Store at 4 °C.

12. 10× Rinse buffer, proprietary composition (Grace Bio-Labs). Store at 4 °C. Prior to use, prepare 400 mL of 1× Rinse Buffer by adding 40 mL of 10× Rinse Buffer to 360 mL of dH_2O. Mix thoroughly. Store at 4 °C.
13. Qdot nanocrystals conjugated to secondary antibodies: 800 nm goat anti-rabbit and 655 nm goat anti-mouse. Store at 4 °C. Do not freeze.
14. Detection reagent diluent, proprietary composition (Grace Bio-Labs). Store at 4 °C.

2.2 Automated Dual-Color Qdot Immunostaining

1. Reverse phase protein microarrays printed with whole cell lysates, serum, vitreous, laser capture microdissected cell lysates, peripheral blood mononuclear cells, or other protein containing body fluids.
2. Automated slide stainer with open-source reagents (e.g., Dako Autostainer).
3. Orbital shaker.
4. Plastic trays/petri dishes with lids.
5. Pipettes (1 μL–50 mL delivery volume).
6. Type 1 reagent grade water (dH_2O).
7. Antigen retrieval solution 10×, 10% sodium hydroxide in water. Store at 4 °C. Dilute to 1× prior to use as described in **item 7** of Subheading 2.1.
8. Q-Block blocking buffer, proprietary composition (Grace Bio-Labs). Store at 4 °C.
9. Primary antibodies (anti-rabbit and/or anti-mouse, unconjugated) (*see* **Note 1**).
10. Antibody diluent, proprietary composition (Grace Bio-Labs). Store at 4 °C.
11. 10× Wash Buffer Q, proprietary composition (Grace Bio-Labs). Store at 4 °C. Dilute to 1× prior to use as described in **item 11** of Subheading 2.1.
12. 10× Rinse buffer, proprietary composition (Grace Bio-Labs). Store at 4 °C. Dilute to 1× prior to use as described in **item 12** of Subheading 2.1.
13. Qdot nanocrystals conjugated to secondary antibodies: 800 nm goat anti-rabbit and 655 nm goat anti-mouse. Store at 4 °C. Do not freeze.
14. Detection reagent diluent, proprietary composition (Grace Bio-Labs). Store at 4 °C.

2.3 ArrayCAM Scanning of Reverse Phase Protein Microarrays

1. ArrayCAM imager with ArrayCAM software (Grace Bio-Labs), computer (Windows 7 or 8).
2. Printed and stained reverse phase protein microarray. RPPAs can be stained with Sypro Ruby Protein Blot stain, Fast Green, Qdot nanocrystals (800 nm, 655 nm, and/or 585 nm) conjugated to rabbit or mouse immunoglobulin, or colorimetric (e.g., diaminobenzidine (DAB)) detection molecules [18].

3 Methods

Personalized medicine requires analysis of single patient samples in a timely manner. RPPAs can accommodate many samples, standards, and controls on a single array (single nitrocellulose pad) or can be constructed with one or more samples on a series of smaller nitrocellulose pads (sectors) on a single slide. Regardless of the array format, analyte detection strategies must be specific, sensitive, and unaltered by contaminating proteins [21]. To compare analyte levels across samples on an array, a normalizer molecule, usually total protein, single stranded DNA or a housekeeping protein, is measured on a separate array. However, a more precise normalization procedure would be to measure the normalizer molecule in the same spot, on the same array, in which the analyte is measured. Multiplex detection of more than one target per slide is not possible using colorimetric catalyzed signal amplification methods [31–34]. Non-amplified signal detection capitalizes on the ability to combine two or more quantum nanocrystal (Qdot) fluorescent molecules in a single detection solution. This dual-color detection system permits two different analytes to be measured within the same array spot (Fig. 1). Qdots have narrow emission spectra, permitting Qdots with different wavelengths to be combined in a single detection cocktail. Qdots provide high signal:noise ratios, resist photobleaching and quenching, and provide a stable signal. Herein we describe manual and automated slide stainer detection methods for RPPAs using non-amplified signal detection with intra-spot, multicolor analysis.

3.1 Manual Dual-Color QDot RPPA Immunostaining

1. Retrieve RPPA slides from storage (−20 °C). Allow the slides to reach room temperature. The number of slides to stain equals the number of primary antibodies plus 1 for a secondary antibody only control [12, 18].
2. Wash RPPA slides 4 × 15 min in fresh dH_2O. Place slides with the nitrocellulose film facing up, in a plastic dish. Add adequate amounts of water to completely immerse the slides. Place the dish on an orbital rocker with gentle rocking for 15 min. Remove and discard the water. Repeat the wash steps three times. Do not allow the slides to dry out at any point during the staining.

3. Attach a ProPlate chamber to the slide. Apply one ProPlate chamber at a time. Fill the chamber with fresh dH_2O.
4. Remove and discard the dH_2O. Add 2 mL of 1× Antigen Retrieval buffer. Incubate for 15 min at room temperature.
5. Wash 3 × 5 min each with 2 mL of fresh dH_2O for each wash with gentle rocking on an orbital rocker.
6. Discard the water after each wash.
7. Add 2 mL 1× Wash Buffer Q. Wash for 5 min with gentle rocking on an orbital rocker.
8. Remove and discard the Wash Buffer Q. Add 2 mL Q-Block Blocking Buffer. Incubate for 10 min at room temp with gentle rocking on an orbital rocker.
9. While the slides are blocking, prepare the primary antibody dilutions in the Antibody Diluent. Each array will be probed with a cocktail of two antibodies: the anti-mouse antibody and the anti-rabbit antibody. Prepare an antibody cocktail for each primary antibody of interest (one cocktail per array). One slide will not have any primary antibody added—this is the secondary antibody only control. This slide will have antibody diluent only added in place of the primary antibody cocktail.
10. Remove and discard the Q-Block Blocking Buffer. Wash twice for 2 min with 2 mL 1× Wash Buffer Q, with gentle rocking on an orbital rocker.
11. Remove and discard the blocking buffer. Add the appropriate primary antibody cocktail to each slide. Add antibody diluent only to the secondary antibody only control slide. Incubate at room temperature for 90 min with gentle rocking on an orbital rocker.
12. Remove and discard the antibody cocktail/antibody diluent.
13. Wash 3 × 5 min with 2 mL 1× Wash Buffer Q with gentle rocking on an orbital rocker.
14. While the slides are washing, prepare the Detection cocktail (Qdot conjugated anti-IgG, mouse and rabbit). For each slide, prepare 2100 μL of a 1:500 dilution of each Qdot conjugated anti-IgG (800 and 655 nm) in 1× Detection Reagent Diluent. Example: for one slide, add 4.2 μL of 500× Qdot 800 nm anti-rabbit IgG and 4.2 μL of 500× Qdot 655 nm anti-mouse IgG to 2091.6 μL of 1× Detection Reagent Diluent. Mix thoroughly.
15. Remove and discard the wash buffer. Wash 2 × 2 min with 2 mL 1× Rinse Buffer.
16. Remove and discard the Rinse Buffer. Add 2100 μL of Detection cocktail. Incubate at room temperature for 90 min with gentle rocking on an orbital rocker.
17. Wash 3 × 5 min with 2 mL 1× Rinse Buffer with gentle rocking on an orbital rocker.

18. Remove and discard the Rinse Buffer. Remove the ProPlate chamber from each array slide.
19. Rinse the slides in copious volumes of fresh dH_2O.
20. Allow the slides to dry at room temperature, protected from light.

3.2 Automated Dual-Color Qdot RPPA Immunostaining (Fig. 3)

1. Program the Dako Autostainer according to Table 2. Determine the volume of Wash Buffer Q and water required for the staining run. Typically five slides require approximately 2000 mL of Wash Buffer Q in the buffer carboy.
2. Add the diluted Wash Buffer Q to the Autostainer buffer carboy.
3. Retrieve RPPA slides from storage (−20 °C). Allow the slides to reach room temperature. The number of slides to stain equals the number of primary antibodies plus 1 for a secondary antibody only control [12, 18].
4. Wash RPPA slides 2 × 5 min in fresh dH_2O. Place slides with the nitrocellulose film facing up, in a plastic dish. Add adequate amounts of water to completely immerse the slides. Place the dish on an orbital rocker with gentle rocking for 15 min. Remove and discard the water. Repeat the wash steps three times. Do not allow the slides to dry out at any point during the staining.

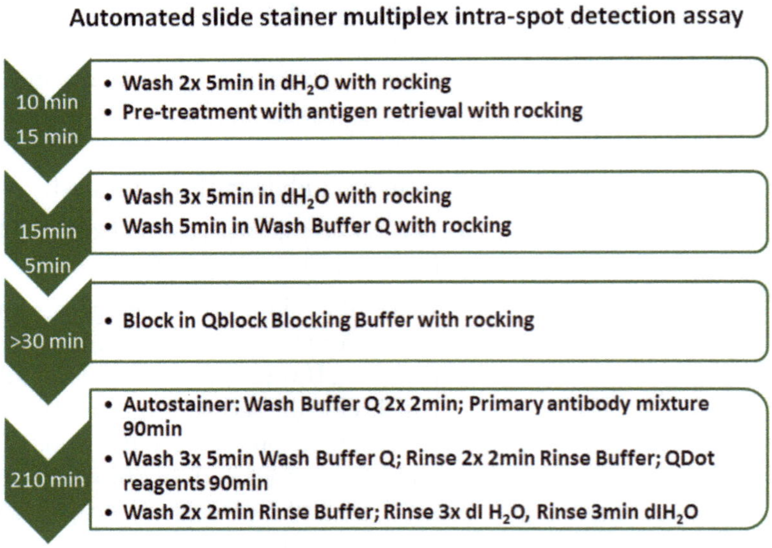

Fig. 3 Workflow for automated dual-color, non-amplified staining of reverse phase protein microarrays. The slides are rinsed, washed, and blocked prior to loading them on the Autostainer. On the Autostainer, the slides are rinsed, incubated with primary antibody cocktails, rinsed, incubated with secondary antibody detection reagents, with final rinses in buffer and water. The total staining time on the Autostainer is 285 min

Table 2
Autostainer program for multiplex, two-color Qdot RPPA staining

Reagent category	Reagent	Time (min)
Rinse	Wash buffer Q	
Auxiliary	Wash buffer Q	2
Rinse	Wash buffer Q	
Auxiliary	Wash buffer Q	2
Rinse	Wash buffer Q	
Primary Antibody	Primary Antibodies (species: rabbit and mouse)	90
Rinse	Wash buffer Q	
Auxiliary	Wash buffer Q	5
Rinse	Wash buffer Q	
Auxiliary	Wash buffer Q	5
Rinse	Wash buffer Q	
Auxiliary	Wash buffer Q	5
Rinse	Wash buffer Q	
Tertiary reagent	Rinse buffer	2
Rinse	Wash buffer Q	
Tertiary reagent	Rinse buffer	2
Rinse	Wash buffer Q	
Substrate	Qdot 655 nm + Qdot 800 nm	90
Rinse	Wash buffer Q	
Tertiary reagent	Rinse buffer	2
Rinse	Wash buffer Q	
Tertiary reagent	Rinse buffer	2
Rinse	dH_2O	
Rinse	dH_2O	
Rinse	dH_2O	
Rinse	dH_2O	
Auxiliary	dH_2O	3
Rinse	dH_2O	

5. Remove and discard the dH_2O. Add 2 mL of 1× Antigen Retrieval buffer. Incubate for 15 min at room temperature.
6. Wash 3×5 min each with 2 mL of fresh dH_2O for each wash with gentle rocking on an orbital rocker.
7. Discard the water after each wash.
8. Add 2 mL 1× Wash Buffer Q. Wash for 5 min with gentle rocking on an orbital rocker.
9. Remove and discard the Wash Buffer Q. Add 2 mL QBlock Blocking Buffer. Incubate for 30 min at room temp with gentle rocking on an orbital rocker.
10. While the slides are blocking, prepare the primary antibody dilutions in the Antibody Diluent. Each array will be probed with a cocktail of two antibodies: the anti-mouse antibody and the anti-rabbit antibody. Prepare an antibody cocktail for each primary antibody of interest (one cocktail per array). One slide will not have any primary antibody added—this is the secondary antibody only control. This slide will have antibody diluent added in place of the primary antibody cocktail.
11. Prepare the Detection cocktail (Qdot conjugated anti-IgG, mouse and rabbit). For each slide, prepare 800 μL of a 1:500 dilution of each Qdot conjugated anti-IgG (800 and 655 nm) in 1× Detection Reagent Diluent. Example: for one slide, add 1.6 μL of 500× Qdot 800 nm anti-rabbit IgG and 1.6 μL of 500× Qdot 655 nm anti-mouse IgG to 796.8 μL of 1× Detection Reagent Diluent. Mix thoroughly.
12. Remove the slides from the QBlock Blocking Buffer. Place the slides in the Autostainer slide rack and start the Autostainer program (Table 2). Do not allow the slides to dry out. 1× Wash Buffer Q can be poured on to the slides to keep them wet while they are on the Autostainer.
13. Program the Autostainer to dispense 800 μL of reagent per slide. Prime the water and buffer on the Autostainer. Start the Autostainer.
14. Remove the Autostainer rack with the stained arrays.
15. Allow the slides to dry at room temperature, protected from light.

3.3 ArrayCAM Imaging of Reverse Phase Protein Microarrays (Fig. 4)

Image acquisition for fluorescent or colorimetric detection systems can be conducted on the ArrayCAM imager. All Qdots can be excited with a common laser wavelength which contributes to simplified and lower-cost instrumentation for fluorescence detection. The ArrayCAM uses a violet semiconductor diode laser as the excitation source. Streamlining and standardizing the data analysis steps can be achieved using an ArrayCAM high-resolution, dual mode chromogenic/fluorescent array imager. The ArrayCAM design facilitates adoption of RPPA technology in any size laboratory.

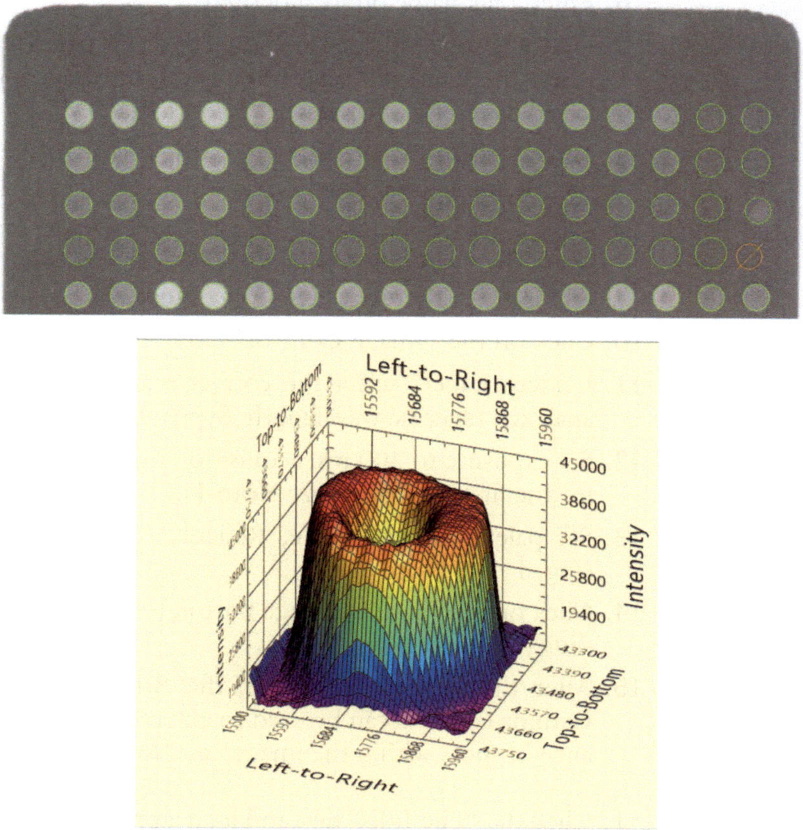

Fig. 4 ArrayCAM spot finding features. The ArrayCAM spot finding software depicts spots that meet the spot search parameters as *green circles* and those spots that fail the search parameters are depicted as *red circles* with lines through them. The 3-dimensional features of the spot can be readily visualized with a 3-D spot inspection tool

3.3.1 ArrayCAM Setup

1. Make sure ArrayCAM is connected to the computer.
2. Turn on ArrayCAM, blue power light will indicate that the machine is operational.
3. Click on ArrayCAM shortcut icon to launch program.
4. After program has opened a green "Imager Ready…" status bar should be visible at the top right corner of the program to confirm that the computer is connected to the ArrayCAM.

3.3.2 Image Acquisition

1. Load a slide by pushing the ArrayCAM lid backwards to expose the slide stage.
2. Orient the slide face down with the label on the left (*see* **Note 2**).
3. Secure the slide on the stage by pushing the two clasps backwards and close the lid.
4. Click on "File Info" tab (Fig. 1).

5. Under "Select Image Type", select TIFF.
6. Enter a file name prefix if desired.
7. Under "Image Storage Directory Path", select the file destination where the scanned images will be sent, by navigating to the preferred directory and pressing "Select folder."
8. Under "Misc File Items", check the "Save Images?" box and the "Use universal time?" box.
9. Click on the "Configure Imager" tab. Under image profiles select the "full slide" tab and Select All. The slide icon will be highlighted in green (Fig. 1) (*see* **Note 3**).
10. Click on the "Image Control" tab.
11. Under "Label" select the correct setting for the slide being analyzed (specific wavelength, Sypro Ruby, or colorimeteric).
12. Set exposure to 200 ms, acquisition time to 1 s, and gain at 50 for the first array scan (*see* **Note 4**).
13. Set palette to gray scale, Reticule off and invert color off (Fig. 1) (*see* **Note 5**)
14. Click on the Capture Image button to acquire an image of the slide (*see* **Note 6**).
15. After the image is obtained, the Brightness, Contrast, and Gamma controls can be used to enhance or reduce the appearance of spots within the image (*see* **Note 7**).

3.3.3 Image Analysis

1. Click the "File Info" tab, and load a previously created navigational .gal file by clicking on the "Load .gal File" button.
2. Select the "1 × 1 Sub-array" and click ok, then choose the appropriate .gal file.
3. Click on the "Image Control" tab. Click on the "Invert Color" button if the image represents a colorimetric assay.
4. Under the "Auto tab", set values as (*see* **Note 8**):

• Intensity Measure: Volume	• Annular Ring Thk %: 75
• Bkg Method: Annular Ring	• Intensity Correction: Sig-Bkg
• Bkg Diameter: 2.0	• Avg Bkg Reference: Fiducial (optional)

5. Click the "Params" button and set the values as (*see* **Note 9**):

• Spot Search Diameter: 1.5 • Spot Evaluation Diameter: 1.25 • Spot Detection Sensitivity: 25 % • Spot Diameter Flag (Lower Limit): 0.5 • Max Spot Offset: 1.5 • Normalize values: Yes • Search mode: One Time	• Spot Diameter Flag (Upper Limit): 2.0 • Spot Circularity (%): 75 • All "Spot Types to Analyze" options should be highlighted green • Printer Vertical Drift: 1.0 (*see* **Note 10**)

6. Click the "Identify" button, then the "Manual" button. Select "Continue" to create a gal file. A navigational gal file may be created in the ArrayCAM software, or can be copied and pasted from another source (*see* **Note 11**). The ArrayCAM software creates the navigational gal file in the correct format when the array parameters are entered. Enter information for the array: # of columns, # of rows, column pitch, row pitch, spot diameter, left origin, and top origin. The arrays are analyzed in a vertical position, with the barcode label at the bottom of the image (Fig. 2).
7. The key stroke sequences for pasting columns of text into the sample ID and spot type columns in the gal file template are: Type in your value, click enter, highlight the text in the first cell that you wish to copy, hold the shift key, scroll down, click in the last box (while still holding the shift key), then click Fill Down.
8. Click Set, click Done, then save the gal file.
9. On the File Info tab, load the gal file.
10. Click on the Image Control tab. Ensure that the correct image is loaded by verifying the image file name in the image box. Click Verify.
11. Use the rectangle box tool to draw a rectangle that fits closely around the first row of spots (e.g. draw the rectangle so it is just touching them at their perimeters) and click "ok" to analyze the spots. Use the magnifying glass tool to zoom in and out when drawing the box, left mouse click will zoom in, left mouse click + shift will zoom out.
12. A new window will appear with the spot boundaries marked as circles (*see* **Note 12**).
13. Use the ellipse tool button in the newly created window to readjust any spot boundaries whose locations have been identified incorrectly. Draw the new circle where the spot location is intended and click the "Set" button. A new, adjusted circle will appear and the old circle will disappear. If many circles require adjustment, the spot evaluation parameters should be adjusted in "Params" window (*see* **Note 13**).
14. After readjustments have been completed, click the "recalculation" button to reanalyze the slide (*see* **Note 14**).
15. Data from analysis will be transferred to previously selected destination folder (*see* **Note 15**).

4 Notes

1. The antibodies directed against the analytes of interest should be from a different species than the antibody used to detect the

protein to be used as the normalizer. For example, if β-actin is selected as the endpoint to be used for normalization between spots, select an anti-mouse β-actin antibody, which will be detected with Qdot 655 nm conjugated to anti-mouse IgG. All other antibodies for detecting individual analytes should be anti-rabbit. Primary antibodies must be validated by western blotting to verify specificity to the protein of interest. A dominant band on the western blot, at the specified molecular weight, provides a means of determining antibody specificity.

The primary antibody concentration (dilution) with the non-amplified staining method is approximately fourfold greater than with catalyzed signal amplification (CSA) methods. For example, if the recommended primary antibody dilution for the CSA method is 1:100, use the antibody at 1:400 with the non-amplified method.

2. Viewing the microarray slide as a roadmap, the narrow edge of the slide is west-to-east (left-to-right) and the long edge is north-to-south (top-to-bottom); the label or barcode is on the "south" side. When placing the slide in the ArrayCAM carriage, the film slide is always oriented so that the nitrocellulose surface is downward and the label is on the left.
3. Slide configuration settings are compatible with single pad or multi-pad nitrocellulose film slides. Single-pad film slides are imaged as five separate images and the resulting images are stitched to form a montage.
4. The optimal exposure, acquisition, and gain settings must be determined empirically for each slide. Typical settings for DAB (colorimetric) stained arrays are 20 ms exposure, 1 s acquisition, 20 gain.
5. Images can be viewed with various palette settings. Monochrome and grayscale will provide lesser contrast. Rainbow and Spectrum will provide greater variation in color contrast for highlighting detail. Adjustments to these controls do not affect the actual image data, only the displayed image.
6. The acquisition progress will be indicated with a progress bar. Acquisition time is less than 1 min for a full slide and approximately 10 s for a single region of a slide. When the acquisition is finished, an image will be displayed. If several sections of the slide are being imaged, the last image acquired will be displayed.
7. Contrast between spots and background can be enhanced by adjusting the Image Enhance and Image Diminish controls. Adjustments to these controls do not affect the actual image data, only the displayed image.
8. Intensity Measure is normally set to Volume or Median. Volume provides slightly better sensitivity while Median is more tolerant to outlier pixels. The ArrayCAM software allows

the user to change the type of background measurement, methods for determining the local area background, and signal intensity corrections for the local area background. Annular Ring background is highly tolerant to spot bleed into the background area.

9. Parameter settings allow the user to define the stringency of the spot finding algorithms. Of these, the Detection Sensitivity has the most effect, as it determines the discrimination of signal vs. background in defining the spot boundary. Spot Search Diameter, Spot Evaluation Diameter, and Maximum Spot Offset all limit the search algorithm to the immediate region of the expected spot location. These settings reduce the probability that spot centers will need to be adjusted manually. Definitions of spot search parameters:

   Spot Search Diameter: This sets the maximum distance about the expected spot center to search for the actual spot "centroid". The centroid location enables the spot location algorithm to narrow the search to an area close to the centroid. The units are in spot diameters. For example, a setting of 1.5 will permit searching within 1.5 diameters of the expected spot location.

   Spot Evaluation Diameter: This sets the maximum distance about the located spot centroid to locate the actual spot using the spot location algorithm. The units are in spot diameters. For example, a setting of 1.5 will permit searching within 1.5 diameters of the spot centroid.

   Spot Detection Sensitivity: This tells the algorithm how sensitive to be in finding the spot. A high-sensitivity will locate spots of very low intensity but will sometimes react to "noise".

   Spot Diameter Flag (Lower Limit): Flags spots in the results text file when the evaluated diameter is less than the flag limit. Units are in diameters.

   Spot Diameter Flag (Upper Limit): Flags spots in the results text file when the evaluated diameter is greater than the flag limit. Units are in diameters.

   Spot Circularity: Flags spots in the results text file when the circularity (aspect ratio) exceeds the percentage set.

   Max Spot Offset: Limits the location algorithm to finding spot within the limits set. For example, a limit of 1.5 will limit the offset relative to the expected location to 1.5 diameters. If a spot is not found within the setting limit, a spot circle is drawn at the expected location and is flagged with a slash.

   Search Mode: Performs the spot location once or twice, depending on whether the switch is set to "One Time" or "Iterative."

10. During printing, any variation in the position of the slide, printer pins, or sample plates can cause the deposited sample spots to vary slightly in their horizontal/vertical position.

Printer Vertical Drift compensates for this drift. If the ArrayCAM circles are lower than the actual printed spots, set the Printer Vertical Drift to <1 (i.e., 0.993). If the ArrayCAM circles are higher than the actual printed spots, set the Printer Vertical Drift to >1 (i.e., 1.005).

11. A microarray block configuration is normally printed according to a .gal file that controls the printing process. ArrayCAM™ uses a "navigational gal file" for auto analysis that is similar to a normal gal file and obeys the rules for creating gal files. The gal file created for printing the arrays can simply be modified and saved as the navigational gal for the purpose of the automated image analysis. However, the format of the navigational gal file must follow a specific format as described below.

    The following example demonstrates the configuration of the navigational .gal file. In this example, only the header and the first several rows are included in the list. The navigational gal file must conform to the general rules for gal file construction. In addition to the basic rules, a few modifications are required as indicated below.

    - Add a new line in the header that indicates the slide type and include the corresponding numeric code.

Slide type	Numeric
Four-pad	0
16-pad	1
Full slide pad	2

    - Add column #6 to indicate the *type* of spot.

    The number of header rows can be unlimited. However, the numerical value in the first column of the second row must be equal to the total number of header rows minus three. In the example below, there are eleven (11) header rows and the numerical value is eight (8):

    11 header rows − 3 required rows = 8 ← enter this number

    Additionally, the numerical value in the second column of the second row must be "6" as there are exactly six columns. The block designator row must be enclosed in double quotes. For example:

    "Block1 = 4400,1400,200,13,433,13,433"

    More than one block designator row can be provided as long as the entire list is consistent with the individual block designators.

    The "Type" designator in column 6 can have any of the following values:

    fiducial
    analyte
    positive

negative
buffer
blank
control
content
blocker

Each type designator can be of all lower case letters and the first letter can be capitalized. No other case options are permitted. Each must be spelled exactly as shown.

The user may construct any navigational .gal file as long as it obeys the general rules for .gal file construction and includes the additional information as described above.

Save the navigational .gal file to the directory C:\GBL\Gal\Controls

ATF	1				
8	6				
Type = GenePix ArrayList V1.0					
Supplier = Grace Bio-Labs					
ArrayerSoftwareName = None					
ArrayerSoftwareVersion = None					
BlockCount = 1					
BlockType = 0					
SlideType = 1					
"Block1 = 4400,1400,200,13,433,13,433"					
Block	Column	Row	ID	Name	Type
1	1	1	ID	FIDUCIAL	fiducial
1	2	1	ID	BLANK	blank
1	3	1	ID	BLANK	blank
1	4	1	ID	FIDUCIAL	fiducial
1	5	1	ID	BLANK	blank
1	6	1	ID	BLANK	blank
1	7	1	ID	BLANK	blank
1	8	1	ID	FIDUCIAL	fiducial
1	9	1	ID	BLANK	blank
1	10	1	ID	BLANK	blank
1	11	1	ID	BLANK	blank
1	12	1	ID	BLANK	blank
1	13	1	ID	FIDUCIAL	fiducial
1	1	2	ID	Content	analyte

(continued)

1	2	2	ID	Content	analyte
1	3	2	ID	Content	analyte
1	4	2	ID	Content	analyte
1	5	2	ID	Content	analyte
1	6	2	ID	Content	analyte
1	7	2	ID	Content	analyte
1	8	2	ID	Content	analyte
1	9	2	ID	Content	analyte
1	10	2	ID	Content	analyte
1	11	2	ID	Content	analyte
1	12	2	ID	Content	analyte
1	13	2	ID	Content	analyte

12. Green circles indicate satisfactorily located spot boundaries. Orange circles with horizontal slash indicate spots are either not found or have been found with peculiar shapes. Regardless of the spot fidelity, all spots will be listed in the results text file. Those with peculiar shapes or diameters out of limit will be flagged accordingly in the text file.
13. Spot boundaries that have been manually adjusted will be displayed in red.
14. If multiple circles require adjustment, the adjusted circle layout may be saved as a "map" and loaded onto subsequent images, obviating the need to manually adjust the circles for each image. Adjust the circles as desired, then click "save map". Name the map file for future use. To use the saved map, open the image and click "verify" to start the analysis. After the analysis is complete, click "load map", choose the saved map file. The image reanalysis proceeds automatically. Click OK to save the data analysis.
15. The data file is linked to the image name. The existing data file will be over-written with the most current data when the image is reanalyzed/recalculated. If you wish to analyze an image with different spot analysis parameters and to save both the original and modified data analysis, make a copy of the image and save it with a different file name *prior* to repeating the analysis.

Acknowledgments

This work was funded in part by George Mason University and Grace Bio-Labs, Inc.

References

1. Holmes FA, Espina V, Liotta LA, Nagarwala YM, Danso M, McIntyre KJ (2013) Pathologic complete response after preoperative anti-HER2 therapy correlates with alterations in PTEN, FOXO, phosphorylated Stat5, and autophagy protein signaling. BMC Res Notes 6:507
2. Korf U, Derdak S, Tresch A, Henjes F, Schumacher S, Schmidt C (2008) Quantitative protein microarrays for time-resolved measurements of protein phosphorylation. Proteomics 8:4603–4612
3. Liotta LA, Espina V, Mehta AI, Calvert V, Rosenblatt K, Geho D (2003) Protein microarrays: meeting analytical challenges for clinical applications. Cancer Cell 3:317–325
4. MacBeath G, Schreiber SL (2000) Printing proteins as microarrays for high-throughput function determination. Science 289:1760–1763
5. Paweletz CP, Charboneau L, Bichsel VE, Simone NL, Chen T, Gillespie JW (2001) Reverse phase protein microarrays which capture disease progression show activation of pro-survival pathways at the cancer invasion front. Oncogene 20:1981–1989
6. Petricoin EF III, Espina V, Araujo RP, Midura B, Yeung C, Wan X (2007) Phosphoprotein pathway mapping: Akt/mammalian target of rapamycin activation is negatively associated with childhood rhabdomyosarcoma survival. Cancer Res 67:3431–3440
7. Gallagher RI, Espina V (2014) Reverse phase protein arrays: mapping the path towards personalized medicine. Mol Diagn Ther 18:619–630
8. Huels C, Muellner S, Meyer HE, Cahill DJ (2002) The impact of protein biochips and microarrays on the drug development process. Drug Discov Today 7:S119–S124
9. Jameson JL, Longo DL (2015) Precision medicine—personalized, problematic, and promising. N Engl J Med 372:2229–2234
10. Mueller C, Liotta LA, Espina V (2010) Reverse phase protein microarrays advance to use in clinical trials. Mol Oncol 4:461–481
11. Cremona M, Espina V, Caccia D, Veneroni S, Colecchia M, Pierobon M (2014) Stratification of clear cell renal cell carcinoma by signaling pathway analysis. Expert Rev Proteomics 11:237–249
12. Espina V, Liotta LA, Petricoin EF III (2009) Reverse-phase protein microarrays for theranostics and patient tailored therapy. Methods Mol Biol 520:89–105
13. Jameson GS, Petricoin EF, Sachdev J, Liotta LA, Loesch DM, Anthony SP (2014) A pilot study utilizing multi-omic molecular profiling to find potential targets and select individualized treatments for patients with previously treated metastatic breast cancer. Breast Cancer Res Treat 147:579–588
14. Pierobon M, Silvestri A, Spira A, Reeder A, Pin E, Banks S (2014) Pilot phase I/II personalized therapy trial for metastatic colorectal cancer: evaluating the feasibility of protein pathway activation mapping for stratifying patients to therapy with imatinib and panitumumab. J Proteome Res 13:2846–2855
15. Robertson FM, Petricoin EF III, Van Laere SJ, Bertucci F, Chu K, Fernandez SV (2013) Presence of anaplastic lymphoma kinase in inflammatory breast cancer. Springerplus 2:497
16. Wulfkuhle JD, Speer R, Pierobon M, Laird J, Espina V, Deng J (2008) Multiplexed cell signaling analysis of human breast cancer applications for personalized therapy. J Proteome Res 7:1508–1517
17. Xia W, Petricoin EF III, Zhao S, Liu L, Osada T, Cheng Q (2013) An heregulin-EGFR-HER3 autocrine signaling axis can mediate acquired lapatinib resistance in HER2+ breast cancer models. Breast Cancer Res 15:R85
18. Gallagher RI, Silvestri A, Petricoin EF III, Liotta LA, Espina V (2011) Reverse phase protein microarrays: fluorometric and colorimetric detection. Methods Mol Biol 723:275–301
19. Tibes R, Qiu Y, Lu Y, Hennessy B, Andreeff M, Mills GB (2006) Reverse phase protein array: validation of a novel proteomic technology and utility for analysis of primary leukemia specimens and hematopoietic stem cells. Mol Cancer Ther 5:2512–2521
20. VanMeter AJ, Rodriguez AS, Bowman ED, Jen J, Harris CC, Deng J (2008) Laser capture microdissection and protein microarray analysis of human non-small cell lung cancer: differential epidermal growth factor receptor (EGPR) phosphorylation events associated with mutated EGFR compared with wild type. Mol Cell Proteomics 7:1902–1924
21. Chiechi A, Mueller C, Boehm KM, Romano A, Benassi MS, Picci P (2012) Improved data normalization methods for reverse phase protein microarray analysis of complex biological samples. Biotechniques 1–7
22. Mannsperger HA, Gade S, Henjes F, Beissbarth T, Korf U (2010) RPPanalyzer: analysis of reverse-phase protein array data. Bioinformatics 26:2202–2203
23. Stanislaus R, Carey M, Deus HF, Coombes K, Hennessy BT, Mills GB (2008) RPPAML/

RIMS: a metadata format and an information management system for reverse phase protein arrays. BMC Bioinformatics 9:555

24. Troncale S, Barbet A, Coulibaly L, Henry E, He B, Barillot E (2012) NormaCurve: a SuperCurve-based method that simultaneously quantifies and normalizes reverse phase protein array data. PLoS One 7:e38686
25. von der Heyde S, Sonntag J, Kaschek D, Bender C, Bues J, Wachter A (2014) RPPanalyzer toolbox: an improved R package for analysis of reverse phase protein array data. Biotechniques 57:125–135
26. Michalet X, Pinaud FF, Bentolila LA, Tsay JM, Doose S, Li JJ (2005) Quantum dots for live cells, in vivo imaging, and diagnostics. Science 307:538–544
27. Resch-Genger U, Grabolle M, Cavaliere-Jaricot S, Nitschke R, Nann T (2008) Quantum dots versus organic dyes as fluorescent labels. Nat Methods 5:763–775
28. Shao L, Gao Y, Yan F (2011) Semiconductor quantum dots for biomedicial applications. Sensors (Basel) 11:11736–11751
29. Nichkova M, Dosev D, Davies AE, Gee SJ, Kennedy IM, Hammock BD (2007) Quantum dots as reporters in multiplexed immunoassays for biomarkers of exposure to agrochemicals. Anal Lett 40:1423–1433
30. Goldman ER, Clapp AR, Anderson GP, Uyeda HT, Mauro JM, Medintz IL (2004) Multiplexed toxin analysis using four colors of quantum dot fluororeagents. Anal Chem 76:684–688
31. Bobrow MN, Harris TD, Shaughnessy KJ, Litt GJ (1989) Catalyzed reporter deposition, a novel method of signal amplification. Application to immunoassays. J Immunol Methods 125:279–285
32. Bobrow MN, Litt GJ, Shaughnessy KJ, Mayer PC, Conlon J (1992) The use of catalyzed reporter deposition as a means of signal amplification in a variety of formats. J Immunol Methods 150:145–149
33. Bobrow MN, Shaughnessy KJ, Litt GJ (1991) Catalyzed reporter deposition, a novel method of signal amplification. II. Application to membrane immunoassays. J Immunol Methods 137:103–112
34. King G, Payne S, Walker F, Murray GI (1997) A highly sensitive detection method for immunohistochemistry using biotinylated tyramine. J Pathol 183:237–241

Chapter 13

Quantitative Proteomics Using SILAC

Kian Kani

Abstract

The ability to enumerate all of the proteins in a cell is quickly becoming a reality. Quantitative proteomics adds an extra dimension to proteome-wide discovery experiments by enabling differential measurements of protein concentrations, characterization of protein turnover, increased stringency of co-immunoprecipitation reactions, as well as many other intriguing applications. One of the most widely used techniques that enable relative protein quantitation is stable isotope labeling by amino acids in cell culture (SILAC) (Ong et al., Mol Cell Proteomics 1(5):376–386, 2002). Over the past decade, SILAC has become the preferred approach for proteome-wide quantitation by mass spectrometry. This approach relies on the metabolic incorporation of isotopically enriched amino acids into the proteome of cells—the proteome of "light" (^{1}H, ^{12}C, ^{14}N) cells can then be compared to "heavy" (^{2}H, ^{13}C, ^{15}N) cells as the isotopically labeled proteins and peptides are easily distinguished in a mass spectrometer. Since cellular uptake and response to isotopically different amino acid(s) is naïve, it is without impact on cell physiology. We provide a detailed step-by-step procedure for performing SILAC-based experiment for proteome-wide quantitation in this chapter.

Key words SILAC, AMT, Quantitative, Proteomics, Multiplex

1 Introduction

Recent advances in mass spectrometers and pre-fractionation techniques have enabled near complete coverage of various proteomes [2]. Researchers are confronted with the challenge of determining the biological significance of long lists of proteins. Quantitative proteomics enables researchers to extract more impactful data from their data-sets. A number of different techniques are available to researchers to implement quantitative proteomic experiments. The determining factor for choosing the right technique is based on the type of sample, the number of controls and cases, and the type of mass spectrometer (Fig. 1). The most widely used technique for quantitative proteomics with cell line models is the SILAC approach. The reliability and reproducibility of this technique is superior to other methods of quantitation [3, 4].

Lucio Comai et al. (eds.), *Proteomics: Methods and Protocols*, Methods in Molecular Biology, vol. 1550,
DOI 10.1007/978-1-4939-6747-6_13, © Springer Science+Business Media LLC 2017

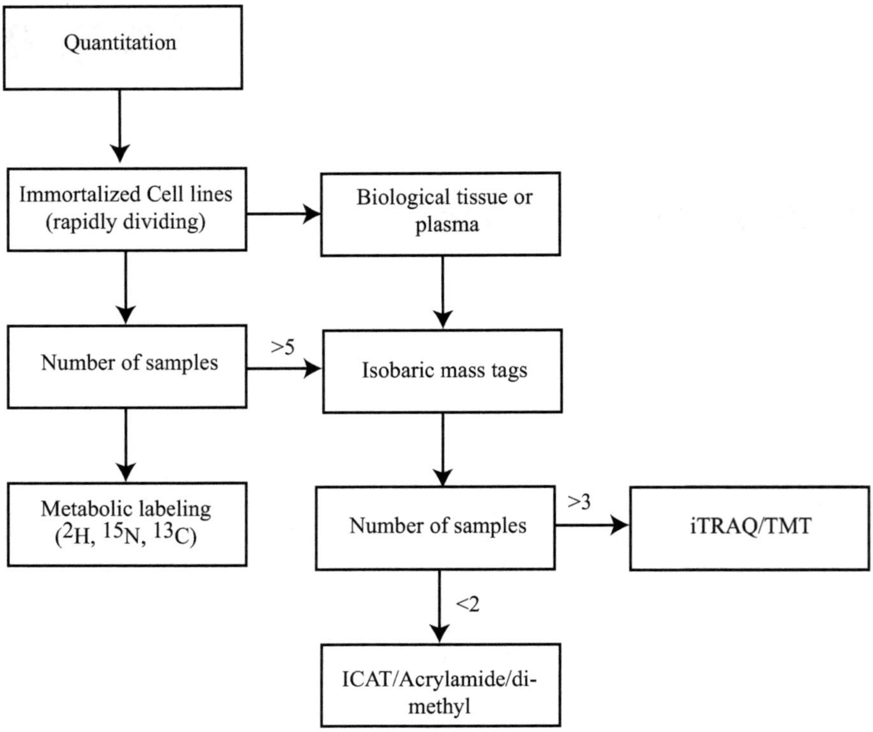

Fig. 1 Quantitative proteomics flowchart for either metabolic labeling or chemical addition of isobaric mass tags

SILAC is ideal for experiments that compare the differential expression of proteins in up to five samples [5], however, the majority of SILAC based experiments utilize no more than three samples [6]. The main principle for metabolic labeling involves the incorporation of amino acids with unusual elemental isotopes (e.g., one or more ^{13}C replacing the more natural ^{12}C) into proteins in cells that cannot endogenous produce that specific amino acid. The reliance of exogenous amino acids into biochemical pathways is termed auxotrophy. For this reason, modern SILAC experiments for human cell lines rely on the incorporation of arginine and lysine. Most two-plex SILAC experiments utilize ^{13}C versus ^{12}C labeled lysine for a mass shift of 6 Da. In general, we recommend using combinations of "light" and "heavy" peptides with an expected mass shift greater than 4 Da. Metabolically labeling with arginine offers greater flexibility in the mass due to the number of "heavy" carbons, deuterium and/or nitrogen atoms. For experiments that require the comparison of three conditions, the most widely used combination of amino acids is ^{12}C lysine, ^{13}C lysine, and ^{13}C/^{15}N arginine which allows for a mass shift of 6 and 10 Da from the control.

One of the complications of this technique is the unintended interconversion of isotopic amino acids during the labeling process into other byproducts. This conversion alters MS/MS signals from one amino acid and introduces various artifacts that affect quantification by increasing the number of unwanted peptide ion peaks. Of particular concern is the biosynthesis of proline from excess arginine [7]. This impacts all SILAC experiments that utilize arginine. The best workaround for this complication is to supplement SILAC media with standard L-proline, thereby suppressing the endogenous conversion enzymes [8]. We suggest that all SILAC media contain 100 mg/L of L-proline in order to prevent the conversion of isotope-coded arginine to proline.

The application of the SILAC approach is based on the research aims of the investigators. Some of the early works focused on differential expression of proteins in various model systems (yeast, *E. coli*, *C. elegans*, *Arabidopsis*, etc.) [9–11]. The combination of SILAC based quantitation with enrichment strategies that enable measurement of protein posttranslational modifications has been a powerful application of this technique. For example, Bose et al. discovered novel downstream phosphorylation targets of the oncogene HER2 [12]. A number of other reports have applied SILAC based approaches to measure changes in glycosylation [13], ubiquitinylation [14], methylation [15], and histone modification [16]. Global changes in various host systems may also be monitored by the SILAC approach. For example, Kruger et al. metabolically labeled a mouse with heavy lysine and demonstrated the application of quantitative proteomics with knockout mice [17]. A number of other experimental strategies involving SILAC have been developed. For instance, Schwanhausser et al. utilized pulsed incorporation of amino acids to measure kinetics of protein turnover (pulsed-SILAC) [18]. SILAC may also be used to enhance the reliability of absolute protein quantitation [19] and facilitate improvement in reducing the false positive identification of contaminant proteins in CO-IP LC-MS/MS experiments [20, 21]. In the following section, we describe a step-by-step procedure for a typical two-plex based SILAC experiment (Fig. 2).

2 Materials

All cell culture material should be sterile and cell lines should be tested for mycoplasma contamination prior to amino acid incorporation with a suitable kit (Mycoprobe Mycoplasma detection kit, R&D Systems, Minneapolis, MN). All buffers and solvents should be prepared with Milli-Q water (Millipore, Billerica, MA).

1. Appropriate media for your cell line deficient in arginine, lysine, and glutamine (*see* **Note 1**). For example, RPMI 1640 Medium deficient in arginine, lysine, and glutamine (Life Technologies, Carlsbad, CA).

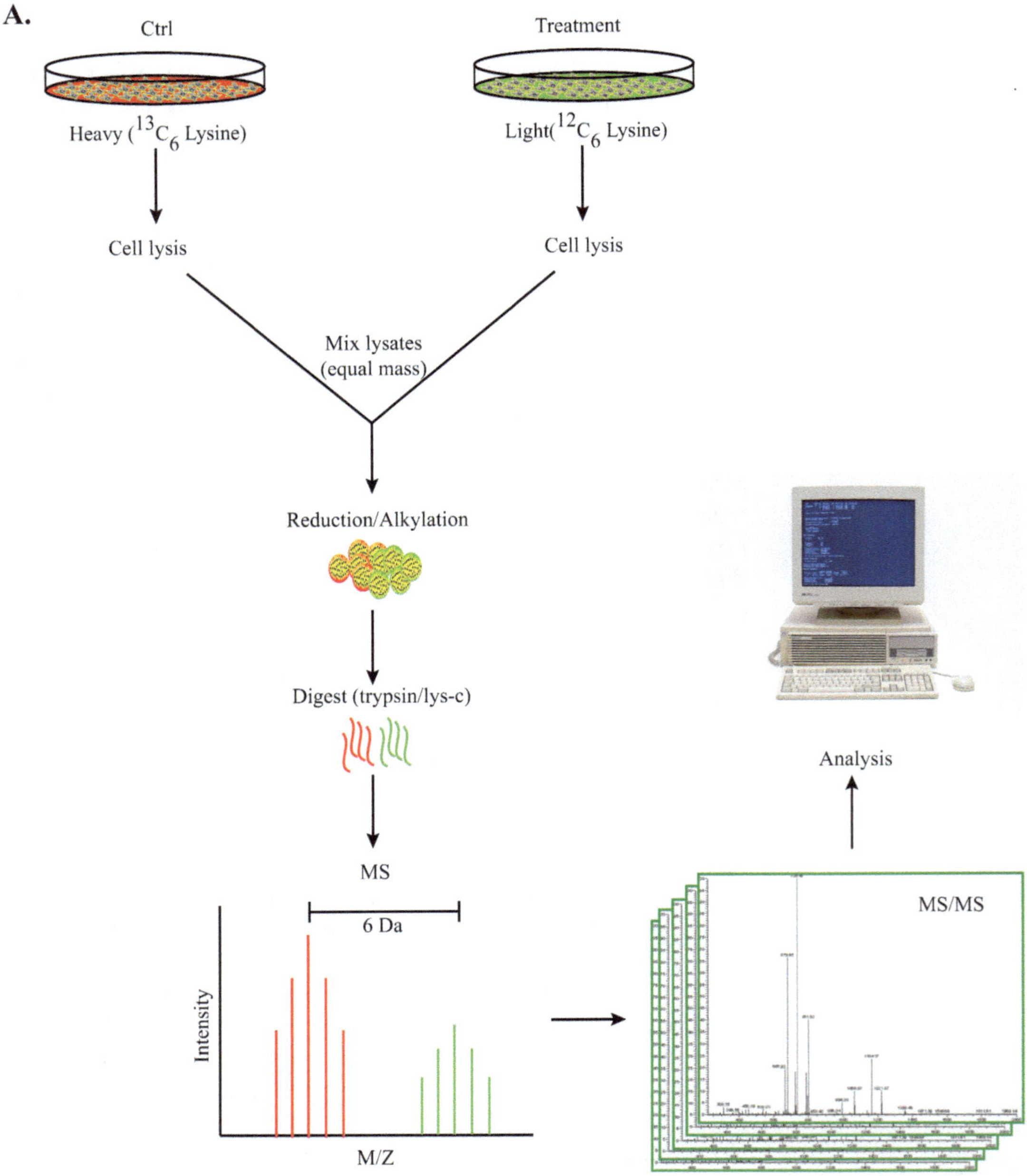

Fig. 2 Schematic of a typical two-plex SILAC experiment. Metabolic labeling of cells occurs in rapidly dividing cultures. Cell are passaged six times in SILAC media. Cells are lysed in an appropriate buffer and subject to protein concentration determination by Bradford or BCA. Equal masses of proteins are mixed and subject to reduction and alkylation. Lysates are treated with trypsin or lys-c and analyzed by LC-MS/MS. The area of the "light" and "heavy" peptides are used for quantitation

2. Dialyzed Fetal Bovine Serum (FBS) (Life Technologies, Carlsbad, CA) (*see* **Note 2**).
3. Select "Heavy" and "Light" amino acids based on the level of multiplexing (Table 1). For example, for a two-plex experiment obtain "light" lysine and arginine as well as the isotopically "heavy" versions.

Table 1
Recommended amino acids for SILAC experiments

	Sample 1	Sample 2	Sample 3	Sample 4	Sample 5
2 plex	$^{12}C/^{14}N$ light amino acids	$^{13}C_6$ Lysine and $^{13}C_6$, $^{15}N_4$ arginine			
3 plex	$^{1}H/^{12}C/^{14}N$ light amino acids	D_4 lysine and $^{13}C_6$ arginine	$^{13}C_6$ $^{15}N_2$ Lysine and $^{13}C_6$, $^{15}N_4$ arginine		
4 plex	$^{1}H/^{12}C/^{14}N$ light amino acids	D_4 lysine and $^{13}C_6$ arginine	$^{13}C_6$ $^{15}N_2$ Lysine and $^{13}C_6$, $^{15}N_4$ arginine	$^{13}C_6$ $^{15}N_2$ D_9 Lysine and D_7 arginine	
5 plex	$^{12}C_6$, $^{14}N_4$ arginine	$^{12}C_6$, $^{15}N_4$ arginine	$^{13}C_6$, $^{14}N_4$ arginine	$^{13}C_6$, $^{15}N_4$ arginine	D_7 $^{13}C_6$, $^{15}N_4$ arginine

(a) "Heavy" and "Light" lysine (Cambridge Isotope Laboratories, Tewksbury, MA), if needed, should be dissolved to a final concentration of 146 mg/mL in media to create a 1000× stock (*see* **Note 3**).

(b) "Heavy" and "Light" arginine (Cambridge Isotope Laboratories, Tewksbury, MA), if needed, should be dissolved to a final concentration of 100 mg/mL in media to create a 1000× stock (*see* **Note 3**).

4. Unlabeled (i.e., "normal" or "light") L-proline should be dissolved to a final concentration of 100 mg/mL in media to make a 1000× stock.
5. Unlabeled L-glutamine should be dissolved in media to a final concentration of 200 mM to yield a 100× stock solution.
6. Penicillin/streptomycin (10,000 U/10,000 μg stock solution).
7. 1 M HEPES buffer.
8. Trypsin–EDTA solution (trypsin, 500 mg/L) (Life Technologies, Carlsbad, CA).
9. Phosphate buffered saline (PBS).
10. 100 mm (10 cm) tissue culture plates.
11. Sterile Vacuum Filter Units, 1 L (EMD Millipore Stericup, 0.22 mm, Fisher Scientific, St. Louis, MO).
12. Microcentrifuge tubes, 1.5 mL.
13. Spin filter, 0.22 μm (e.g., 0.22 μm Corning Costar Spin-X centrifuge tube filters , Sigma-Aldrich, Pittsburg, PA).
14. BCA protein analysis kit.

15. Dithiothreitol (DTT) 1 M stock made fresh in water (100×).
16. Iodoacetamide (IAA) 1 M stock made fresh in water (18.18×) (*Note: stock solutions of iodoacetamide at 1 M may require sonication in order to obtain complete solubility*).
17. Sequencing grade Trypsin and Lys-C (optional).
18. Glacial acetic acid.
19. Trifluoroacetic acid (TFA).
20. Formic acid.
21. Urea lysis buffer: 6 M urea, 2 M thiourea in 100 mM ammonium bicarbonate (pH 8) (*Note: must be made fresh*).

3 Methods

One of the major concerns for accurate proteome quantitation is to ensure that metabolic incorporation of amino acids is complete. We outline a sample protocol that may be adapted to both smaller and larger scale experiments. It is highly recommended to use low passage cells for any SILAC experiments. This is especially important because the cells need to be cultured in the SILAC media for at least six passages.

3.1 Cell Culture and Labeling

1. Seed one 10 cm plate per condition in normal media at 25–50 % confluence the day before starting the SILAC experiment (Day 0, *see* **Note 4** for comments on experiments requiring a larger quantity of cells for final analysis).
2. Obtain one (1 L) Sterile Vacuum Filter for each growth condition (e.g., “Heavy” and “Light”) inside a sterile laminar biologic safety bench.
3. Prepare SILAC media by adding the following items to the top chamber of the vacuum filter:
   (a) 860 mL of media.
   (b) 1 mL of 1000× proline (to suppress arginine conversion).
   (c) 1 mL of heavy or light 1000× amino acids as appropriate (*see* **Note 5**).
   (d) 10 mL 100× glutamine.
   (e) 18 mL 1 M HEPES.
   (f) 10 mL 100× of Penicillin/streptomycin.
   (g) 100 mL dialyzed FBS.
4. Allow components to filter through the bottom chamber using vacuum, discard the upper section. Label the media appropriately (e.g., “Heavy” or “Light”).
5. On Day 1, aspirate the media and wash cells with PBS thrice.

6. Add 10 mL of SILAC media to each 10 cm plate.
7. Monitor the growth kinetics of the cells during the course of metabolic labeling to ensure that the SILAC media does not interfere with cell physiology (*see* **Note 6**).
8. Passage cells at appropriate times.
   (a) Aspirate media off plates.
   (b) Wash off residual media with 10 mL of PBS.
   (c) Remove residual PBS by vacuum aspirator.
   (d) Add 1 mL of trypsin and incubate cells for 5 min in a temperature controlled incubator at 37 °C (5 % CO_2).
   (e) Add 9 mL of appropriate SILAC media to quench the trypsin and transfer 1 mL to a new 10 cm plate.
   (f) Add 9 mL of fresh SILAC media.
9. Repeat until the cells have gone through five successive passages (six total passages).
10. Confirm full incorporation of amino acids by performing a small LC-MS/MS experiment.

3.2 Cell Lysis

Each lysis condition is application and protocol dependent. We will describe a simple urea lysis procedure that is applicable to many subsequent analysis pipelines.

1. Seed the final passage of cells such that they reach 85 % confluence 48 h after passaging.
2. For Adherent cells, place 10 cm plates on ice and remove by aspiration.
3. Wash cells thrice with 10 mL of ice cold PBS. Aspirating PBS between washes (*Note: Do not wash cells off the plate*).
4. After the last wash, tilt plate at 45° angle for 1 min thereby allowing the remaining PBS to drain to bottom of plate.
5. Completely aspirate remaining PBS.
6. Add 1 mL of urea lysis buffer.
7. Scrape cells of plate with a Nunc Cell scraper (Fisher Scientific, Pittsburg, PA).
8. Remove cells and pipette into a prechilled 1.5 mL microcentrifuge.
9. Freeze cells at −20 °C for at least 2 h (*Note: cells can be kept frozen for later analysis*).
10. Thaw cells on ice.
11. Prepare cell lysate by passing cell material through a 27 gauge needle at least ten times (BD Sciences, Thermo Scientific, Pittsburg, PA).

12. Centrifuge cell lysate at 4 °C for 10 min at 14,000 × *g*.
13. Apply supernatant to 0.22 μm Corning Costar Spin-X centrifuge tube filters (Sigma-Aldrich, Pittsburg, PA) and centrifuge at 5,000 × *g* for 10 min.
14. Place filtrate in a prechilled Eppendorf microcentrifuge tube.
15. Determine protein concentration using the BCA (*Note: most lysates must be diluted 1:4 to 1:10 in order to fall into the linear range of the assay*).
16. Mix equal mass of "heavy" and "light" labeled cells in a new Eppendorf microcentrifuge tube.
    (a) Example: 1 mg of "light" labeled lysates should be mixed with 1 mg of "heavy" labeled lysates.
17. Samples must be reduced with 10 mM final concentration of DTT (Sigma-Aldrich, St. Louis, MO). Incubate samples at 65 °C for 45 min.
18. Allow samples to cool to room temperature and add iodoacetamide (Sigma-Aldrich, St. Louis, MO) at a final concentration of 55 mM for 30 min in the dark.
19. Protein digestion is accomplished by addition of sequencing grade trypsin (Promega, Madison, WI) and/or Lys-c (Sigma-Aldrich, St. Louis, MO). Efficient protein digestion occurs at 1 M urea, diluted with 55 mM ammonium bicarbonate supplemented with 1:100 (mass ratio) trypsin. Incubate at 37 °C for 16 h with gentle agitation or rotation.
20. Add additional 1:100 (mass ratio) trypsin for 4 h at 37 °C.
21. Trypsin digest should be quenched by addition of 1 μL of glacial acetic acid (Sigma-Aldrich, St. Louis, MO).
22. Lyophilize samples in a Speed-Vac.
23. Resuspend lyophilized peptides in 20 μL of 0.5 % trifluoroacetic acid (TFA) in 5 % acetonitrile (ACN) (Sigma-Aldrich, St. Louis, MO).
24. Desalt samples by applying peptides through a C18 spin column (Life Technologies, Carlsbad, CA). Wash columns with 400 μL of 0.5 % TFA, 5 % acetonitrile thrice. Elute peptides in 50 % methanol, 50 % acetonitrile, 0.1 % formic acid.
25. Proceed to LC-MS/MS.

3.3 Practical Example

The following section will outline some of the practical considerations that are involved in a typical two-plex SILAC based experiment. This example can be modified to fit various experimental conditions, cell lines, and controls. One of the key challenges in clinical oncology is to determine which patient is going to benefit from a particular therapeutic regime. Biomarkers that indicate therapy response are useful clinical tools for patient stratification.

We will utilize the cellular dependence of the A431 cell line (ATCC, Manassas, VA) to EGFR signaling as a surrogate for lung cancer patients that harbor EGFR activating mutations in their tumors. These patients have dramatic response rates to tyrosine kinase inhibitors (TKIs) that target EGFR. The goal of the following experiment is to identify proteins that are either upregulated or downregulated in the EGFR dependent A431 cell line upon treatment with an EGFR TKI (Iressa).

1. Two-plex SILAC experiments can be done with various combinations of "light" and "heavy" amino acids. In this example, utilize SILAC with $^{13}C_6$ lysine which results in a mass shift of 6 Da.
2. Maintain two cultures of A431 cells in either heavy or light SILAC media and passage them for at least six generations (*see* **Note 7**).
3. Perform technical replicates by dosing "heavy" labeled cells with drug (in this case Iressa) in order to comparing the proteome with "light" labeled cells treated with vehicle. The reciprocal experiment is "heavy" labeled cells treated with vehicle and compared to "light" labeled cells treated with drug.
4. Two 15 cm plates of "heavy" labeled A431 cells and two 15 cm plates of "light" labeled A431 cells are seeded at 50% confluence (*Note: the confluence and general health of cells is an important component to SILAC based in vitro experiments. Proper care should be placed on cell health*).
5. The following day, dose cells "heavy" labeled cells with 100 nM Iressa and "light" labeled cells with vehicle control.
6. Repeat the dosing on the reciprocal labeled cells.
7. After 16 h, proceed to cell lysis step as detailed in Subheading 3.2.
8. Perform LC-MS/MS analysis; in general, peptides identifications are based on the MS/MS spectra and the *m/z* of the monoisotopic peak.
9. Peptide quantitation is obtained from the extracted ion chromatogram (XIC) of two differently labeled versions of the peptide. The XIC is the contribution of a given *m/z* to the total ion chromatogram (TIC). The main components of the XIC are the *m/z* value, the elution time profile, and the intensity of the peak. A number of different software packages are available including Proteowizard [22], MSQuant [23], MaxQuant [24, 25], and Census [26, 27].
10. In order to assess the robustness of each quantitative peptide, manual inspection of the XIC may be performed (Fig. 3). Peptides with poor XIC overap and or low intensity should be excluded.

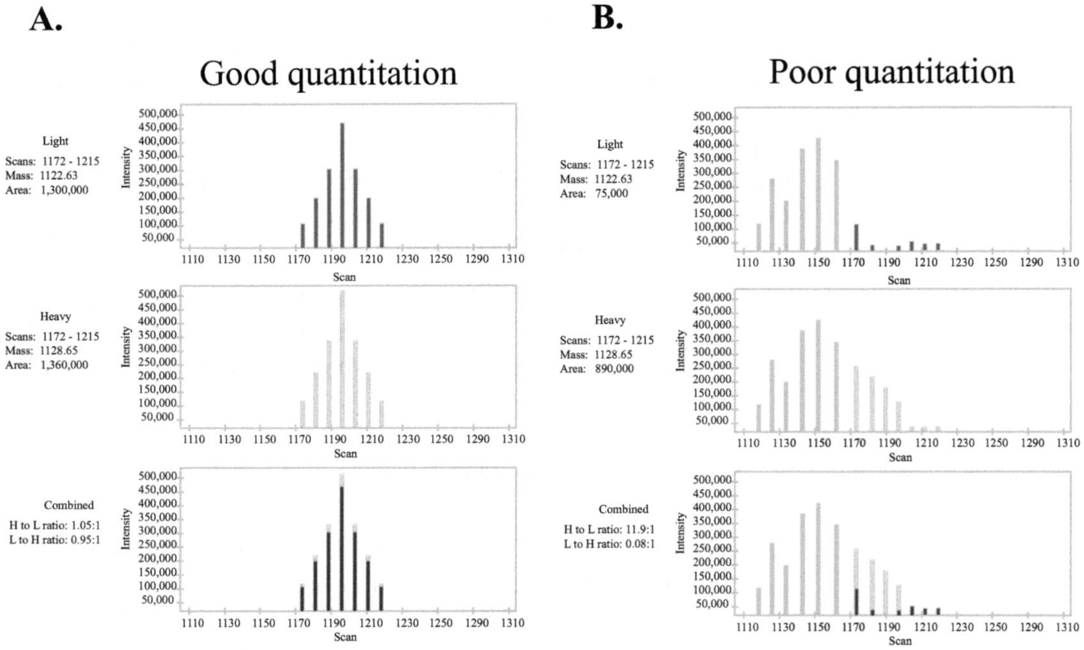

Fig. 3 (**a**) Accurate quantitation is obtained when the elution profile of the light and heavy peptide overlap in time but differ in *m/z*. (**b**) In this example, the elution profile of the heavy and light peptide show traces of a secondary peak which results in poor overlap of the XIC and unreliable quantitation

11. The final ratio for the protein expression reflects the expression change as a function of Iressa dosing. The SILAC ratio of any given protein can be determined by averaging the area of the "light" and "heavy" peptides. The statistical significance of the SILAC ratio must include the following parameters:
    (a) The number of quantitative peptides (at least two).
    (b) The intensity of the quantitative peptides should be greater than 1000.
    (c) The mean ratio for all peptides from this protein.
    (d) The standard deviation (lower the better).
    (e) Error determined by the student's *t*-test (less than 0.05).
12. Lists of proteins may be exported to a suitable spreadsheet for further analysis.
13. The initial step is to convert the "heavy" and "light" fold changes into a $\log_2$ scale (Fig. 4a).
14. In order to compare results from different proteomic experiments (reciprocal mixing) and eliminate to systematic errors in the mixing step the data must be normalized using the median of the $\log_2$ ratio. This may be achieved by dividing the ratio of each protein by the median value.

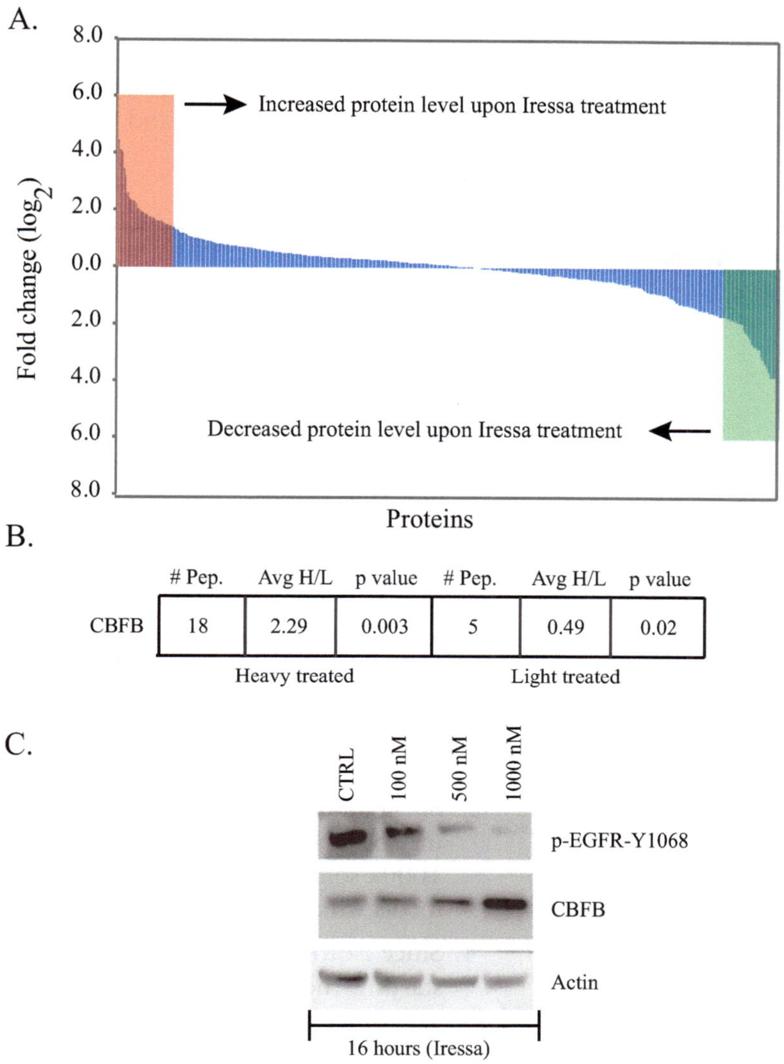

Fig. 4 (**a**) Distribution of protein fold change for the heavy treated sample. (**b**) Example of several experimental parameters used to determine robustness of data. (**c**) Western blot validation of protein fold change with loading control

15. We assess biological significance by performing a mock proteomic experiment with untreated "heavy" mixed with untreated "light" cells. Based on the distribution of proteins in this experiment, we suggest a cutoff of at least 1.3. Most proteomic publications utilize a threshold of 1.5 in SILAC ratios.
16. The results for the "heavy" Iressa treated sample should be compared to the "light" Iressa treated sample. In an ideal experiment, the ratios of the two experiments will be inversed (Fig. 4b).

17. Validation of proteomic experiments is a key step in assuring reproducibility of proteomic experiments. Techniques that rely on protein fold change are suitable for verification of proteomic results (ELISA, and Western blots being the most common) (Fig. 4c).

4 Notes

1. Cell lines often require a unique formulation of media for optimum survival. We recommend checking with an online depository (http://www.atcc.org/) for the selection of correct media. Most immortalized cell lines may be used with this application, but, try to avoid cells that divide slower than once in a 48 h period.
2. Some cell lines require a unique combination of growth factors that are either lost during the dialysis step or not present in sufficient concentrations to promote cell growth. In these conditions, we suggest exogenous addition of growth factors or dialysis of any preferred FBS with a lower molecular weight cutoff. In this case, we suggest dialysis with a 3500 Da cutoff Snakeskin tubing (Life Technologies, Carlsbad, CA) with Phosphate Based Saline (PBS).
3. Small differences in the molecular mass of the "heavy" isotopes may be ignored when preparing the stock solutions. As these are 1000× solutions, 1 mL of stock is sufficient to make 1 L of final media which is the usual amount of media to make at one time.
4. Since isotopically "heavy" reagents are expensive, we suggest starting the labeling procedure in 10 cm dishes. Large-scale experiments may increase the number of plates after the third passage.
5. For a two condition differential proteomics experiment, one condition will be grown in all light media and the other condition will be grown in media supplemented with heavy lysine and arginine. Sequential protein digestion with Lys-c and trypsin provide increased peptide and protein coverage compared to Lys-c alone; while single protease digestion with Lys-c increases quantification of lysine-labeled proteins [28]. Media will have to be supplemented with the heavy or light versions of all amino acids absent in the media. That is, you will add 1 mL of 1000× lysine (heavy or light) and 1 mL of 1000× arginine (heavy or light) to every bottle of media used in the experiment.
6. Changes in glucose, pyruvate, growth factors, and steroids may be optimized if growth kinetics become altered during labeling.
7. Prior to the start of any large-scale proteomics experiment, it is highly advisable to plan out positive and negative controls. In this example, the use of the A431 cell line was motivated by their dra-

matic response to Iressa treatment. In order to optimize drug dosage and incubation time we performed a number of mock experiments to ensure our experimental strategy was robust. Based on these results, we dosed A431 cells for 16 h with 100 nM of the drug (Note: we utilized the phosphorylation of EGFR as a readout to optimize dosing regimens. A similar approach should be used to ensure validity of any large scale proteomics experiment).

References

1. Ong SE, Blagoev B, Kratchmarova I, Kristensen DB, Steen H, Pandey A, Mann M (2002) Stable isotope labeling by amino acids in cell culture, SILAC, as a simple and accurate approach to expression proteomics. Mol Cell Proteomics 1(5):376–386
2. Beck S, Michalski A, Raether O, Lubeck M, Kaspar S, Goedecke N, Baessmann C, Hornburg D, Meier F, Paron I, Kulak NA, Cox J, Mann M (2015) The impact II, a very high-resolution quadrupole time-of-flight instrument (QTOF) for deep shotgun proteomics. Mol Cell Proteomics 14(7):2014–2029. doi:10.1074/mcp.M114.047407, M114.047407 [pii]
3. Zhang G, Fenyo D, Neubert TA (2009) Evaluation of the variation in sample preparation for comparative proteomics using stable isotope labeling by amino acids in cell culture. J Proteome Res 8(3):1285–1292. doi:10.1021/pr8006107
4. Bantscheff M, Schirle M, Sweetman G, Rick J, Kuster B (2007) Quantitative mass spectrometry in proteomics: a critical review. Anal Bioanal Chem 389(4):1017–1031. doi:10.1007/s00216-007-1486-6
5. Tzouros M, Golling S, Avila D, Lamerz J, Berrera M, Ebeling M, Langen H, Augustin A (2013) Development of a 5-plex SILAC method tuned for the quantitation of tyrosine phosphorylation dynamics. Mol Cell Proteomics 12(11):3339–3349. doi:10.1074/mcp.O113.027342, O113.027342 [pii]
6. Hilger M, Mann M (2012) Triple SILAC to determine stimulus specific interactions in the Wnt pathway. J Proteome Res 11(2):982–994. doi:10.1021/pr200740a
7. Van Hoof D, Pinkse MW, Oostwaard DW, Mummery CL, Heck AJ, Krijgsveld J (2007) An experimental correction for arginine-to-proline conversion artifacts in SILAC-based quantitative proteomics. Nat Methods 4(9):677–678. doi:10.1038/nmeth0907-677, nmeth0907-677 [pii]
8. Bendall SC, Hughes C, Stewart MH, Doble B, Bhatia M, Lajoie GA (2008) Prevention of amino acid conversion in SILAC experiments with embryonic stem cells. Mol Cell Proteomics 7(9):1587–1597. doi:10.1074/mcp.M800113-MCP200, M800113-MCP200 [pii]
9. de Godoy LM, Olsen JV, de Souza GA, Li G, Mortensen P, Mann M (2006) Status of complete proteome analysis by mass spectrometry: SILAC labeled yeast as a model system. Genome Biol 7(6):R50. doi:10.1186/gb-2006-7-6-r50, gb-2006-7-6-r50 [pii]
10. Gruhler A, Schulze WX, Matthiesen R, Mann M, Jensen ON (2005) Stable isotope labeling of Arabidopsis thaliana cells and quantitative proteomics by mass spectrometry. Mol Cell Proteomics 4(11):1697–1709. doi:10.1074/mcp.M500190-MCP200, M500190-MCP200 [pii]
11. Gruhler A, Olsen JV, Mohammed S, Mortensen P, Faergeman NJ, Mann M, Jensen ON (2005) Quantitative phosphoproteomics applied to the yeast pheromone signaling pathway. Mol Cell Proteomics 4(3):310–327. doi:10.1074/mcp.M400219-MCP200, M400219-MCP200 [pii]
12. Bose R, Molina H, Patterson AS, Bitok JK, Periaswamy B, Bader JS, Pandey A, Cole PA (2006) Phosphoproteomic analysis of Her2/neu signaling and inhibition. Proc Natl Acad Sci U S A 103(26):9773–9778. doi:10.1073/pnas.0603948103, 0603948103 [pii]
13. Boersema PJ, Geiger T, Wisniewski JR, Mann M (2013) Quantification of the N-glycosylated secretome by super-SILAC during breast cancer progression and in human blood samples. Mol Cell Proteomics 12(1):158–171. doi:10.1074/mcp.M112.023614, M112.023614 [pii]
14. Dhungana S, Merrick BA, Tomer KB, Fessler MB (2009) Quantitative proteomics analysis of macrophage rafts reveals compartmentalized activation of the proteasome and of proteasome-mediated ERK activation in response to lipopolysaccharide. Mol Cell Proteomics 8(1):201–213. doi:10.1074/mcp.M800286-MCP200, M800286-MCP200 [pii]
15. Ong SE, Mittler G, Mann M (2004) Identifying and quantifying in vivo methylation sites by heavy methyl SILAC. Nat Methods 1(2):119–126. doi:10.1038/nmeth715, nmeth715 [pii]

16. Zhang K, Li L, Zhu M, Wang G, Xie J, Zhao Y, Fan E, Xu L, Li E (2015) Comparative analysis of histone H3 and H4 post-translational modifications of esophageal squamous cell carcinoma with different invasive capabilities. J Proteomics 112:180–189. doi:10.1016/j.jprot.2014.09.004, S1874-3919(14)00419-9 [pii]
17. Kruger M, Moser M, Ussar S, Thievessen I, Luber CA, Forner F, Schmidt S, Zanivan S, Fassler R, Mann M (2008) SILAC mouse for quantitative proteomics uncovers kindlin-3 as an essential factor for red blood cell function. Cell 134(2):353–364. doi:10.1016/j.cell.2008.05.033, S0092-8674(08)00695-8 [pii]
18. Schwanhausser B, Gossen M, Dittmar G, Selbach M (2009) Global analysis of cellular protein translation by pulsed SILAC. Proteomics 9(1):205–209. doi:10.1002/pmic.200800275
19. Hanke S, Besir H, Oesterhelt D, Mann M (2008) Absolute SILAC for accurate quantitation of proteins in complex mixtures down to the attomole level. J Proteome Res 7(3):1118–1130. doi:10.1021/pr7007175
20. Rees JS, Lilley KS, Jackson AP (2015) SILAC-iPAC: a quantitative method for distinguishing genuine from non-specific components of protein complexes by parallel affinity capture. J Proteomics 115:143–156. doi:10.1016/j.jprot.2014.12.006, S1874-3919(14)00559-4 [pii]
21. Tackett AJ, DeGrasse JA, Sekedat MD, Oeffinger M, Rout MP, Chait BT (2005) I-DIRT, a general method for distinguishing between specific and nonspecific protein interactions. J Proteome Res 4(5):1752–1756. doi:10.1021/pr050225e
22. Kessner D, Chambers M, Burke R, Agus D, Mallick P (2008) ProteoWizard: open source software for rapid proteomics tools development. Bioinformatics 24(21):2534–2536. doi:10.1093/bioinformatics/btn323, btn323 [pii]
23. Mortensen P, Gouw JW, Olsen JV, Ong SE, Rigbolt KT, Bunkenborg J, Cox J, Foster LJ, Heck AJ, Blagoev B, Andersen JS, Mann M (2010) MSQuant, an open source platform for mass spectrometry-based quantitative proteomics. J Proteome Res 9(1):393–403. doi:10.1021/pr900721e
24. Cox J, Mann M (2008) MaxQuant enables high peptide identification rates, individualized p.p.b.-range mass accuracies and proteome-wide protein quantification. Nat Biotechnol 26(12):1367–1372. doi:10.1038/nbt.1511, nbt.1511 [pii]
25. Cox J, Matic I, Hilger M, Nagaraj N, Selbach M, Olsen JV, Mann M (2009) A practical guide to the MaxQuant computational platform for SILAC-based quantitative proteomics. Nat Protoc 4(5):698–705. doi:10.1038/nprot.2009.36, nprot.2009.36 [pii]
26. Moore RE, Young MK, Lee TD (2002) Qscore: an algorithm for evaluating SEQUEST database search results. J Am Soc Mass Spectrom 13(4):378–386. doi:10.1016/S1044-0305(02)00352-5
27. Qiao Y, Zhang H, Bu D, Sun S (2011) PI: an open-source software package for validation of the SEQUEST result and visualization of mass spectrum. BMC Bioinformatics 12:234. doi:10.1186/1471-2105-12-234, 1471-2105-12-234 [pii]
28. Ma J, Li W, Lv Y, Chang C, Wu S, Song L, Ding C, Wei H, He F, Jiang Y, Zhu Y (2013) A new insight into the impact of different proteases on SILAC quantitative proteome of the mouse liver. Proteomics. 13(15):2238–2242

Chapter 14

Relative Protein Quantification Using Tandem Mass Tag Mass Spectrometry

Lichao Zhang and Joshua E. Elias

Abstract

Measuring protein changes over time or following stimuli is one of the important tasks of proteomics. In the past decade, several strategies have been developed for the relative quantification of proteins using mass spectrometry (MS). Isobaric labeling strategies for relative quantitative proteomics allow for parallel multiplexing of quantitative experiments. With this technique, multiple peptide samples are chemically labeled with isobaric chemical tag variants and each variant has the same molecular structure and mass. Each variant, however, is designed to produce a unique "reporter ion" when fragmented inside a mass spectrometer. Once peptide samples are labeled, combined, and analyzed using MS, differentially labeled peptides are indistinguishable in a first, MS spectrum of intact peptides. However, since each tag variant contains a labile component with different mass, "reporter ions" can be generated and recorded in a subsequent MS^2 spectrum. Intensities from each variant are recorded to represent the relative abundances of the peptide in each sample. Isobaric tags for relative and absolute quantitation (iTRAQ) and tandem mass tags (TMT) are commercially available reagents for performing this technique. Here, we describe the general workflow of relative quantification of proteins using TMT by MS^2, or an additional MS^3 spectrum.

Key words Mass spectrometry, Quantification, Tandem mass tag

1 Introduction

Several labeling techniques measure relative abundances of proteins in different biological samples by mass spectrometry. Two general labeling strategies—in vivo and in vitro labeling—have been widely applied to quantitative protein analysis. In vivo labeling requires heavy isotopes to be incorporated into the proteins of living organisms by feeding them isotopically labeled nutrients in the growth medium or food [1]. This approach has limitations, such as the requirement for large amounts of costly growth medium, and difficulties in controlling of labeling efficiency [2]. In vitro labeling overcomes these limitations by chemically labeling any peptide sample with high efficiency. Isotope-coded affinity tag (ICAT) introduced the concept of in vitro, quantitative peptide

Lucio Comai et al. (eds.), *Proteomics: Methods and Protocols*, Methods in Molecular Biology, vol. 1550,
DOI 10.1007/978-1-4939-6747-6_14, © Springer Science+Business Media LLC 2017

labeling for MS-based detection [3]. With this method, chemically identical probes with distinct masses are utilized to label two peptide samples. Their relative abundances are then determined from the each form's MS^1 peak intensity. The identity of each peptide is inferred from a MS^2 spectrum derived from at least one of the labeled forms.

The introduction of Isobaric tags such as TMT and iTRAQ enables more accurate and multiplexed quantification by using MS^2 spectra for quantitative analysis [2, 4]. These reagents are typically composed of a mass reporter, a mass normalizer and an amine reactive group (Fig. 1a). The mass reporter and mass normalizer moieties incorporate stable isotopes in multiple configurations such that each mass reporter's mass can be resolved in a MS^2 spectrum. The intact mass of each isobaric tag variant, however, is the same. With this labeling strategy, digested peptides from multiple samples are first labeled in parallel with as many as ten tag variants (TMT). They are then mixed and analyzed using reversed phase high performance liquid chromatograph (HPLC) coupled with a mass spectrometer capable of tandem MS analysis (Fig. 2). The mass reporter can be cleaved off the labeled peptide through collision-induced dissociation (CID) prior to detection with a time-of-flight (TOF) mass analyzer, or through higher energy collision dissociation (HCD) prior to detection with an orbitrap mass analyzer. Intensities measured in the resulting MS^2 spectrum report relative peptide quantification (Fig. 3). The isobaric and chemically identical nature of isobaric tags ensure that identical peptides labeled with different tag variants will have the same chromatographic elution profile, and experience identical ionization processes in the mass spectrometer. This leads to a more accurate quantification in comparison to the original ICAT strategy. In addition, this strategy often improves the signal-to-noise ratio of quantitative measurements by enabling quantification at the MS^2 level, which is inherently less noisy than MS^1 scans of intact peptides.

However, in highly complex peptide mixtures, co-isolation and co-fragmentation of multiple ions is increasingly likely to occur with MS^2, resulting in distorted ratios of isobaric tag reporter ions [5]. In this case, distorted reporter ion ratios no longer reflect the true proportions of a selected peptide precursor's components. Combining ion trap and orbitrap technologies allows reporter signals to be measured following a third stage of ion isolation and fragmentation (MS^3). This technique was shown to greatly reduce ratio distortion resulting from interfering ions [5]. Moreover, a "MultiNotch MS^3" method (also referred to as "synchronous precursor selection") was demonstrated to increase detection sensitivity by simultaneously selecting multiple fragment ions from MS^2 to enhance reporter ion intensities at the MS^3 level [6].

This chapter describes a protocol for protein quantification using commercially available TMT reagents, which include 2plex,

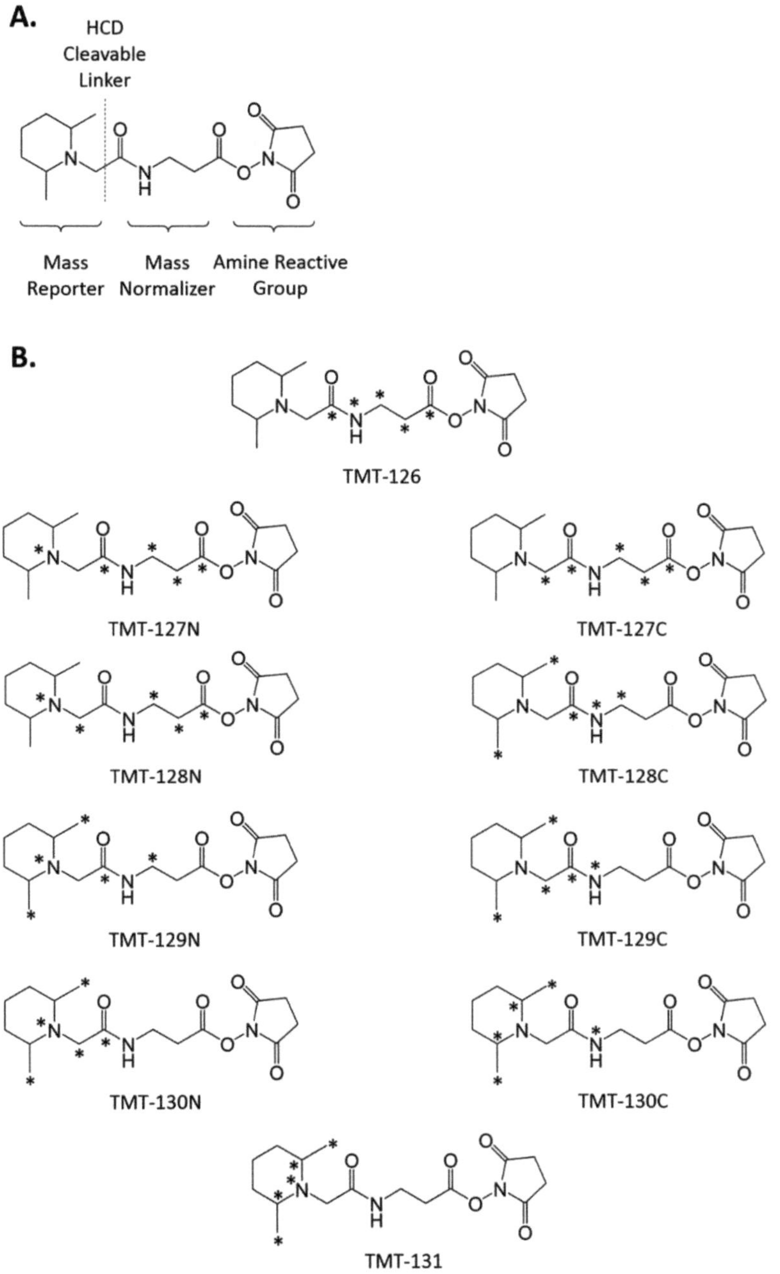

Fig. 1 Structure of TMT reagents. (**a**) TMT reagent contains mass reporter, mass normalizer and amine reactive group. A HCD cleavable linker is between the mass reporter and mass normalizer. (**b**) 10plex TMT. Heavy labeled C12 and N13 are indicated by *asterisks*. Figures are adapted from Thermo Scientific instructions for TMT10plex mass tag labeling kits and reagents

6plex, and 10plex (Fig. 1b). The TMT 10plex technique is used to describe the analysis procedure, including protein digestion, TMT reaction, sample purification, fractionation, HPLC-MS, and data analysis.

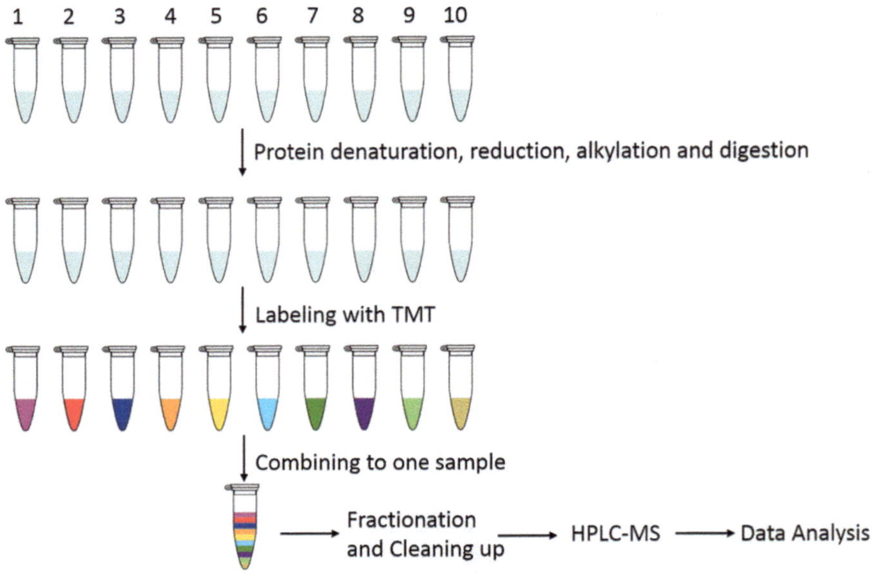

Fig. 2 Workflow for relative quantification using 10plex TMT. Ten different protein samples are denatured, reduced, alkylated and digested into peptide samples. These peptide samples are then labeled with 10plex TMT reagent kit and then combined into one sample. The combined sample is fractionated using HPRP liquid chromatography and the fractions are concatenated into 12 samples. After cleaning up, each sample is subjected to HPLC-MS analysis and the data are then analyzed to obtain the identity and relative quantification of proteins

2 Materials

2.1 Reduction, Alkylation and Protein Digestion

1. Ammonium bicarbonate.
2. HPLC grade water.
3. Dithiothreitol.
4. Iodoacetamide.
5. Sequencing grade modified trypsin.
6. Formic acid, 99%.

2.2 TMT Labeling

1. Acetonitrile (ACN), anhydrous, 99.8%.
2. HEPES sodium salt (Na-HEPES), minimum 99.5% titration.
3. TMT 10plex Mass Tag Labeling Kit (Thermo Scientific, Rockford, IL).
4. Hydroxylamine solution 50 wt%.

2.3 Sample Purification and Fractionation

1. Sep-Pak C18 1 cc vac cartridge, 50 mg sorbent per cartridge (Waters, Milford, MA).
2. Vacuum manifold (Varian, Palo Alto, CA).
3. Vortex mixer.

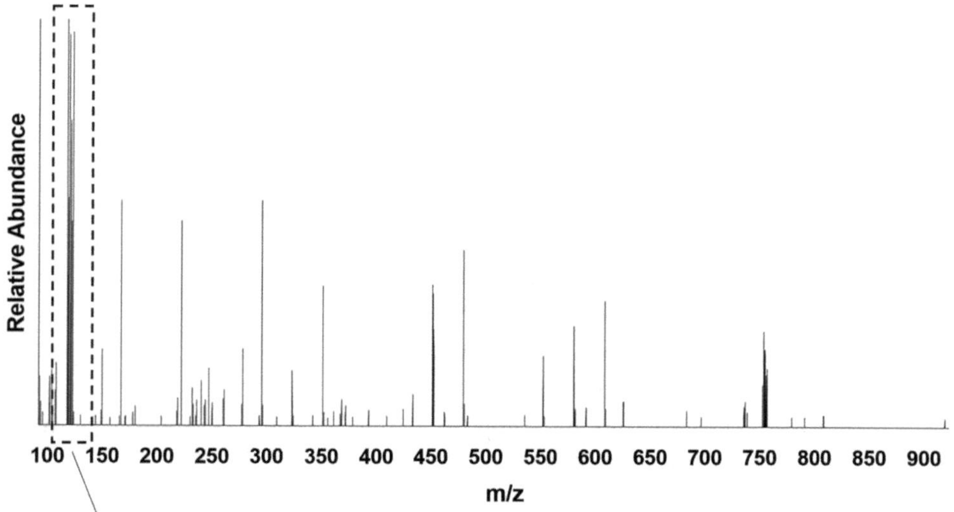

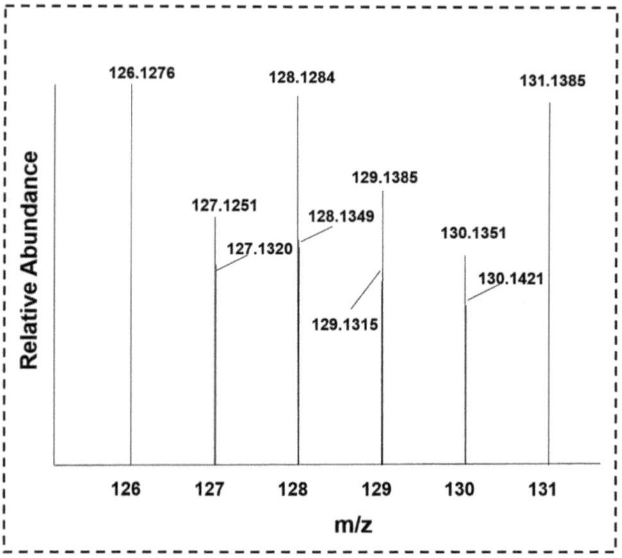

Reagent	HCD Monoisotopic Reporter Ion Mass
126	126.127725
127N	127.124760
127C	127.131079
128N	128.128114
128C	128.134433
129N	129.131468
129C	129.137787
130N	130.134822
130C	130.141141
131	131.138176

Fig. 3 MS analysis of 10plex TMT samples for peptide sequencing and quantification. *Top figure* shows a typical MS^2 HCD spectrum. *Bottom left figure* shows the zoomed spectrum of the reporter ions region. Theoretical HCD monoisotopic reporter ion masses for 10plex TMT reagent are listed at *bottom right*

4. Centrifuge.
5. Methanol.
6. Acetonitrile.
7. Acetic acid, glacial.
8. Ammonium formate for HPLC ≥ 99.0.
9. Ammonium hydroxide (Fisher Scientific, Fair Lawn, NJ).
10. Centrivap concentrator (Labconco, Kansas City, MO).

11. Empore™ SPE Disks, C18, 47 mm (Sigma-Aldrich, St. Louis, MO).
12. ZORBAX Extend-C18 column (Agilent technologies, Santa Clara, CA).
13. HPLC binary pump.

2.4 HPLC-MS Analysis

1. Nano-flow C18 HPLC columns are commercially available such as Acclaim PepMap100 C18 column of 75 μm in i.d. and 250 mm in length, packed with 3 μm 100 Å beads (Fisher Scientific, Fair Lawn, NJ).
2. The following equipment are needed for in-house packed nano-flow C18 column:

   Laser-based micropipette puller P-2000 (Sutter Instrument, Novato, CA).

   Fused Silica Capillary Tubing, 100 μm (Polymicro Technologies, Phoenix, AZ).

   Reprosil-Pur 120 C18-AQ, 5 μm and 3 μm (Dr. Maisch GmbH, Ammerbuch, Germany).
3. Ekspert nanoLC 425 (AB Sciex, Redwood City, CA).
4. LTQ-Obitrap Velos/Elite Mass spectrometer, or Obitrap Fusion Tribrid Mass Spectrometer (Thermo Scientific, San Jose, CA).

3 Methods

3.1 Reduction, Alkylation and Protein Digestion

1. Suspend 100 μg of protein in 100 mM ammonium bicarbonate (pH ~ 8) (*see* **Notes 1–4**).
2. Prepare 0.5 M DTT in HPLC grade water. Add 0.5 M DTT to the protein sample to obtain a final concentration of 5 mM and incubate for 60 min at 37 °C to reduce disulfide bonds.
3. Allow sample to cool to room temperature. Prepare 0.5 M iodoacetamide in HPLC grade water and add to the sample to obtain a final concentration of 14 mM. Incubate for 45 min at room temperature in the dark to alkylate cysteines.
4. Add trypsin or LysC (*see* **Note 5**) to the sample at an enzyme to protein ratio of 1:20 to 1:50 and digest for 6 h to overnight at 37 °C (*see* **Note 4**).
5. Acidify the sample to pH 3 with formic acid to quench digestion (*see* **Note 3**).
6. Desalt the digest using C18 solid-phase extraction (Sep-Pak). The C18 columns are conditioned and equilibrated with 1 mL of methanol (three times), 1 mL of 70% ACN in 0.5% Acetic acid (one time), 1 mL of 40% ACN in 0.5% acetic acid, and 1 mL of 0.5% Acetic acid (three times). Add the sample, wash

with 1 mL of 0.5 % acetic acid (three times) and elute the peptides with 1 mL of 40 % ACN in 0.5 % acetic acid. The eluted samples are dried in a centrivap concentrator and stored in −20 °C (*see* **Note 6**).

3.2 TMT Labeling (Using TMT 10plex as Example)

1. Resuspend ten samples in 100 μL of 50 mM Na-HEPES (pH 8).
2. Add 30 μL of anhydrous ACN to each sample.
3. Equilibrate TMT reagent at room temperature for 10–15 min before opening lids of reagent vials (*see* **Note 7**). Quickly spin down the reagent vial to collect any residue to the bottom of the tube and then resuspend the TMT reagent in 40 μL of anhydrous ACN. Allow reagent to dissolve for 5 min with occasional vortexing, and spin down briefly at the end.
4. Add 10 μL of TMT reagent to each corresponding peptide sample and mix with a very gentle and quick vortex (*see* **Notes 8** and **9**).
5. Incubate the samples for 1 h at room temperature.
6. Prepare 5 % hydroxylamine solution by diluting 50 % hydroxylamine with 50 mM Na-HEPES. Add 10 μL of 5 % hydroxylamine to each sample and incubate for 15 min at room temperature to quench the reaction.
7. Acidify the samples to pH 2 with 60 μL of 25 % formic acid.

3.3 Sample Preparation for Ratio and Labeling Efficiency Check (See Note 10)

1. Combine 1 % of each labeled sample and dry down in a centrivap concentrator.
2. Dry down the rest of the labeled samples and store in −20 °C.
3. Purify the combined sample using STAGE-Tip [7]. Cut out four cookies (2 mm o.d.) from the Empore C18 Disks and pack into a P200 pipette tip. The STAGE-Tip column is conditioned and equilibrated with 80 μL of methanol, 40 μL of 40 % ACN in 5 % formic acid, 40 μL of 5 % formic acid (two times). Resuspend the dried sample in 40 μL of 0.1 % formic acid. Add the sample, wash with 40 μL of 5 % formic acid (two times) and elute the peptides with 40 μL of 40 % ACN in 5 % formic acid. The eluted samples are dried in a centrivap concentrator and stored in −20 °C.
4. Perform HPLC-MS analysis described in Subheading 3.5 and data analysis in Subheading 3.6.
5. Verify if the reporter ion distributions follow desired overall ratio (e.g., 1:1:1:1:1:1:1:1:1:1). Calculate sample ratio adjustments, if necessary (*see* **Note 10**).

3.4 Sample Purification and Fractionation See Note 11)

1. Resuspend the samples in 50 μL of 0.1 % formic acid and combine the samples with the adjusted ratio (Subheading 3.3, **step 5**).
2. Purify the combined sample using Sep-Pak.

3. Fractionate the sample using offline high pH reversed phase (HPRP) liquid chromatography (*see* **Note 11**). Buffer A is 10 mM ammonium formate and buffer B is 90% acetonitrile/10% water (pH for both buffers is adjusted to 10 with ammonium hydroxide). Solubilize the labeled and mixed peptides in buffer A and separate on an Agilent ZORBAX 80 Å Extend-C18 column (5 μm particles, 4.6×250 mm) with an Agilent 1200 binary pump equipped with a photodiode array detector. Use an 84 min linear gradient from 0 to 40% buffer B in 66 min to separate the peptide mixture into 96 fractions. Concatenate the 96 fractions into 12 samples by combining eight fractions with equal time interval [8]. Acidify each sample with formic acid to pH 3, and dry the samples in a centrivap concentrator.
4. Redissolve each sample in 0.1% formic acid and purify with Sep-Pak. Vacuum-dry and reconstitute the samples in 0.1% formic acid for HPLC-MS analysis.

3.5 HPLC-MS Analysis

1. Prepare the following buffers: buffer A is 0.2% formic acid in water; buffer B is 0.2% formic acid in ACN.
2. Peptides are separated on a 100 μm inner diameter microcapillary column. The tip of the column is pulled in-house with laser puller and the column is packed with Reprosil C18 resin (0.5 cm of 5 μm resin close to the tip followed by 18 cm of 3 μm resin). There are also commercially available nano-flow C18 columns that can be used.
3. Run the HPLC gradient as follows: 98% A to 60% A in 130 min, to 40% A in 20 min, and then to 2% A in 18 min. The total gradient time is 180 min.
4. The TMT reporter ions can be analyzed at either MS^2 or MS^3 level on a LTQ-Orbitrap instrument (Velos or Elite), and the "synchronous precursor selection" MS^3 method is available with Obitrap Fusion Tribrid Mass Spectrometer (*see* **Note 12**) [5, 6].
   (a) MS^2 method. The MS^1 scans are performed in the Orbitrap in the mass range of 300–1500 m/z and the resolution is set to 60,000. Ten most intense ions (intensity above 500 counts) are selected for HCD fragmentation. The precursor isolation width is set to 2 m/z. The AGC settings are 1E6 and 5E4 for $FTMS^1$ and FT MSn scans, respectively. Maximum ion times for $FTMS^1$ and $FTMSn$ are both 250 ms. The normalized collision energy for HCD is 45% at 30 ms activation time. Monoisotopic precursor selection and charge state rejection are enabled. Singly charged ion species and ions with no unassigned charge states are excluded from MS^2 analysis. Ions within ±10 ppm m/z window around ions selected for MS^2 are excluded from further selection for fragmentation for 90 s.

(b) MS^3 method. The MS^1 scans are performed in a similar way. Ten most intense ions (intensity above 500 counts) are selected for CID fragmentation in the ion trap with precursor ion isolation width of 2 *m/z*, AGC setting of 5E3, maximum ion time of 150 ms and normalized collision energy of 35 % at activation time of 20 ms. Following each MS^2 analysis, the most intense fragment ion (intensity above 500 counts) is selected for HCD MS^3 analysis with isolation width of 2 *m/z*, normalized collision energy of 60 % at activation time of 50 ms and resolution of 60,000. Charge state screening is disabled to allow fragment ions to be selected for MS^3 (*see* **Note 13**).

(c) MS^3 method with synchronous precursor selection. This method can be selected from the available predefined workflows. Similarly, the full MS scans are performed in the Orbitrap in the mass range of 300–1500 *m/z* and the resolution is set to 120,000. The AGC settings is 4E5 and maximum ion time for $FTMS^1$ is 50 ms. Data dependent mode is set to top speed and precursor priority is set to most intense. Precursors with charge states 2–7 for MS^2 CID are isolated in quadrupole with isolation width of 0.7 *m/z*, AGC setting of 1E4 and maximum ion time of 50 ms. CID collision energy is set to 35 % and the mass analysis is performed in ion trap. Ions within ±10 ppm *m/z* window around ions selected for MS^2 are excluded from further selection for fragmentation for 90 s. Following each MS^2 CID, a MS^3 HCD fragmentation is performed with synchronous precursor selection enabled (the number of precursors set to 5), AGC setting of 1E5, and maximum ion time of 105 ms. HCD collision energy is set to 65 % and the fragment ions are detected in Orbitrap in the scan range of 120–500 *m/z* with resolution of 60,000 (*see* **Note 14**).

3.6 Data Analysis

1. Several software packages such as Proteome Discoverer, Mascot, and MaxQuant, are available to process TMT data for protein identification and quantification. Here we describe a protocol using Proteome Discoverer 2.1.
2. Create a workflow including "Spectrum Files", "Spectrum Selector", "Sequest HT" or "Mascot", "Percolator", and "Reporter Ions Quantifier". Trypsin or LysC is selected as the enzyme with at most two missed cleavage sites. Precursor mass tolerance is set to ±10 ppm and fragment mass tolerance is set to ±0.6 Da. At most four dynamic modifications are allowed per peptide. Carbamidomethylation of cysteine (+57.021 Da) is set as static modification, oxidation of methionine (+15.995 Da) is set as differential modification, TMT-labeled N-terminus and lysine (+229.163) are set as differential

modifications or static modifications (*see* **Note 15**). Percolator is applied to filter out the false MS^2 assignments at a strict false discovery rate of 1 % and relaxed false discovery rate of 5 %. The maximum delta Cn is set to 0.05. For quantification, a mass tolerance of ±20 ppm window is applied to the integration of report ions using the most confident centroid method.

3. Create a consensus workflow (*see* **Note 16**). In the "peptide and protein quantification" node, set "apply quan value corrections" to "true" to apply correction for the isotopic impurity of each reporter ions (values provided by the manufacturer in the product insert). Reporter abundances can be present using S/N values in Proteome Discoverer 2.1. Peptides are normally filtered based on average reporter S/N and isolation efficiency. The threshold for average reporter S/N can be set to 10 and the threshold for co-isolation can be set to 30%.
4. Data normalization is often necessary for more accurate quantification. Proteome discoverer provides the function of normalization and scaling in the "peptide and protein quantifier" node in the consensus workflow. The normalization mode can be set to "total peptide amount" or "specific protein amount" (*see* **Note 17**). The scaling mode is normally set to "on channels average" (*see* **Note 18**). For ratio check analyses (Subheading 3.3) normalization is not needed but the scaling can still be performed, then the average values of all the scaled reporter ion abundances for each channel are calculated and compared. The rest of the samples should be combined with adjusted ratio.
5. Protein abundances can be calculated as the sum, mean, or median of peptide abundances. Proteome Discoverer 2.1 uses the sum of peptide abundances as protein abundances. Proteins are then evaluated by the extent to which they increased or decreased in one sample relative to another. A scatter plot comparing the magnitude of a log-transformed ratio (*x*-axis) versus the significance of the ratio (*y*-axis) can be a useful way to intuitively select meaningful changes between any two experimental variables. Significance can be calculated based on the variance across biological replicates if TMT channels encode such replicates. Alternatively, significance can be estimated from the variance in ratios recorded from each peptide that contributed to a protein's quantification.

4 Notes

1. It is recommended to measure protein amount using a Bradford or BCA protein assay, or SDS-PAGE with standard amount of protein before protein digestion.

2. This protocol is based on labeling 100 μg of peptide sample per condition. If more than 100 μg of protein is digested, an aliquot of 100 μg of sample can be prepared for TMT labeling.
3. Digestion protocols often use primary amine-containing buffers such as ammonium bicarbonate or 50 mM Tris–HCl (pH ~ 8.8). Such buffers will partially or completely quench the TMT reagent, leading to incomplete peptide labeling. Thus, this protocol desalts peptides by solid phase extraction, prior to TMT labeling. If no such cleanup step is possible or desired, non-primary amine-based buffers (such as TEAB, HEPES) should be used as the digestion buffer. With this modification, there is no need to quench digestion before labeling.
4. If the protein sample is difficult to solubilize, prepare 8 M urea in 100 mM ammonium bicarbonate as the denaturing buffer instead. After reduction and alkylation, dilute urea to 1 M with 100 mM ammonium bicarbonate for trypsin digestion. However, exposure of protein to urea may lead to carbamylation on peptide N-terminus and lysine residues [9]. To avoid carbamylation, the sample in urea buffer should not be heated to above 37 °C. Normally room temperature or 37 °C is efficient for reduction and digestion.
5. Digestion with LysC instead of trypsin is recommended for TMT quantification at the MS^3 level [5]. Since TMT is reactive to peptide N-terminus and lysine residues, digestion with LysC ensures all *b*- and *y*-type ions contain at least one TMT reporter. Therefore when performing conventional MS^3 analysis, the most abundant ion from an MS^2 spectrum always contains TMT and its MS^3 fragment ions can always provide quantitative information. However, since LysC generates longer peptides with higher charge states, the number of peptides identified using collision-induced dissociation (CID) will decrease. If "synchronous precursor selection" MS^3 analysis is applied, tryptic digestion becomes an option again since multiple *b*- and/or *y*-type ions will be selected for MS^3 analysis and there is large chance that one or more of the selected ions contain TMT.
6. It is recommended to perform quality control on the digested sample using HPLC-MS to make sure the digestion is complete and there are no chemical modifications on N-terminus and lysine (such as carbamylation) that could affect TMT labeling. Also, the concentration of protein digest can be measured using quantitative peptide assay such as Thermo Scientific Pierce Quantitative Florescent Peptide Assay or Thermo Scientific Pierce Quantitative Colorimetric Peptide Assay.
7. TMT reagents are sensitive to moisture. Equilibrate the reagents to room temperature before opening lid to avoid forming condensation inside the reagent vial.

8. Thermo scientific instructions for TMT10plex mass tag labeling kits and reagents suggests that 25–100 μg of protein can be labeled with one vial of reagent. However one vial of reagent is actually enough to label up to 400 μg of peptides. Therefore in this protocol when labeling 100 μg of peptides, one quarter of the reagent is used. If more TMT reagent is desired, the volume of anhydrous ACN added to the sample should be decreased in order to maintain the same the final concentration of ACN in the labeling reaction.
9. If any TMT reagent remains once the vials have been opened, it can be stored in anhydrous ACN at −20 °C for up to 1 week. After 1 week, the reagent will degrade and the labeling efficiency will decrease.
10. After TMT labeling, it is highly recommended to combine the same small amount of sample from each condition and perform a HPLC-MS run to check the ratio of different conditions as well as TMT labeling efficiency. Since the measurement of protein amount may be inaccurate, the digestion may have different efficiencies and the post-digestion purification may have different peptide recoveries, the actual ratio of labeled peptide samples may be not at 1:1. Normally ratio variation of 1.5-fold or less is acceptable and can be normalized later. However if the ratio variation is too large, the channels with low reporter ion intensities will be skewed towards zero. For final TMT sample, samples of different conditions can be combined with the ratio adjusted based on ratio check result. Deviant test ratios could also reflect material loss affecting one or more samples prior to labeling. It could also reflect inaccurate protein quantification (*see* **Note 1**), or nonuniform protein abundance distributions across the experimental samples (e.g., one dominant protein in one sample). To resolve these possibilities, SDS-PAGE analysis of each sample should be considered. The ratio check run is also helpful to evaluate the TMT labeling efficiency. Labeling efficiency is defined as the percentage of fully labeled unique peptides in all the peptides identified. Using this protocol 99% of labeling efficiency can be achieved routinely. If the labeling efficiency is lower than 90%, contaminants in the sample or in the buffer used for TMT labeling could have quenched the TMT reagent. Samples can be relabeled with fresh TMT reagent in newly made buffers after being purified. Also if the amount of peptides is suspected to be higher than expected, larger amount of TMT reagents should be used.
11. Two-dimensional liquid chromatography is necessary for complex samples to further reduce sample complexity before MS analysis so that the dynamic range and proteome coverage can be improved, and the co-eluting/co-fragmented signal interferences for quantification can be reduced. Offline, concatenated

HPRP liquid chromatography is recommended for its ability to maximize the quantity of identifications a mass spectrometer can make per minute when combined with standard HPLC-MS [8]. Alternative fractionation techniques include strong cation exchange, isoelectric focusing, and hydrophilic interaction liquid chromatography.

12. The "MultiNotch MS^3" method from ref. 6 was implemented on an Orbitrap-Elite mass spectrometer with experimental software. Essential aspects of this method were implemented commercially on the Obitrap Fusion platform and called "synchronous precursor selection".
13. With charge state screening disabled, ions with charge state +1 and no charge state assigned can still be excluded for MS^2 analysis. Charge state exclusion is only applied to precursor selection for MS^2 but not for MS^3.
14. For TMT 10plex reporter ions detection, resolution of 60,000 is recommended. However for TMT 6plex or less, lower resolution can be applied for orbitrap mass analysis or low resolution analysis in ion trap can be performed.
15. TMT-labeled N-terminus and lysine (+229.163) can be set as differential modifications for the purpose of checking labeling efficiency.
16. An established workflow template for TMT quantification is named "CWF_Comprehensive_Enhanced Annotation_Quan" in Proteome Discoverer.
17. If data is normalized with total peptide amount, it is based on the assumption that the majority of peptides are essentially invariant across samples while only a small portion of peptides increase or decrease among different samples. This assumption is not necessarily valid for all experiments. After normalization with total peptide amount, the total abundance of peptides is the same for all channels. This step is to correct different amounts of sample input in each channel resulting from individual sample preparation and final mixing.
18. In this mode the abundances of each channel are scaled so that for every peptide and protein the average of all channels is 100. This step is to correct the differences of reporter ion abundance resulting from different amount of peptides detected.

Acknowledgments

The authors would like to thank Dr. Lihua Jiang from Stanford University and Dr. Xiaoyue Jiang from Thermo Scientific for helpful discussions about TMT methods on Obitrap Fusion Tribrid Mass Spectrometer and data processing using Proteome Discoverer.

References

1. Oda Y, Huang K, Cross F, Cowburn D, Chait B (1999) Accurate quantitation of protein expression and site-specific phosphorylation. Proc Natl Acad Sci U S A 96:6591–6596
2. Thompson A, Schafer J, Kuhn K, Kienle S, Schwarz J, Schmidt G, Neumann T, Hamon C (2003) Tandem mass tags: a novel quantification strategy for comparative analysis of complex protein mixtures by MS/MS. Anal Chem 75:1895–1904
3. Gygi S, Rist B, Gerber S, Turecek F, Gelb M, Aebersold R (1999) Quantitative analysis of complex protein mixtures using isotope-coded affinity tags. Nat Biotechnol 17:994–999
4. Ross P, Huang Y, Marchese J, Williamson B, Parker K, Hattan S, Khainovski N, Pillai S, Dey S, Daniels S, Purkayastha S, Juhasz P, Martin S, Bartlet-Jones M, He F, Jacobson A, Pappin D (2004) Multiplexed protein quantitation in Saccharomyces cerevisiae using amine-reactive isobaric tagging reagents. Mol Cell Proteomics 3:1154–1169
5. Ting L, Rad R, Gygi SP, Haas W (2011) MS3 eliminates ratio distortion in isobaric multiplexed quantitative proteomics. Nat Methods 8:937–940
6. McAlister GC, Nusinow DP, Jedrychowski MP, Wuehr M, Huttlin EL, Erickson BK, Rad R, Haas W, Gygi SP (2014) MultiNotch MS3 enables accurate, sensitive, and multiplexed detection of differential expression across cancer cell line proteomes. Anal Chem 86:7150–7158
7. Rappsilber J, Mann M, Ishihama Y (2007) Protocol for micro-purification, enrichment, pre-fractionation and storage of peptides for proteomics using StageTips. Nat Protoc 2:1896–1906
8. Wang Y, Yang F, Gritsenko MA, Wang Y, Clauss T, Liu T, Shen Y, Monroe ME, Lopez-Ferrer D, Reno T, Moore RJ, Klemke RL, Camp DG II, Smith RD (2011) Reversed-phase chromatography with multiple fraction concatenation strategy for proteome profiling of human MCF10A cells. Proteomics 11:2019–2026
9. Stark G, Stein W, Moore S (1960) Reactions of cyanate present in aqueous urea with amino acids and proteins. J Biol Chem 235:3177–3181

Chapter 15

Pathway-Informed Discovery and Targeted Proteomic Workflows Using Mass Spectrometry

Caroline S. Chu, Christine A. Miller, Andy Gieschen, and Steve M. Fischer

Abstract

Recent advancements in mass spectrometry (MS) and data analysis software have enabled new strategies for biological discovery using proteomics. Proteomics has evolved from routine discovery and identification of proteins to integrated multi-omics projects relating specific proteins to their genes and metabolites. Using additional information, such as that contained in biological pathways, has enabled the use of targeted protein quantitation for monitoring fold changes in expression as well as biomarker discovery. Here we discuss a full proteomic workflow from discovery proteomics on a quadrupole Time-of-Flight (Q-TOF) MS to targeted proteomics using a triple quadrupole (QQQ) MS. A discovery proteomics workflow encompassing acquisition of data-dependent proteomics data on a Q-TOF and protein database searching will be described which uses the protein abundances from identified proteins for subsequent statistical analysis and pathway visualization. From the active pathways, a protein target list is created for use in a peptide-based QQQ assay. These peptides are used as surrogates for target protein quantitation. Peptide-based QQQ assays provide sensitivity and selectivity allowing rapid and robust analysis of large batches of samples. These quantitative results are then statistically compared and visualized on the original biological pathways with a more complete coverage of proteins in the studied pathways.

Key words Mass spectrometry, Proteomics, QTOF, QQQ, Time-of-flight, Quadrupole, Proteins, Informatics, Data-dependent acquisition, Protein quantitation

1 Introduction

Here we discuss a full proteomic workflow using MS from discovery proteomics LC/MS on a quadrupole Time-of-Flight (Q-TOF) MS to targeted proteomics using a triple quadrupole (QQQ) LC/MS. For the discovery proteomics on the Q-TOF both ultra-high performance liquid chromatography (UHPLC) and nano-LC conditions will be included in the discussion depending on the user's sample matrix. The multistep discovery proteomics workflows (Fig. 1) encompass acquisition of the data-dependent proteomics data on the Q-TOF, protein database searching using Spectrum

Lucio Comai et al. (eds.), *Proteomics: Methods and Protocols*, Methods in Molecular Biology, vol. 1550,
DOI 10.1007/978-1-4939-6747-6_15,

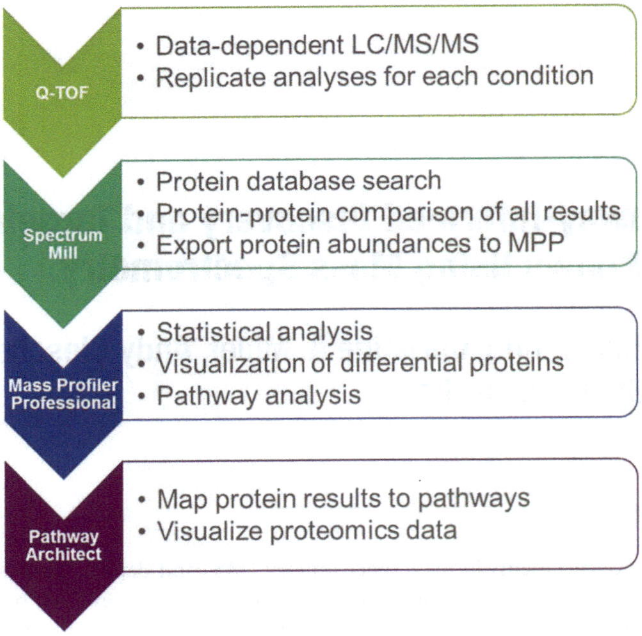

Fig. 1 Workflow for discovery proteomics

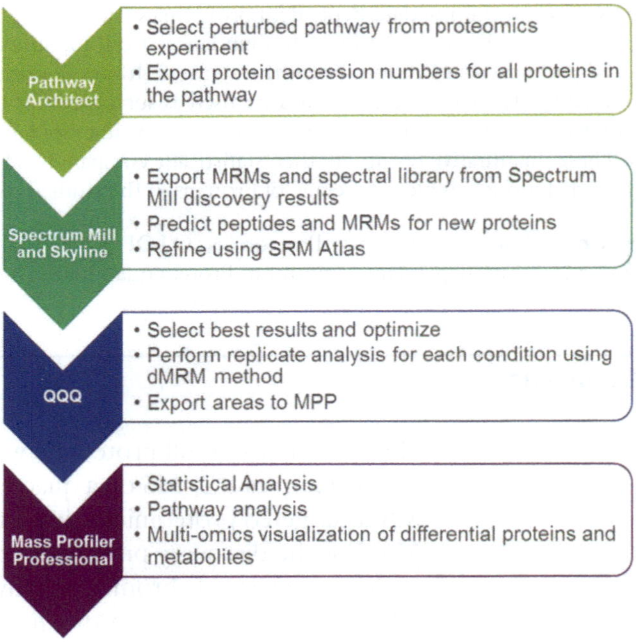

Fig. 2 Targeted proteomics workflow

Mill to the statistical comparison using Mass Profiler Professional and visualization of the experimental results in biological pathways in Pathway Architect used for the development of the targeted proteomic experiments. For the targeted proteomics workflows

(Fig. 2) the targeted proteins obtained from Pathway Architect are transformed to a targeted peptide method using Spectrum Mill or Skyline for use on the QQQ, and the data is then statistically compared using Mass Profiler Professional.

2 Materials

Prepare all solutions in water of 18 MΩ-cm resistivity or better and free of organics. For formic acid, we use LC/MS grade and prefer to buy small volume bottles to maintain contamination-free acid.

2.1 Protein Digest Reagents

1. Protein Lysate: As a quality assessment sample, we use either a *Pyrococcus furiosus* (Pfu, Agilent p/n 400510) [1, 2] or *E. coli* protein lysate (BioRad p/n 163-2110).
2. Ammonium bicarbonate (100 mM): add 100 mL water to 0.7906 g ammonium bicarbonate.
3. Dithiothreitol (DTT, 200 mM) >99+%: add 1 mL water to 0.031 g DTT in a 1.5 mL Eppendorf tube.
4. Iodoacetamide (IAM, 200 mM) 97%: add 1 mL water to 0.037 g IAM in a 1.5 mL Eppendorf tube.
5. Trypsin Stock Solution: Sequencing-grade trypsin and resuspension buffer (Agilent p/n 204310): Prepare a 1 μg/μL solution in a) 50 mM acetic acid if not all the solution will be used (store remainder in freezer) or b) 50 mM ammonium bicarbonate if it will all be used immediately.

2.2 Discovery Proteomics Standard-Flow HPLC Mass Spectrometry

1. AdvanceBio Peptide Mapping Column, 2.1×250 mm, 2.7 μm (Agilent p/n 651750-902) for 60, 90, 120, and 150 min LC gradients (Tables 1, 2, 3, and 4)
2. Mobile Phase Buffer A: 0.1% formic acid in water, for a 1 L solution add 1 mL formic acid to a clean bottle and 999 mL of water and mix thoroughly.
3. Mobile Phase Buffer B: 0.1% formic acid in 90% acetonitrile in water, for a 1 L solution add 1 mL of formic acid to a clean bottle, 900 mL acetonitrile, 99 mL of water, mix thoroughly.
4. Needle Wash: 0.1% formic acid in 50% (v/v) methanol in water, for a 1 L solution, add 1 mL formic acid to a clean bottle, add 500 mL of methanol, 499 mL water, and mix thoroughly.

2.3 Discovery and Targeted Proteomics Nanoflow HPLC-Chip Mass Spectrometry

1. Agilent 1260 Infinity HPLC-Chip/MS for use with all Agilent 6000 series Mass Spectrometers.
2. Q-TOF Internal Reference Mass (IRM) Low Mass Solution (1000 μg/mL methyl stearate in acetonitrile): Put 10 mg methyl stearate into a 15 mL Falcon conical tube, add 10 mL of acetonitrile, vortex to get methyl stearate into solution.

Table 1
60 min Gradient with the Agilent 1290 UPLC

LC conditions	
Column	AdvanceBio Peptide Mapping, 2.1 × 250 mm, 2.7 μm (Agilent p/n 651750-902)
Column temperature	50 °C
Injection volume	20 μL
Autosampler temp	4 °C
Needle wash	10 s in wash port (50:50 water:methanol with 0.1 % formic acid)
Mobile phase	A = 0.1 % formic acid in water B = 0.1 % formic acid in 90 % acetonitrile in water
Flow rate	0.40 mL/min
Gradient program	Time, mins % B 0.0 3 52.0 35 55.0 70 57.0 70 58.0 3
Stop time	60.0 min
Post time	5.0 min

Table 2
90 min Gradient with the Agilent 1290 UPLC

LC conditions	
Column	AdvanceBio Peptide Mapping, 2.1 × 250 mm, 2.7 μm (Agilent p/n 651750-902)
Column temperature	50 °C
Injection volume	20 μL
Autosampler temp	4 °C
Needle wash	10 s in wash port (50:50 water:methanol with 0.1 % formic acid)
Mobile phase	A = 0.1 % formic acid in water B = 0.1 % formic acid in 90 % acetonitrile in water
Flow rate	0.40 mL/min
Gradient program	Time, mins %B 0.0 3 82.0 35 85.0 70 87.0 70 88.0 3
Stop time	90.0 min
Post time	5.0 min

Table 3
120 min Gradient with the Agilent 1290 UPLC

LC conditions		
Column	AdvanceBio Peptide Mapping, 2.1 × 250 mm, 2.7 μm (Agilent p/n 651750-902)	
Column temperature	50 °C	
Injection volume	20 μL	
Autosampler temp	4 °C	
Needle wash	10 s in wash port (50:50 water:methanol with 0.1 % formic acid)	
Mobile phase	A = 0.1 % formic acid in water B = 0.1 % formic acid in 90 % acetonitrile in water	
Flow rate	0.40 mL/min	
Gradient program	Time, min	%B
	0.0	3
	110.0	40
	115.0	70
	117.5	70
	118.0	3
Stop time	120.0 min	
Post time	5.0 min	

Table 4
150 min GRADIENT with the Agilent 1290 UPLC

LC conditions		
Column	AdvanceBio Peptide Mapping, 2.1 × 250 mm, 2.7 μm (Agilent p/n 651750-902)	
Column temperature	50 °C	
Injection volume	20 μL	
Autosampler temp	4 °C	
Needle wash	10 s in wash port (50:50 water:methanol with 0.1 % formic acid)	
Mobile phase	A = 0.1 % formic acid in water B = 0.1 % formic acid in 90 % acetonitrile in water	
Flow rate	0.40 mL/min	
Gradient program	Time, mins	%B
	0.0	3
	140.0	40
	145.0	70
	147.5	70
	148.0	3
Stop time	150.0 min	
Post time	5.0 min	

3. Q-TOF IRM, High Mass Solution: Add 1 mL acetonitrile to a 1.5 mL Eppendorf tube. Add 20 μL HP-1221 (Agilent p/n G1982-85001) high mass compound, mix on a vortex mixer. Both solutions should be refrigerated when not in use.
4. For nanoflow, typically a short column HPLC-Chip (ProtID-Chip-43 (II), Agilent p/n G4240-62005, 43 mm 300 Å C18 chip with 40 nL trap column) is used for simple samples such as 2D-gel spots or single protein digests. The Polaris-HR-Chip 3C18 (Agilent p/n G4240-62030, High Resolution Chip 150 mm 180 Å 3 μm C18 Chip with 360 nL trap column), offers improved resolutions and peak capacities for peptides in complex protein digests.
5. As an alternative the HPLC-Chip interface and columns, the Agilent G1992A Nanospray Ion Source with gas distributor (Agilent p/n G1964-20303) and spray shield kit (Agilent p/n G1988-60007) can be used to interface any nanocolumn to Agilent mass spectrometers.

2.4 Targeted Proteomics Standard-Flow HPLC Mass Spectrometry

1. AdvanceBio Peptide Mapping Column, 2.1 × 100 mm, 2.7 μm (Agilent p/n 655750-902) for 60, 90, 120 min LC gradients. (Tables 1–4)
2. Mobile Phase Buffer A: 0.1 % formic acid in water, for a 1 L solution add 1 mL formic acid to a clean bottle and 999 mL of water and mix thoroughly.
3. Mobile Phase Buffer B: 0.1 % formic acid in 90 % acetonitrile in water, for a 1 L solution add 1 mL of formic acid to a clean bottle, 900 mL acetonitrile, 99 mL of water, mix thoroughly.
4. Needle Wash: 0.1 % formic acid in 50 % (v/v) methanol in water, for a 1 L solution, add 1 mL formic acid to a clean bottle, add 500 mL of methanol, 499 mL water, and mix thoroughly.

2.5 Mass Spectrometry: 6500 Series Q-TOF and 6400 Series QQQ

1. ESI-L Low Concentration Tuning Mix (Agilent p/n G1969-8500) (for Dual ESI or ESI with Jet Stream Technology).
2. Glass Calibrant Delivery System (CDS) bottle (Agilent p/n 9300-2576) and cap (Agilent p/n 9300-2575).
3. Tuning and calibrating with the ESI with Jet Stream Technology on the 6500 series iFunnel Q-TOFs in positive ion mode, add 10 mL ESI-L to a clean CDS bottle, add 88.5 mL acetonitrile, 1.5 mL water, and 5 μL of 0.1 mM HP-0.321 (included in the Biopolymer Reference Mass Kit (Agilent p/n G1969-85003). Mix thoroughly and place in position Bottle B in the CDS.
4. Tuning and calibrating with the dual ESI on the 6500 series iFunnel Q-TOFs in positive ion mode, add 25 mL ESI-L to a clean CDS bottle, add 71.25 mL acetonitrile, and 3.75 mL water. Mix thoroughly and place in position Bottle B in the CDS.

5. Internal Reference Mass Solution for the ESI with Jet Stream Technology on the 6500 Series Q-TOFs; ES-TOF Reference Mass Solution Kit (Agilent p/n G1969-85001) containing two ampoules (2.2 mL/ampoule) of the following reference ions: 100 mM ammonium trifluoracetate (TFA-NH_4) in 90:10 acetonitrile:water, 5 mM purine in 90:10 acetonitrile:water, and 2.5 mM hexakis(1H, 1H, 3H-tetrafluoropropoxy)phosphazine (HP-0921) in 90:10 acetonitrile:water (*see* **Note 1**). To a 1 L Nalgene bottle, add 950 mL acetonitrile, 50 mL water, 0.4 mL purine, and 1.0 mL HP-0921. Cap and invert the bottle several times to mix the reference solution. Pour 100 mL into a CDS bottle and place onto Bottle A in the CDS. Alternatively an isocratic pump can be used with the 1 L stock solution with a 1:100 splitter (Agilent p/n G1607-60000) connected to the reference nebulizer.
6. Internal Reference Mass Solution for the ESI on the 6500 Series Q-TOFs; ES-TOF Reference Mass Solution Kit (Agilent p/n G1969-85001) containing two ampoules (2.2 mL/ ampoule) of the following reference ions: 100 mM TFA-NH_4 in 90:10 acetonitrile:water, 5 mM purine in 90:10 acetonitrile:water, and 2.5 mM HP-0921 in 90:10 acetonitrile:water. To a 1 L Nalgene bottle, add 950 mL acetonitrile, 50 mL water, 0.5 mL TFA-NH_4, 1.0 mL purine, and 0.45 mL HP-0921 (*see* **Note 2**) Tuning and calibrating with the ESI with Jet Stream Technology and ESI on the 6400 series QQQs in positive ion mode, add 100 mL ESI-L to a clean CDS bottle and place in position Bottle B in the CDS.

2.6 Peptide Quantitation Checkout on the 6400 Series QQQ

1. Bovine Serum Albumin (BSA) Stock: Prepare a 1 pmol/μL stock of trypsinized bovine serum albumin, BSA digest (Agilent p/n G1900-85000) by adding 500 μL of 15% acetonitrile/85% water with 0.1% formic acid to the standard (500 pmol per vial). This can be aliquoted into 0.5 mL Eppendorf tubes and frozen.
2. BSA Dilution Solution: Prepare a 10 fmol/ μL solution by diluting the BSA Stock 1:100 with 15% acetonitrile/85% water with 0.1% formic acid. For the levels shown below, you should make 1 mL (10 μL of BSA Stock plus 990 μL of solvent).
3. Human Serum Albumin (HSA) Stock: Prepare the stock solution of human serum albumin, HSA, peptides standard (Agilent p/n G2455-85001) by adding 500 μL of the BSA Dilution Solution. Vortex well to completely dissolve the standard. The resulting stock solution is 1 pmol/μL and contains seven peptides. Only one peptide will be used for the quantitation checkout; however, optimization information for the other six is shown in the appendix.

4. Prepare dilutions as shown in the table below by adding specified volumes of HSA standard and BSA Dilution Solution to vials. (Best practice is to put the BSA Dilution Solution in the vial first, then add the standard volume.) For convenience, dilutions can be prepared directly in the conical bottom polypropylene autosampler vials (Agilent p/n 5190-3155) and sealed with the appropriate caps (Agilent p/n 5182-0541).
5. For HPLC-Chip systems, use a range appropriate to the 6400 Series QQQ model. Typically your samples should cover six orders of linear dynamic range (*see* **Note 3**).
6. Prepare a vial containing 60% acetonitrile/40% water with 1% trifluoroacetatic acid, TFA (or 50% 2,2,2-trifluorethanol,TFE in water) to clean the injector before doing low-level samples. This solution works well for solubilizing hydrophobic peptides. If a clean blank cannot be achieved after running several injections of this solution, the best action is to remove the needle seat (with seat capillary) and place it seat-side down in a beaker with this solution, then sonicate for 5–10 min.
7. For the HPLC Chip Cube interface, the ProtID-Chip-43 (II) (Agilent p/n G4240-62005) 43 mm×0.075 mm chip with 40 nL enrichment column is used. For standard flow, use the Eclipse Plus EC-C18 RRHD 2.1×50 mm, 1.8 μm column (Agilent p/n 857750-902).
8. The mobile phases used for Channel A: 0.1% formic acid in 3% acetonitrile in water (v/v) and Channel B: 0.1% formic acid in 90% acetonitrile in water (v/v) (*see* **Note 4**)
9. Commercial Targeted Proteomics Kits:
   - PeptiQuant™ MRM-MS Workflow Performance Kit: LC/MRM/MS PEPTIQUANT WORKFLOW PERFORM KIT FOR AGILENT 6495 1 RUN (Cambridge Isotope Laboratories, Inc., Item #: WFPK-A6495-1).
   - PeptiQuant™ LC/MS Platform Performance Kit: LC/MRM/MS PEPTIQUANT PLATFORM PERFORM. KIT FOR AGILENT 6495/UPLC 1290 1WK SUPPLY (Cambridge Isotope Laboratories, Inc., Item #: LCMSP-D-A6495-1).

2.7 Data Analysis

1. Agilent G2721AA/G2733AA Spectrum Mill MS Proteomics Workbench.
2. Agilent G3835AA MassHunter Mass Profiler Professional Software.
3. Agilent G6825AA Pathway Architect.
4. Skyline(https://brendanx-uw1.gs.washington.edu/labkey/project/home/software/Skyline/begin.view).

2.8 Instrumentation

1. 1290 UPLC, consisting of
   - G4220A 1290 Infinity Binary Pump.
   - G1330B 1290 Thermostat.
   - G4226A 1290 Infinity Autosampler.
   - G1316C 1290 Infinity Thermal Column Compartment.
2. 1260 Infinity HPLC-Chip/MS System, consisting of
   - G4225A 1260 HiP Degasser (quantity 2).
   - G1376A 1260 Infinity Capillary Pump.
   - G2226A 1260 Infinity Nanopump.
   - G1377A Infinity Autosampler.
   - G1330B 1290 Thermostat.
   - G4240A 1260 HPLC Chip Cube.
3. G6550AA 6550 iFunnel Q-TOF LC/MS.
4. G6495AA 6495 QQQ LC/MS.

3 Methods

All procedures are performed under room temperature. Biological samples and hazardous chemicals should be handled with caution.

3.1 Protein Digestion

1. Dissolve sample in 50% TFE in 50 mM ammonium bicarbonate buffer to yield a 1.35 mg/mL solution (*see* **Note 5**) For larger amounts of sample (such as the *E. coli* lysate), aliquot 100 μL per tube into 1.5 mL Eppendorf tubes (*E. coli* sample yields 18 tubes; 135 μg per tube).
2. Add 2.5 μL DTT stock solution (200 mM) to each tube and vortex to mix. Heat at 60 °C for 45 min.
3. Add 10 μL IAM stock solution (200 mM). Vortex briefly. Allow to stand at room temperature for 45 min in the dark (foil covered rack).
4. Add 2.5 μL DTT stock solution (200 mM) to remove excess IAM. Allow to stand at room temperature for 30 min in the dark.
5. Add 600 μL water and 200 μL ammonium bicarbonate to each vial (Note: pre-mix this to reduce pipetting).
6. Add 6 μL trypsin stock solution at 1:20 or 1:50 enzyme:substrate. Vortex briefly. Incubate overnight at 37 °C.
7. Add 4 μL neat formic acid or TFA to stop trypsin activity. Vortex briefly. Digest is ready to analyze by LC/MS. Final concentration should be ~150 ng/μL. For larger amounts of sample, such as the quality assessment samples, it is convenient

to mix all the digest vials together (15 mL Falcon tube), then aliquot to Eppendorf tubes (100 μL per vial) and store at −80 °C (or −20 °C if you don't have a −80 °C freezer).

3.2 Discovery Proteomics: Data-Dependent Acquisition (DDA) on an Agilent 6500 Series Q-TOF

1. With the ESI with Jet Stream Technology source [3] on the Q-TOF, in MassHunter Acquisition, go to the Tune page. Tune and calibrate the Q-TOF in the Extended Mass Range, 2 GHz at mass range m/z 3200.
2. Select Quadrupole Tune and perform a Quad Tune.
3. Once the tune is completed, change the mass range to m/z 1700 and select "Apply".
4. Once the 20 min equilibration is complete, select m/z 118 to m/z 1622, and select calibrate the TOF.
5. Save the Tune File before switching back to acquisition window.
6. In the acquisition window, generate the following acquisition method for standard flow with the ESI with Jet Stream Technology source:

   Liquid Chromatography Gradient: Table 2.

   Source: Table 5.

   MS Acquisition: Table 6.
7. Save method as "DDA_90min_AJS.m".
8. In the acquisition window, generate the following acquisition method for nanoflow with the HPLC-Chip source [4, 5]:

   Liquid Chromatography Gradient: Table 7.

Table 5
Source conditions for the 6500 series QTOF with the ESI with Agilent Jet Stream source

Ion source/mode	Agilent Jet Stream, positive
Gas temperature	250 °C
Drying gas flow	14 L/min
Nebulizer gas	35 psi
Sheath gas temperature	250 °C
Sheath gas flow	11 L/min
Capillary voltage	3500 V
Nozzle voltage	0 V
Fragmentor	360 V
Reference mass	m/z 322.0481 and 1221.9906

Table 6
Discovery proteomics acquisition method for the 6500 series QTOF

Parameter	Setting
Acquisition mode	Extended dynamic range (2 GHz), high sensitivity, low mass range *m/z* 1700
Mass range	*m/z* 300–1700
Acquisition rate/time	8 spectra/s
Auto MSMS range	*m/z* 50-1700
MSMS acquisition rate/time	3 spectra/s (max)
Isolation width	Narrow (~1.3 Hz)
Precursors/cycle	Top 20
Collision energy	3.6*(*m/z*)/100–4.8
Threshold for MSMS	1000 counts and 0.001
Dynamic exclusion	On; 1 repeat then exclude for 0.2 min
Precursor abundance-based scan speed	Yes
Target	25,000
Use MS/MS accumulation time limit	Yes
Purity	100% stringency, 30% cutoff
Isotope model	Peptides
Sort precursors	By abundance only; +2, +3, >+3

Source: Table 8.

MS Acquisition: Table 9.

9. Save method as "DDA_130min_Chip.m.

3.3 Discovery Proteomics: Data Analysis Using Spectrum Mill (Fig. 3)

1. Move your DDA data from the QTOF to a new folder under the "smdata" folder on your Spectrum Mill server. Typically data is organized in subfolders under smdata to organize projects (*see* **Note 6**).
2. Open Internet Explorer and load Spectrum Mill, navigate to the Data Extractor page.
3. In the **Data Directories** section, click the **Select**... button to select the folder or folders that contain your files.
4. For Agilent Q-TOF data and most extractions, adjust the parameters as outlined below:
   - Click the **Choose**... button to select the "Carbamidomethylation (C) "modification".

Table 7
Source and LC gradient conditions for the 6500 series QTOF with the HPLC Chip

LC conditions	HPLC-Chip, positive
Column	Agilent Polaris-HR-Chip-3C18 with a 360 nL enrichment and 150 mm × 0.075 mm analytical column
Injection volume	Adjusted to load 1 μg of total protein per injection
Autosampler temp	4 °C
Needle wash	10 s in wash port (50:50 water:methanol with 0.1 % formic acid)
Mobile phase	A = 0.1 % formic acid in water B = 0.1 % formic acid in 90 % acetonitrile in water
Flow rate	Loading: 2 μL/min with 3 % B Analytical: 300 nL/min
Gradient program	Time, min %B 0.0 3 90.0 25 120.0 40 125.0 90 130.0 90 130.1 3
Stop time	133.0 min
Post time	5.0 min

Table 8
Source conditions for the 6500 series QTOF with HPLC Chipcube

Ion source/mode	HPLC ChipCube, positive
Gas temperature	250 °C
Drying gas flow	11 L/min
Capillary voltage	1800–1950 V
Fragmentor	360 V
Reference mass	299.2944 and 1221.9906

- MS/MS Spectral Feature Finding:

  MH+ 600.0-6000.0 Da.

  Scan Time Range: 0–300 min.

  Sequence tag length >0.

  Ignore spectra with dissociation mode: Disable both CID and ETD.

- Merge nearby MS^n scans with same precursor *m/z*:

Table 9
Discovery proteomics acquisition method for the 6500 series QTOF with HPLC-Chip

Parameter	Setting
Acquisition mode	Extended Dynamic Range (2GHz), High Sensitivity, Low Mass Range *m/z* 1700
Mass range	*m/z* 275–1700 (MS) and 50–1700 (MS/MS)
Acquisition rate/time	8 spectra/s
Auto MSMS range	*m/z* 50–1700
MSMS acquisition rate/ time	3 spectra/s (max)
Isolation width	Narrow (~1.3 Hz)
Precursors/cycle	Top 20 precursors per cycle using precursor abundance-based acquisition rate with accumulation time limit enabled; active exclusion after one spectrum for 0.5 min
Collision energy	3.6*(*m/z*)/100–4.8
Threshold for MSMS	1000 counts and 0.001
Dynamic Exclusion	On; 1 repeat then exclude for 0.2 min
Precursor abundance-based scan speed	Yes
Target	25,000
Use MS/MS accumulation time limit	Yes
Purity	100% stringency, 30% cutoff
Isotope model	Peptides
Sort precursors	By abundance only; +2, +3, >+3

Retention time & *m/z* tolerance: ±45 s, ±1.4 *m/z*

General MS/MS Merging Constraints: Select "Spectral Similarity & RT & *m/z*" from the pull down menu.

- Precursor *m/z* and Charge Assignment:

  Precursor Charge: Select "Find".

  Maximum (z): 7.

  Minimum MS1 S/N: 25.

  Find ^{12}C precursor *m/z*: Enabled.

  MS Noise threshold: 400.

5. Navigate to MS/MS Search. Adjust the parameters as outlined below:

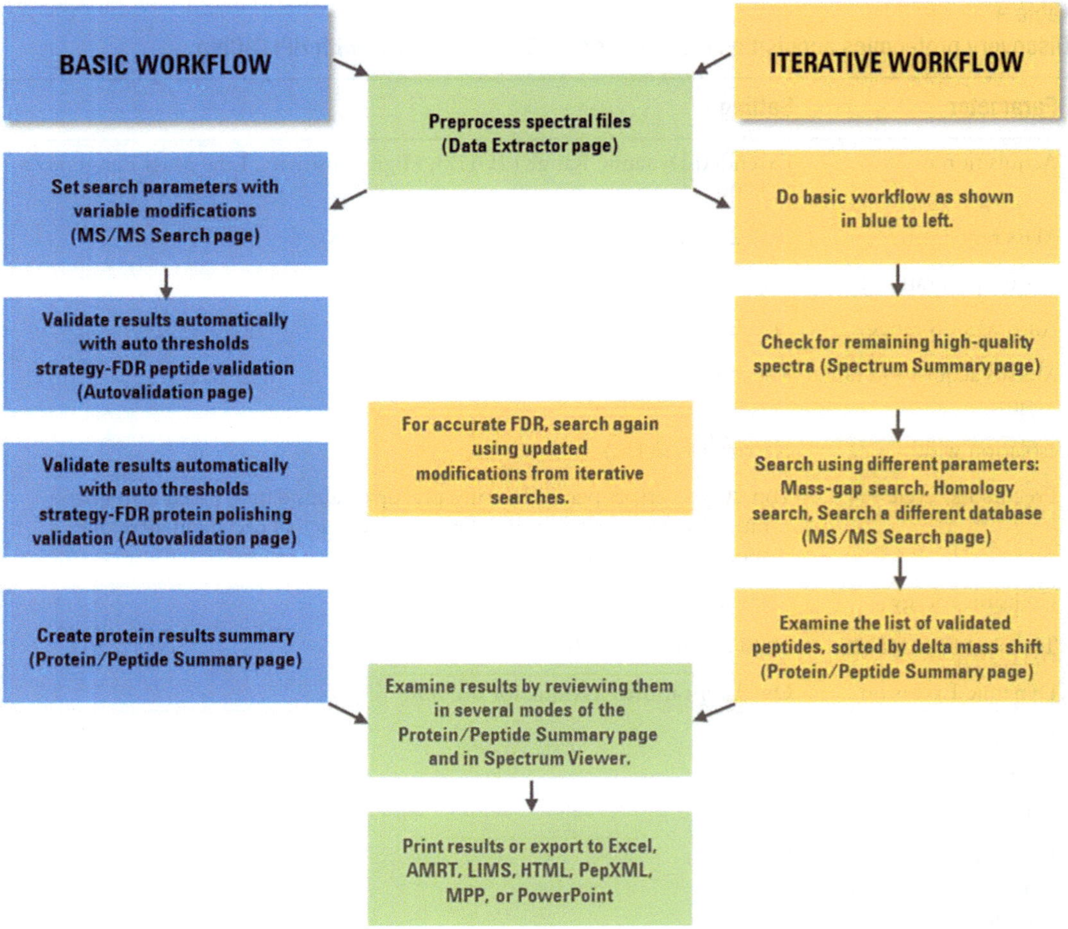

Fig. 3 Spectrum Mill workflow

- Select the data directory where the Data Extraction was performed.
- Search Parameters:

  Validation filter: Select "spectrum-not marked-sequence-not-validated".

  Batch size: 150.

  Search previous hits: Disabled.

  Max reported hits: 5.

  Database: Select the appropriate database for your sample.

  Digest: Select "Trypsin".

  Species: Select "All".

  Maximum # missed cleavages: 2.

  Modifications: Select "Carbamidomethylation (C)".

Search Criteria: Matching Tolerances.

Minimum matched peak intensity: 50 %.

Masses are: Select "Monoisotopic".

Precursor mass tolerance: ±20 ppm.

Product mass tolerance: ±40 ppm.

Maximum ambiguous precursor charge: 3.

Search Criteria: Spectral Quality.

Sequence tag length: Disabled.

Minimum detected peaks: 4.

Search Criteria: Search Mode.

Calculate revered database scores: Enabled.

Dynamic peak thresholding: Enabled.

Discriminant scoring: Select "Disable (same as Score).

Search mode: Select "Variable modifications" and add the appropriate modifications for your sample such as phosphorylated S, T, or Y. Typical sample handling modifications can be included such as oxidized M and deamidated N.

Search Criteria: Data Files.

Fragmentation mode: Select "All".

Spectrum files (./cpick_in/): Enter "*.pkl".

6. It is recommended that you require your peptide spectral matches meet a specified false discovery rate (FDR) by navigating to MS/MS Autovalidation, then selecting the following parameters:
   - Select the data directory where the Data Base Search was performed.
   - Strategy: Auto thresholds.
   - Mode: Peptide.
   - Optimize score and R1-R2 score thresholds with max FDR: 1.2 % across each LC Run.
   - Precursor charge range: 2–4 (see Note 7).
   - Min sequence length: 6.
   - Required AAs: any.
   - Disallowed AAs: none.
   - Filtering: None (ppm) and None (SC/pI).
7. To view your search results and prepare for cross-sample comparison, navigate to Protein/Peptide Summary.
8. Adjust the parameters as outlined below:

- Select the data directory or directories where the Data Base Search was performed.
- Summarize Results for Review:

  Select MPP Generic Export from pulldown menu.

  Mode: Select "Protein-Protein Comparison".

  Validation and Sorting:

  Filter results by: Select "valid".

  Validation preset: Select "none".

  Protein grouping method: Select "1 shared, expand subgroups".

  Sort proteins by: Select "Score".

  Filter by protein score: Select "0".

  Sort proteins by: Select "Score".

  Filter peptides by:

  Score: Select ">0".

  % SPI: Select ">0".

  Required AAs: Select "any".

  Disallowed AAs: Select "none".
- Review Fields:

  Enable the following fields for review:

  Filename.

  Score.

  Subgroup specific (*see* **Note 8**).

  Accession #.

  Protein name.

  Intensity: Total.

  Protein Quantitation Options: Enable Exclude isotope quality Precursor XICs (<0.85 Chi2 vs. Averagine).
- Select Summarize. The exported file can be found under the data subdirectory (or the first of the subdirectories listed if more than one directory was used). Copy this file to a known location for subsequent analysis.

3.4 Discovery and Targeted Proteomics

Data Analysis Using Mass Profiler Professional (MPP) to perform statistical and correlation analysis between samples groups. Using the optional Pathway Architect module allows the results to be mapped to publically available biological pathways databases along with metabolomics and gene data, if available.

1. Open MPP and create a new project.
2. Create a New Experiment within MPP.

3. Use the following parameters for discovery data:
   - Analysis type: Mass Profiler Professional.
   - Experiment type: Identified.
   - Workflow type: Data Import Wizard.
   - Click OK, then select data source: Generic and specify the organism used.
   - Click Next, then select the generic file exported from Spectrum Mill.
   - Click Next to confirm that the appropriate samples are present, then click Next again.
   - For Experimental Groupings, you need to select "Add Parameter", then specify the Parameter Name. For each sample, specify the group. To perform statistics, a minimum of $n = 3$ is required. Click OK after all information is added. Another parameter may be entered at this time if you have multiple parameters you wish to evaluate.
   - Click Next to align which is done by the protein name across all samples and groups.
   - Click Next to specify Normalization Criteria if desired. Typically this is not used in proteomics.
   - Click Next to select Baselining Options. Z-transform can be used when samples from different sources are used. Baseline to median of all samples works well when the samples are very similar, which is often the case with proteomics samples.
   - Under the Workflow: Experiment Setup, select Create Interpretation. Select the experiment parameter of interest and click Next, then Next to create the Interpretations (Averaged and non-averaged) for that parameter.
   - Under Workflow: Navigate to the Quality Control menu, Select Filter by Abundance. Select All Entities for the Entity List and your non-averaged interpretation. Select Next to input parameters. Change the lower cutoff to 2 as MPP changes all zero values (where a protein is not found) into a 1 when importing the data. Specify the required reproducibility. Depending on the number of samples, you can adjust the % of samples in any group that must have values greater than 1. Select Next to view the results, then Next to save the new filtered entity list.
   - Correlation analysis or statistical analysis can now be performed on this filtered entity list. Data can be viewed using a variety of tools including hierarchical clustering, principal component analysis etc.

- After analysis is completed, the processed entity list can be mapped to available pathways using the optional Pathway Architect module (see Note 9) Under Workflow: Navigate to Pathway Analysis, select Single Experiment Analysis. Select the pathway source of interest and click Next. After selecting the interpretation and entity list, click Next. After reviewing the results, select Next to save the pathway results. Selecting any pathway of interest from the list will display the pathway. The relative abundances of each experimental group will be displayed as a heatstrip next to the protein entry on the pathway.
- From the pathway view, all proteins, metabolites and/or genes in the selected pathway can be exported. To export the SwissProt list, right click in the pathway view and choose Select All Entities. Then right click again and select Copy to Clipboard as SwissProt/UniProt IDs. This can be pasted into Spectrum Mill or Skyline software for pathway directed targeted analysis (*see* **Note 10**).

3.5 Targeted Proteomics: HSA in BSA Checkout on the 6400 Series QQQ using Standard Flow with the AJS

1. Create a method named "HSA_BSA_Checkout_AJS.m" with the following:

Autosampler:

Injection volume:	1 μL
Needle wash:	Enabled for 10 s in flushport
Needle flush solvent:	50 % methanol / 50 % H_2O with 0.1 % formic acid
Bottom sensing:	On
Vial offset:	0
ALS Therm:	On, 4 °C

Binary pump:

Solvents:	A1: water with 0.1 % formic acid
	B1: 90 % acetonitrile / 10 % water with 0.1 % formic acid
Flow:	0.6 L/min
Max pressure:	600 bar
Stop time:	3.5 min
Post time:	0 min

Gradient:	Time	%B
	0	5
	1	25
	1.5	70
	1.6	5

Column temperature: 35 °C

QQQ MRM Method:

Ionization mode:	Positive Agilent Jet Stream, positive
Time filter:	ON
Time Filter Width:	0.03
Dry gas flow:	15 L/min.
Dry gas temp:	150 ⁰ C
Sheath gas flow:	11 L/min
Sheath gas temp:	200 °C
Nebulizer:	30 psig
Nozzle voltage:	0
Capillary Voltage:	3500 V is typical
Time segments:	1
Delta EMV:	0–200 V
MS1 Res/MS2 Res:	Wide/unit
Dwell:	60 ms
Fragmentor:	380 V
Cell Accelerator Voltage:	4 V
Polarity:	Positive

MRM scan segments: (Precursor Ion, Product Ion, Collision Energy)

m/z 575.3111, *m/z* 937.4625, 16 V.

m/z 575.3111, *m/zm/z* 823.4196, 20 V.

m/z 575.3111, *m/z* 694.3770, 20 V.

2. Create a worklist with a BSA Blank as the first entry followed by Sample H to A, where Sample A is the last sample injected. Use the above method for all entries with a 1 μL injection volume. For the sample type, use "Blank" for the BSA blank and "Calibration" for the HSA spiked into BSA. For the Level for

Table 10
HSA in BSA dilution table for 6400 Series QQQ checkout

Sample	Final concentration	Volume of HSA standard	Volume of BSA dilution solution (μL)
A	100 fmol/μL	10 μL of HSA Stock	90
B	10 fmol/μL	10 μL of A	90
C	1 fmol/μL	10 μL of B	90
D	100 amol/μL	10 μL of C	90
E	10 amol/μL	10 μL of D	90
F	5 amol/μL	20 μL of E	20
G	2 amol/μL	10 μL of E	40
H	1 amol/μL	10 μL of E	90

each sample enter 1 for Sample H and go up in increments of 1 till you get to Sample A with at Level 8 (Table 10).

3.6 Targeted Proteomics: HSA in BSA Checkout on the 6400 Series QQQ using Standard Flow with the HPLC-Chip

1. Create a method named "HSA_BSA_Checkout_Chip.m" with the following parameters:

Nanoflow pump:

Solvents:	A1: water with 0.1 % formic acid	
	B1: 90 % acetonitrile / 10 % water with 0.1 % formic acid	
Flow:	600 nL/min (primary flow 500 μL/min)	
Max pressure:	200 bar	
Stop time:	9 min	
Post time:	0 min	
Gradient:	Time	%B
	0	3.0
	7	70.0
	7.1	3.0

Capillary pump:

Solvents:	A1: water with 0.1 % formic acid
	B1: 90 % acetonitrile / 10 % water with 0.1 % formic acid
Flow:	3.00 μL/min (primary flow 200 μL/min)
Max pressure:	200 bar

Micro wellplate sampler:

Injection volume:	1 µL
Needle wash:	Enabled for 10 s in flushport
Needle flush solvent:	20% methanol / 80% water with 0.1% formic acid
Bottom sensing:	On
Vial offset:	0
ALS Therm:	On, 4 °C

HPLC-Chip MS Interface:

Injection flush volume: 4 µL.

Pumps configured so Intelligent Sample Loading is activated.

Chip Cube Timetable: Valve to enrichment at 7.5 min.

QQQ MRM:

Ionization mode:	HPLC-Chip, positive
Time filter:	On
Time Filter Width:	0.07
Dry gas flow:	11 L/min.
Dry gas temp:	150 °C
Capillary Voltage:	1750–1900 V is typical
Time segments:	1
Delta EMV:	200 V
MS1 Res/MS2 Res:	Wide/unit
Dwell:	60 ms
Fragmentor:	380 V
Cell accelerator voltage:	4 V
Polarity:	Positive

MRM scan segments: (Precursor Ion, Product Ion, Collision Energy).

m/z 575.3111, *m/z* 937.4625, 16 V.

m/z 575.3111, *m/z* 823.4196, 20 V.

m/z 575.3111, *m/z* 694.3770, 20 V.

2. Create a worklist with a BSA Blank as the first entry followed by Sample H to A, where Sample A is the last sample injected. Use the above method for all entries with a 1 µL injection vol-

ume. For the sample type, use "Blank" for the BSA blank and "Calibration" for the HSA spiked into BSA. For the Level for each sample enter 1 for Sample H and go up in increments of 1 till you get to Sample A with at Level 8 (Table 10).

3.7 Targeted Proteomics: Commercial Kits for Quality Control Using Standard Flow on the 6400 Series QQQ [6, 7, 8]

1. For initial performance checkout of the QQQ for daily or monthly assessment of the LC/MS platform, use the PeptiQuant™ LC/MS Platform Performance kit. Each Platform Performance kit includes seven tryptic-digested plasma samples spiked with stable-isotope labeled standards (SIS). These samples are rehydrated and ready for use with the LC/MRM/MS. Protocols are included within each kit.
2. For quality control of the bottom-up workflow (from denaturation through to detection), use the PeptiQuant™ MRM-MS Workflow Performance Kit. Each Workflow Performance kit contains: human plasma, trypsin, and a SIS peptide mixture. The protocol is included with the kit providing the detailed procedure for the sample preparation. The protocol describes the sequential reduction, alkylation, and quenching of the 10× diluted plasma prior to digestion overnight with trypsin. Digestion is stopped by the addition of the chilled SIS peptide mixture from 250 to 0.025 fmol/μL for standard samples G to A (A being the lowest concentration) and chilled formic acid solution. The samples are allocated into separate Eppendorf tubes, centrifuged, and the peptide supernatant is removed for desalting. The desalted supernatant is lyophilized and rehydrated in 0.1 % formic acid for a final concentration of 1 μg/μL for LC-MRM/MS analysis.

4 Notes

1. Before you break open each ampoule, invert the ampoule several times to mix. Inspect the ampoule's contents to ensure that all the solution is contained in the lower cylindrical base. Shake the ampoule, if needed, to dislodge any air pocket that may prevent solution from settling in the lower portion of the ampoule.
2. Cap and invert the bottle several times to mix the reference solution. Pour 100 mL into a CDS bottle and place onto Bottle A in the CDS. Alternatively an isocratic pump can be used with the 1 L stock solution with a 1:100 splitter (Agilent p/n G1607-60000) connected to the reference nebulizer.
3. System cleanliness is the biggest challenge as this peptide will exhibit some carryover.

4. Best practice is to use fresh mobile phases for studies. Purge both channels when mobile phases are changed.
5. The solution may appear opaque prior to digestion. Mix well to ensure homogeneity before proceeding.
6. Organizing data in subfolders can be important for subsequent database searching.
7. You should have at least 200 peptide spectra matches for all charge states in the specified range for the FDR calculation.
8. This summarizes the protein abundance only from peptides unique to that protein subgroup thus ensuring that peptides which occur in more than one protein are not used.
9. Pathways must be downloaded from publically available sources using the Annotations: Import Pathways menu prior to doing pathway analysis.
10. Alternatively, protein database search results from Spectrum can also be exported as a peptide spectral library for import into Skyline. The user can then use Skyline to select target proteins with corresponding transitions for direct import into the 6400 series QQQ for validation and optimization on the QQQ. The user should review the manuals and training videos provided by Skyline for familiarization.

References

1. Wong CC, Cociorva D, Miller CA et al (2013) Proteomics of *Pyrococcus Furiosus* (Pfu): identification of extracted proteins by three independent methods. J Proteome Res 12(2):763–770
2. Vaudel M, Burkhart JM, Breiter D et al (2012) A complex standard for protein identification, designed by evolution. J Proteome Res 11(10):5065–5071
3. Yang Y, Bhat V, Miller CA. (2015) Jet Stream proteomics for sensitive and robust standard flow LC/MS. Agilent Technical Overview 5991-5687EN http://www.agilent.com/cs/library/technicaloverviews/public/5991-5687EN.pdf Accessed 31 August 2015
4. Miller CA, Jenkins S, Sana TR, et al. (2013) Proteomics in multi-omics workflows using yeast as a model system. Agilent Application Note 5991-2484EN https://www.agilent.com/cs/library/applications/5991-2484EN.pdf 31 August 2015
5. Buckenmaier S, Mora J, van de GoorT, et al. (2012) Enhanced chromatography with the Agilent Polaris-HR-Chip-3C18 improved LC/MS/MS proteomics results. Agilent Technical Overview 5991-0735EN http://www.agilent.com/cs/library/technicaloverviews/Public/5991-0735EN.pdf Accessed 31 August 2015
6. Percy AJ, Chambers AG, Borchers CH, (2014) Application kits for standardizing MRM-based quantitative plasma proteomic workflows on the Agilent 6490 LC/MS system. Agilent Application Note 5991-3601EN https://www.agilent.com/cs/library/applications/5991-3601EN.pdf Accessed 31 August 2015
7. Percy AJ, Chambers AG, Yang J et al (2012) Comparison of standard-and nano-flow liquid chromatography platforms for MRM-based quantitation of putative plasma biomarker proteins. Anal Bioanal Chem 404(4):1089–1101
8. Percy AJ, Mohammed Y, Yang J, Borchers CH (2015) A standardized kit for automated quantitative assessment of candidate protein biomarkers in human plasma. Bioanalysis, 7(23):2991–3004

Chapter 16

Generation of High-Quality SWATH® Acquisition Data for Label-free Quantitative Proteomics Studies Using TripleTOF® Mass Spectrometers

Birgit Schilling, Bradford W. Gibson, and Christie L. Hunter

Abstract

Data-independent acquisition is a powerful mass spectrometry technique that enables comprehensive MS and MS/MS analysis of all detectable species, providing an information rich data file that can be mined deeply. Here, we describe how to acquire high-quality SWATH® Acquisition data to be used for large quantitative proteomic studies. We specifically focus on using variable sized Q1 windows for acquisition of MS/MS data for generating higher specificity quantitative data.

Key words Mass spectrometry, SWATH acquisition, Data-independent acquisitions, Variable windows, Quantitation, Proteomics

1 Introduction

The goal of quantitative proteomics is to both identify and quantify a broad range of peptides and proteins. The extreme complexity and dynamic range of proteins in cells, tissues, and fluids challenges traditional data-dependent workflows to reproducibly and deeply interrogate complex samples. There has been resurgence in the use of data-independent acquisition (DIA) mass spectrometry strategies to increase the reproducibility and comprehensiveness of data collection, enabled by recent technological innovations in both MS hardware and software. This acquisition strategy can now be routinely applied to proteomic samples to collect high quality quantitative data for high numbers of peptides/proteins. These data-independent acquisitions are widely performed on orthogonal time-of-flight (QqTOF or TripleTOF®) platforms, and the workflow is referred to as SWATH Acquisition [1–4]. However, DIA acquisitions can also be executed on other platforms, such as on Q-Exactive mass spectrometers [5]. In SWATH Acquisition on TripleTOF platforms, Q1 isolation

Lucio Comai et al. (eds.), *Proteomics: Methods and Protocols*, Methods in Molecular Biology, vol. 1550,
DOI 10.1007/978-1-4939-6747-6_16, © Springer Science+Business Media LLC 2017

windows are stepped across the mass range in an LC timescale, transmitting populations of peptides for fragmentation, and high-resolution composite MS/MS spectra are acquired at each step. Gillet et al. [2] described their initial SWATH experiments setting the Q1 quadrupole to transmit Δ25 *m/z* windows, with a 1 *m/z* overlap between windows. The MS/MS spectra generated during a SWATH acquisition are of much greater complexity than the MS/MS spectra in typical data-dependent acquisition (DDA) experiments when using unit resolution (1 *m/z*) Q1 windows. In addition, an MS1 scan is typically collected in each cycle, and this cycle of MS1 and SWATH MS/MS scans is repeated throughout the entire LC-MS acquisition. Fragment ion information from the obtained SWATH MS/MS spectra can be used to uniquely confirm detection of specific peptides, typically by comparisons to reference spectra or spectral libraries as described by Gillet et al. [2] and Röst et al. (Chapter 20).

The complexity of the chimeric SWATH MS/MS spectra and the resulting specificity depends on the number of peptides eluting off the column at the same retention time within the same *m/z* isolation window. Initial approaches used fixed sized Q1 windows of 20–25 *m/z*; however, more recent work has shown that decreasing the Q1 window size, as well as varying the window width as a function of precursor ion density [6] can provide higher specificity, less interferences and improvements in the amount of quantitative data extracted to ultimately increase the depth of coverage.

Data-independent SWATH acquisitions are compatible with several downstream data processing algorithms, and recently a variety of new software tools have emerged. One widely used approach is to match SWATH data against pre-assembled MS/MS spectral libraries, using tools such as OpenSWATH [7], Skyline [5, 8], SWATH 2.0 [9], and Spectronaut [10]. These tools analyze the untargeted DIA acquisition data with targeted data processing taking advantage of existing spectral libraries. Another strategy, MSPLIT-DIA (or mixture-spectrum partitioning using libraries of identified tandem mass spectra) is a spectral matching tool for untargeted peptide identification in DIA data [11]. Lastly, two recently published algorithms DIA-Umpire [12], and Group-DIA [13] apply an untargeted data analysis approach in which SWATH or DIA data is first processed by detecting precursor and fragment ion features and subsequently assembling them into pseudo-tandem MS/MS spectra, which then can be searched by any typical database search engine.

2 Materials

2.1 Samples for Analysis

1. Retention Time Calibration Peptides: The iRT peptides (kit P/N Ki-3003, Biognosys, Switzerland) are spiked into complex samples, typically at dilutions of 1/10 (*see* **Note 1**).
2. LC-MS quality control (QC) sample: Beta-Galactosidase (BGal) Stock Solution. Add 625 μL of buffer (10% acetonitrile/0.1% formic acid) to the BGal vial (kit P/N 4465867, SCIEX) to create a 1 pmol/μL stock solution. Aliquot the stock solution into 10–50 μL volumes and freeze at −20 °C for future use.
3. SWATH system suitability sample: pre-digested cell lysate similar to the study sample described under **step 4** to assess SWATH performance.
4. Pre-digested Cell Lysate samples: a typical study sample used here consists of a tryptic digestion of a yeast cell lysate. Proteomic samples were previously reduced, alkylated, digested, and desalted, providing a solution of ~1 μg/μL digested protein lysate for analysis.

2.2 Chromatography

1. LC System—nanoLC™ 425 System (SCIEX) combined with a cHiPLC® System (*see* **Note 2**).
2. Two analytical cHiPLC columns set up in serial mode—each 75 μm × 15 cm ChromXP™ C18-CL chip, 3 μm, 300 Å (P/N 804-00001, SCIEX).
3. LC Buffer A: 98% water with 0.1% formic acid and 2% acetonitrile;
4. LC Buffer B: 98% acetonitrile with 0.1% formic acid and 2 % water.

2.3 Mass Spectrometry

1. MS analysis is performed using the NanoSpray® Source on the TripleTOF® 6600 System (*see* **Note 3**) operating with Analyst® Software TF 1.7 (SCIEX).

2.4 Data Processing

1. SWATH® Acquisition data is processed with the SWATH Acquisition MicroApp 2.0 in PeakView® Software [9], using a spectral ion library [4] generated from prior data-dependent acquisitions (*see* **Note 4**).

3 Methods

3.1 Set Up LC-MS System

1. Using the nanoLC 425 LC system with a cHiPLC System (SCIEX), set up in standard 'trap elute' mode, to desalt first then separate the peptides from the complex tryptic digestions. Set up the autosampler (AS-3) valve with a 10 μL sample loop, and connect to the loading pump, and the cHiPLC system.

Connect the analytical gradient to the cHiPLC system, which is set up with two analytical C18 chip columns in serial mode (see below and *see* **Note 5** and **6**).

2. Place a trap elute jumper chip (PN 800-00389, SCIEX) in slot 1 in the cHiPLC system. Place two analytical chips in slots 2 and 3 (SCIEX), (*see* **Note** 7). Set the temperature on the cHiPLC system to 35 °C for both to maintain a constant temperature (*see* **Note 8**).
3. Connect the eluant from the column to the nanoflow spray tip (New Objectives FS 360-20-10-N-20) mounted on the NanoSpray Ion Source (SCIEX). The source is connected to a TripleTOF 6600 system operated using Analyst Software TF 1.7 (SCIEX).

3.2 Create a Basic LC-MS Acquisition Method

1. Click *Build Acquisition Method* to start a new method. Set experiment 1 to TOF MS and set the *m/z* range to 400 – 1500 and the *Accumulation Time* to 250 msec. Next, open the *Edit Parameters* and set the source conditions that were optimized for the system, these should typically be in the following ranges. On the *Compound* tab, ensure the *Declustering Potential* (*DP*) is set to between 80 and 100 V. On the *Source/Gas* tab, typical conditions will be 2300–2600 V for the *IonSpray Voltage* (*IVSF*), 3-6 for the *Ion Source Gas 1* (*GS1*), 0 for the *Ion Source Gas 2* (*GS2*), 20–25 for the *Curtain Gas* (*CUR*), and 100–150 for the *Interface Heater Temperature* (*IHT*).
2. Create a trap loading method for desalting and add this to the LC-MS acquisition method; for example using a loading pump flow rate of 0.5 μL/min of 100% Buffer A for 30 min (*see* **Note 6**).
3. Create a μL-pickup method for the autosampler, injecting 2 μL of sample using a 10 μL loop and add this to the LC-MS acquisition method.
4. Create the analytical gradient method and add this to the LC-MS method. An example gradient for a complex sample is to use a flow rate of 300 nL/min and the following gradient: 5–30% (v/v) solvent B with solvent A making up the difference (from 0 to 120 min), 30–80% solvent B (from 120 to 130 min), and at 80% solvent B (from 130 to 136 min) then back to 5% solvent B from 136 to 138 min for mobile phase equilibration, with a total run-time of 155 min (*see* **Note 9**).
5. Set the MS acquisition *Duration* to 150 min. Save the LC-MS acquisition method.

3.3 Convert to a SWATH Acquisition Method

1. Click on the *Create SWATH Exp* button in the top right hand side of the LC-MS acquisition method window. The *Create SWATH Experiment* UI appears (Fig. 1). Click on the 'Manual' Tab of the UI on the right to set up a custom SWATH experiment.

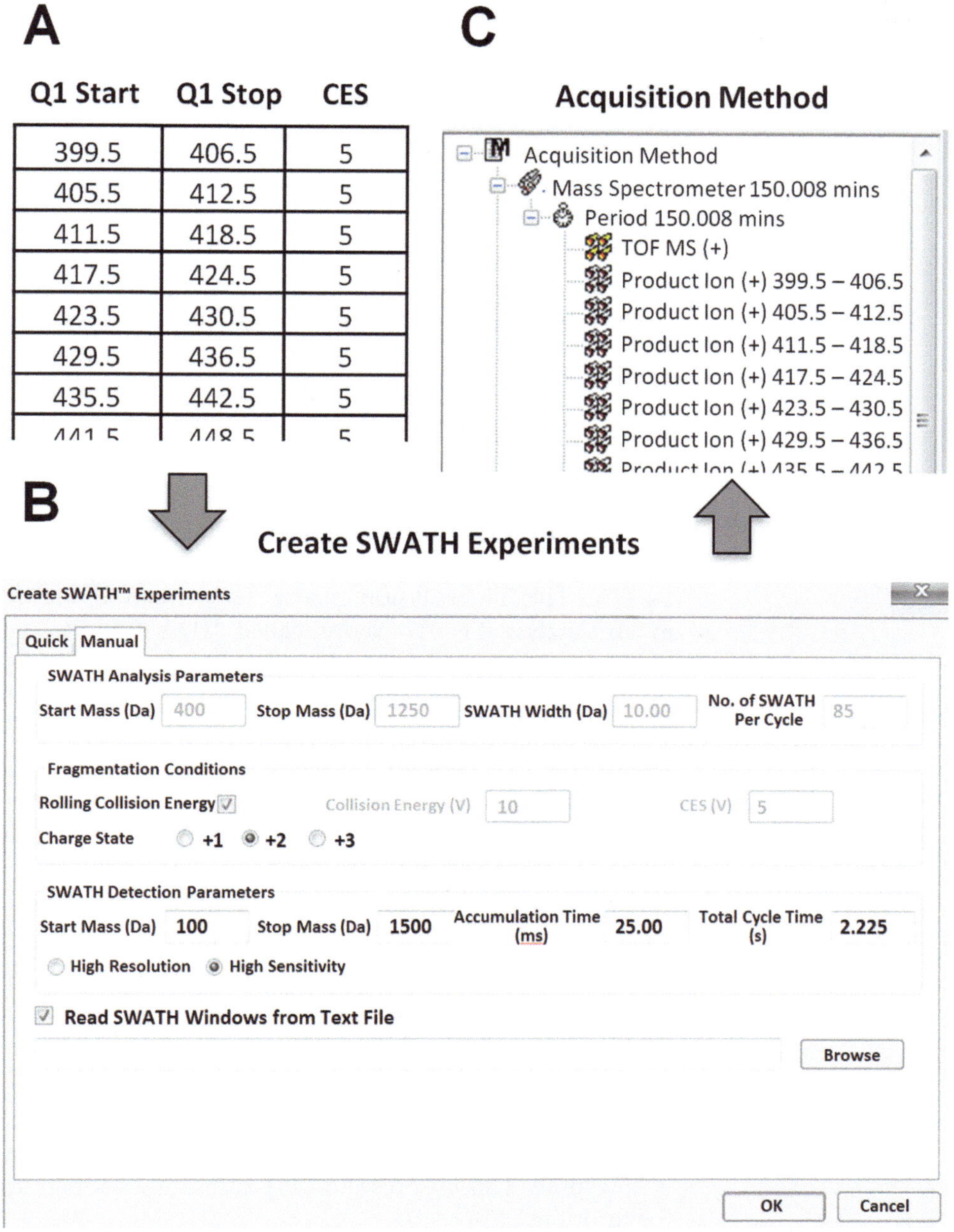

Fig. 1 Build a Variable Window SWATH Method. Any window strategy can be constructed in text file format and loaded into the SWATH Acquisition method editor for automatic method building. (**a**) A text file describing the Q1 isolation window strategy can be constructed. (**b**) Load into SWATH Acquisition method editor and adjust the MS and MS/MS settings as shown. (**c**) Click *OK* to automatically build a SWATH Acquisition method

Prepare the SWATH Variable Window text file that defines the Q1 isolation window strategy as shown in Fig. 1a (*see* **Note 10**). Generate the Variable Window Table manually in Excel or compute from previously acquired MS data on the specific sample of interest (*see* **Note 11**). The Variable Window excel file should be set up to define the varying windows width across the entire *m/z* range with 3 Excel columns 'Q1 Start *m/z*', 'Q1 Stop *m/z*', and 'collision energy spread, CES' (see Fig. 1a). Save the document as a *.txt file. The variable window text file is imported into the LC-MS acquisition method (described below in **step 2** of Subheading 3.3), to build the Variable Window SWATH Acquisition Method in the Analyst software.

2. From the *Manual* tab of the Create SWATH Experiment UI (*see* Fig. 1b) check the *Read SWATH Windows from Text File* option at the very bottom of the UI and load the SWATH Variable Window text file using the *Browse* button. Note that the *SWATH Analysis Parameters* Section (top of window) is grayed out and not used once the *Read SWATH Windows from Text File* option is selected.
3. Next, under the *Fragmentation Conditions* Section click on the *Rolling Collision Energy* option and set the charge state to 2+ (Fig. 1b). The collision energy (CE) equations are defined in an Analyst TF 1.7 Script called 'IDA CE Parameters'. Software defaults can be used or SWATH optimized CE equations can be downloaded (http://sciex.com/community/entity/11856).
4. Finally, set the *SWATH Detection Parameters* section to the following values, MS/MS scans acquired from *m/z* 100–1500 with 25 msec accumulation times in high MS/MS sensitivity mode, as shown in Fig. 1b (*see* **Note 10**). Note the overall cycle time will not be computed until after you click *OK*.
5. Click *OK* at the bottom to build the LC-MS acquisition method (Fig. 1c). The method shows the TOF MS scan (experiment 1), followed by all 100 Product Ion scans (variable Q1 window MS/MS scans—experiments 2–101).
6. After method building, click on the TOF MS experiment (Exp 1) defining the TOF MS1 scan, and confirm that the mass range is set to *m/z* 400–1500 and that the accumulation time is 250 msec. This ensures the MS1 scan is also of high quality for future use [14].
7. The total cycle time in this case is 2.8 s which allows enough measured points across an eluting peak with the given chromatographic setup (*see* **Note 12**). This completes the building of the Variable Window SWATH Acquisition method. For information on creating a fixed window acquisition method (*see* **Note 13**).

3.4 Creating a Practical SWATH Study Acquisition Batch

1. In order to obtain consistent and reliable data from study samples, regular LC-MS tests need to be performed before and during the entire SWATH study. Initially, use the pre-digested BGal standard (*see* **Note 5**) and perform QC acquisitions, or

mass calibration acquisitions, that are typically used in your laboratory.

2. In addition to the simple LC-MS system QC checks discussed above, also use complex standard samples similar to the study samples and test the generated SWATH acquisition method and performance of the system in SWATH mode. Perform before starting the study and also intersperse such SWATH specific tests throughout the entire SWATH study. In addition, retention time calibration (iRT) peptides spiked into the SWATH test samples and the study samples by themselves can be used to monitor retention time stability (*see* **Note 14**).
3. Randomize study samples to avoid systematic errors in the study, block-randomization of biologically different samples often is applied in proteomics studies.
4. Acquire data for the study samples with the acquisition methods generated above, injecting 2 μL of samples at concentrations of ~0.5–1.5 μg/μL yielding an amount of ~1–3 μg of sample on column.

3.5 Data Processing

Once the SWATH data is acquired, several data processing pipelines can be used to process the data (*see* **Note 4**). As an example, a SWATH data set was acquired using the above methods and a number of different Q1 window strategies, then processed using a spectral library approach as previously described for the SWATH 2.0 algorithm by Lambert et al. [9], also referred to as targeted extraction. Figure 2 highlights that higher numbers of peptides are robustly identified and quantified when more, smaller (variable) Q1 windows are used for SWATH acquisitions [6].

4 Notes

1. iRT peptides can also be spiked at 1:20 dilution instead, depending on study sample and preference.
2. Different nano-flow LC systems can be used, such as NanoLC Ultra® 2Dplus (or 1D) HPLC system (Eksigent), or nano-flow LC systems from other vendors.
3. Method setup and acquisitions can similarly be performed on a TripleTOF 5600 or 5600+ system. All steps described will equally apply, as all of the TripleTOF systems use the same acquisition software, Analyst TF Software 1.7.
4. Other SWATH data processing software can be used as discussed in the introduction, i.e., OpenSWATH, Skyline, Spectronaut, DIA Umpire, and others. Please refer to the Chapter by Röst et al. in this book for a detailed description of data processing.

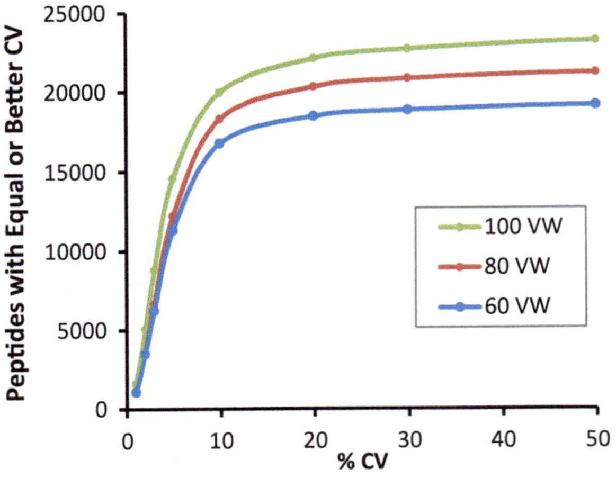

Fig. 2 Increasing Window Numbers/Decreasing Window Size Provides More Robust Peptide Detection. Using a yeast cell lysate (3 μg on column), SWATH data was collected (five replicates) in serial column mode (2 h gradient) using variable Q1 windows. The number of windows used to cover the mass range was increased from 60 to 100 windows and the reproducibility curves were plotted. The number of confident peptide detections (<1 % FDR) with ≤20 % CV further increased as the window size decreases [6]

5. All instructions for operating the LC System and building pump and autosampler methods can be found in the Operators Guide (http://sciex.com/Documents/user%20guides/ekspert-nanolc-400-systems-operator-guide-eng.pdf). All protocols for performing the BetaGalactosidase tests can be found in the System Integration Test (http://sciex.com/Documents/user%20guides/nanolc-system-integration-test-en.pdf).
6. Here, the cHiPLC system setup consists of two analytical column chips in serial mode; in principle this is a "trap–elute" experiment, which uses the first analytical column chip for trapping and desalting of the sample with the Loading solvent, while then upon valve switching the analytical gradient will flow through both analytical column chips (in serial mode) to separate and elute the sample, providing a 30 cm column length. Due to two analytical column chips in serial mode, loading solvent flow rates should be relatively low (~0.5 μL/min) to keep column pressures reasonable (see **step 2** of Subheading 3.2). Alternatively, operators can use a more traditional trap elute set up using a trap chip (with 200 μm × 6 mm ChromXP™ C18-CL chip, 3 μm, 300 Å, P/N 5015841, SCIEX) instead of the first analytical column chip (*see* **Note 5**).
7. The chromatographic configuration described here is one example of a possible configuration. It is possible to use different "trap and elute" configurations and fused silica packed columns for separations. There are different column length

options that can be used, to provide increased peak capacity and potentially increased depth of coverage. Several chromatographic setups are possible and feasible; however, it is essential to perform high quality chromatography to obtain the best separation and best peak shape for SWATH quantitation.

8. Retention time reproducibility is a key consideration when collecting SWATH Acquisition data. It is recommended that column heaters are used to thermostat the analytical column and ideally both the analytical column and trap for highest reproducibility.
9. Sufficient column re-equilibration is important for high retention time reproducibility. Ensure that the column is fully re-equilibrated in the starting conditions of the gradient before injecting the next sample.
10. There are a number of key parameters to consider when deciding on the best variable window strategy to use. First, it is important to determine the average chromatographic peak width, as this will define the cycle time. A typical cycle time estimate would be to take the chromatographic peak width at half height and divide by 6–8 to get an appropriate cycle time and to provide sufficient chromatographic peak sampling for good quantitation (i.e., for a peak width of 15 s at half height therefore a good cycle time would be between 1.8 and 2.5 s). Once the cycle time is determined, the number of windows can be maximized. Accumulation times as fast as 25–30 ms per MS/MS have been shown to produce good quality quantitative data on the TripleTOF systems, so divide the cycle time by 25 ms to get the number of windows (i.e., a cycle time of 2.5 s would allow for 100 SWATH Q1 windows).
11. To easily set up variable windows, the 'variable window calculator' at http://sciex.com/software-downloads-x2110 (SCIEX) or the swathTUNER tool [15] (http://swathtuner.sourceforge.net) can be used to create the variable window text file for loading into Analyst Software. In general, the instrument operator has great flexibility in designing the variable window text file, and can tailor SWATH window sizes either specifically for the type and complexity of the sample measured in the study (Fig. 3), or just using a once established generally applicable window pattern. Within a quantitative study, use the same variable window pattern for all samples. An example of a 100 Q1 variable window text file can be found here (http://sciex.com/community/entity/1217).
12. The number of variable windows, accumulation times, and total cycle times can be further optimized by the individual operator to allow for the optimal number of points across the eluting sample peak to allow >10 points across each peak for

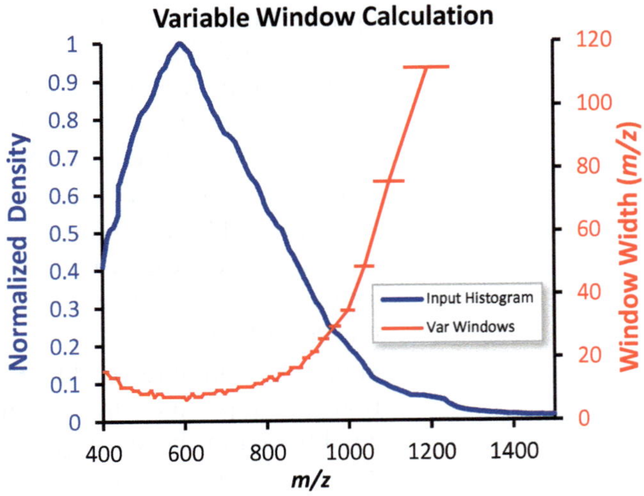

Fig. 3 Variable Q1 Window Widths for SWATH Acquisition. To achieve better specificity in complex matrices, smaller Q1 windows are desirable especially in the *m*/*z* dense regions where many peptide precursors are measured. The *m*/*z* density histograms constructed from the TOF MS data for the proteome of interest (*blue line*) can be used to construct variable sized windows (*red line*), where the density of precursors in each of the isolation windows is equalized across the *m*/*z* range

optimal quantitation. This is very dependent on the chromatography and length of gradient, which can also be optimized for the specific sample type.

13. To build a fixed SWATH acquisition method instead of the variable SWATH window method described above, go to the SWATH Experiment UI and select the 'Manual' tab (see Fig. 1b). First set the *SWATH Analysis Parameters* on the top: set the *Start Mass* to 400, the *Stop Mass* to 1250, and the *SWATH Width* to desired width (typically use 10 *m*/*z*), the # of SWATH MS/MS scans per cycle is then automatically computed to 85 windows. As no text file is imported (as for variable windows), and the box *Read SWATH Windows from text file* is unchecked, the UI can directly calculate SWATH windows under the *SWATH Analysis Parameters*. Set the *Fragmentation Conditions* as defined in **step 3** of Subheading 3.3 and set the *SWATH Detection Parameters* as defined in **step 4** of Subheading 3.3. The cycle time for the MS/MS portion is computed to be 2.2 s. Click *OK*. Confirm the TOF MS settings as defined in **step 6** of Subheading 3.3.
14. Spiked synthetic peptides can also be used to adjust for slight retention time drifts during data processing, and to perform retention time alignments using libraries from other researchers and other LC-MS systems. Endogenous peptides already present in the sample can also be used to perform retention time alignment between library and sample.

Acknowledgments

We acknowledge support from the NIH shared instrumentation grant for the TripleTOF 6600 system at the Buck Institute (1S10 OD016281, B.W.G.).

References

1. Collins BC, Gillet LC, Rosenberger G et al (2013) Quantifying protein interaction dynamics by SWATH mass spectrometry: application to the 14-3-3 system. Nat Methods 10:1246–1253
2. Gillet LC, Navarro P, Tate S et al (2012) Targeted data extraction of the MS/MS spectra generated by data-independent acquisition: a new concept for consistent and accurate proteome analysis. Mol Cell Proteomics 11(O111):016717
3. Liu Y, Buil A, Collins BC et al (2015) Quantitative variability of 342 plasma proteins in a human twin population. Mol Syst Biol 11:786
4. Selevsek N, Chang CY, Gillet LC et al (2015) Reproducible and consistent quantification of the *Saccharomyces cerevisiae* proteome by SWATH-mass spectrometry. Mol Cell Proteomics 14:739–749
5. Egertson JD, MacLean B, Johnson R et al (2015) Multiplexed peptide analysis using data-independent acquisition and Skyline. Nat Protoc 10:887–903
6. Hunter CL, Collins B, Gillet L, et al. (2014) Increasing depth of coverage in data independent acquisition with acquisition improvements and higher sample loads. Proccedings of the 61st Annual ASMS Conference on Mass Spectrometry & Allied Topics, Baltimore, MD, June 15–19, 2014
7. Rost HL, Rosenberger G, Navarro P et al (2014) OpenSWATH enables automated, targeted analysis of data-independent acquisition MS data. Nat Biotechnol 32:219–223
8. MacLean B, Tomazela DM, Shulman N et al (2010) Skyline: an open source document editor for creating and analyzing targeted proteomics experiments. Bioinformatics 26:966–968
9. Lambert JP, Ivosev G, Couzens AL et al (2013) Mapping differential interactomes by affinity purification coupled with data-independent mass spectrometry acquisition. Nat Methods 10:1239–1245
10. Bruderer R, Bernhardt OM, Gandhi T et al (2015) Extending the limits of quantitative proteome profiling with data-independent acquisition and application to acetaminophen-treated three-dimensional liver microtissues. Mol Cell Proteomics 14:1400–1410
11. Wang J, Tucholska M, Knight JD et al (2015) MSPLIT-DIA: sensitive peptide identification for data-independent acquisition. Nat Methods 12(12):1106–1108, Online
12. Tsou CC, Avtonomov D, Larsen B et al (2015) DIA-Umpire: comprehensive computational framework for data-independent acquisition proteomics. Nat Methods 12:258–264, 7 p following 264
13. Li Y, Zhong CQ, Xu X et al (2015) Group-DIA: analyzing multiple data-independent acquisition mass spectrometry data files. Nat Methods 12(12):1105–1106, Online
14. Rardin MJ, Schilling B, Cheng LY et al (2015) MS1 peptide ion intensity chromatograms in MS2 (SWATH) data independent acquisitions. improving post acquisition analysis of proteomic experiments. Mol Cell Proteomics 14:2405–2419
15. Zhang Y, Bilbao A, Bruderer T et al (2015) The use of variable Q1 isolation windows improves selectivity in LC-SWATH-MS acquisition. J Proteome Res 14:4359–4371

Chapter 17

Annotating Mutational Effects on Proteins and Protein Interactions: Designing Novel and Revisiting Existing Protocols

Minghui Li*, Alexander Goncearenco*, and Anna R. Panchenko

Abstract

In this review we describe a protocol to annotate the effects of missense mutations on proteins, their functions, stability, and binding. For this purpose we present a collection of the most comprehensive databases which store different types of sequencing data on missense mutations, we discuss their relationships, possible intersections, and unique features. Next, we suggest an annotation workflow using the state-of-the art methods and highlight their usability, advantages, and limitations for different cases. Finally, we address a particularly difficult problem of deciphering the molecular mechanisms of mutations on proteins and protein complexes to understand the origins and mechanisms of diseases.

Key words Protein–protein interactions, Databases, Mutations

1 Introduction

The era of genome sequencing has unraveled a large number of human genetic variations, as illustrated by the milestone 1000 Genomes project [1]. Mutations and genetic recombinations may occur naturally during the cell division and at the same time may be caused by extrinsic factors. A single nucleotide substitution that results in a codon change encoding a different amino acid is called "missense point mutation" (called "mutation" thereafter). Germline missense mutations are passed to progeny, whereas somatic mutations are not inherited. Due to predominantly neutral effects of many genetic variations, they have accumulated in human population and can be responsible for many individual phenotypic traits in humans and may be used for genetic fingerprinting. Generally, a variant frequently occurring in a population is termed a polymorphisms and single nucleotide polymorphisms (SNP) are one of the most common types of genetic variations.

*Author contributed equally with all other contributors.

Lucio Comai et al. (eds.), *Proteomics: Methods and Protocols*, Methods in Molecular Biology, vol. 1550,
DOI 10.1007/978-1-4939-6747-6_17, © Springer Science+Business Media LLC 2017

Missense mutations can render proteins nonfunctional and may be responsible for many diseases. From the clinical perspective, these non-neutral mutations affecting human health represent the main interest. For some diseases and genes, particularly following the Mendelian inheritance patterns, the causal genotype–phenotype relationship has been already established, while for complex polygenic diseases involving multiple factors it is still unknown. Moreover, genetic variants with low penetrance, weakly associated with disease phenotypes, can only be annotated for large samples and for many diseases their genetic determinants have to be discovered.

2 Materials

2.1 Software

1. Molecular dynamics packages: NAMD (http://www.ks.uiuc.edu/Research/namd/) [2] and CHARMM (http://www.charmm.org/) [3].
2. Structural visualization packages: VMD (http://www.ks.uiuc.edu/Research/vmd/) [4], Chimera (http://www.cgl.ucsf.edu/chimera/) [5] and CN3D (http://www.ncbi.nlm.nih.gov/Structure/CN3D/cn3d.shtml) [6].

2.2 Online Resources

1. FTP site at NCBI (ftp://ftp.ncbi.nih.gov/pub/panch/Mutation_binding/) includes examples of configuration and runfiles for VMD, NAMD, and CHARMM used in protocols of Subheadings 3.5 and 3.6.
2. Databases collecting human genetic variations, mutations, and data on their clinical relevance (Table 1).
3. Web servers for characterization of structural features of mutations (Table 2).
4. Web servers for predicting the phenotypic effects of mutations (Table 3)
5. Web servers and programs for predicting the effects of mutations on protein stability (Table 4).
6. Web servers and programs for predicting the effects of mutations on protein–protein binding affinity (Table 5).

3 Methods

Recent advances in experimental methods reduced the cost of the DNA sequencing and equipped many labs and hospitals with genotyping and sequencing assays, so that these data along with the clinical profiles of patients can be deposited into the central archival facilities. Some of them include the databases of Genotypes and

Table 1
A summary of databases integrating the data on human genetic variations, mutations, and their clinical relevance

Database	Description	URL	Reference
COSMIC	Somatic mutations in cancer	http://cancer.sanger.ac.uk/cosmic	[16]
HGMD	Published gene lesions responsible for human inherited disease	http://www.hgmd.cf.ac.uk/ac/	[52]
TCGA	Cancer Genome Atlas	http://cancergenome.nih.gov/	[17]
OMIM	Human genes, inherited genetic disorders, and germline mutations	http://www.omim.org/	[14]
dbGaP	Clinical studies linking genotypes to disease phenotypes	http://www.ncbi.nlm.nih.gov/gap	[7]
GTR	Genetic Testing Registry	http://www.ncbi.nlm.nih.gov/gtr/	[8]
PheGenI	Phenotype–genotype integrator	http://www.ncbi.nlm.nih.gov/gap/phegeni	[12]
ClinVar	Genomic variations and their relationship to human health	http://www.ncbi.nlm.nih.gov/clinvar/	[15]
dbSNP	Single nucleotide polymorphisms	http://www.ncbi.nlm.nih.gov/SNP/	[9]
dbVar	Large genetic variations	http://www.ncbi.nlm.nih.gov/dbvar	[10]
SwissVar	Disease-related variants in Uniprot, provides structural mapping	http://swissvar.expasy.org	[19]
PharmGKB	Associates genes with drugs. Catalogs genetic variations known to impact drug response	https://www.pharmgkb.org/	[53]
MutDB	Integrates human variations and COSMIC mutations, maps to structural data, KEGG pathways, and includes predictions of effects of variations on phenotype	http://www.mutdb.org/	[18]
CBioPortal	Visualization and analysis of large cancer studies. It is based on TCGA and incorporates the overlapping data from COSMIC	http://cbioportal.org/	[54]

Table 2
Online resources for exploring the structural, cellular, and genomic context of mutated proteins

Database	Description	URL	Reference
HPRD	Integrates information pertaining to domain architecture, posttranslational modifications, interaction networks, and disease association	http://www.hprd.org	[55]
EBI IntAct	Molecular interaction data	http://www.ebi.ac.uk/intact/	[56]
BioSystems	Provides integrated access to genes, proteins, small molecules, and pathways	http://www.ncbi.nlm.nih.gov/biosystems/	[21]
Reactome	Curated and peer-reviewed pathways	http://www.reactome.org	[57]
KEGG	Manually curated pathways	http://www.genome.jp/kegg/pathway.html	[58]
CDD	Annotates functional and binding sites in protein domain families	http://www.ncbi.nlm.nih.gov/Structure/cdd/cdd.shtml	[23]
IBIS	Annotates protein–protein and protein–DNA/RNA/ion/small molecule interactions and binding sites. Identifies conserved binding sites in homologous protein complexes	http://www.ncbi.nlm.nih.gov/Structure/ibis/ibis.cgi	[27]
muPIT	Interactive exploration of mutations and their structural context	http://mupit.icm.jhu.edu	[59]
DMDM	Domain mapping of mutations	http://bioinf.umbc.edu/dmdm/	[60]
PolyDoms	Mapping of mutations to protein domains and prediction of structural and functional impact of mutations	https://polydoms.cchmc.org/polydoms/	[61]

Phenotypes (dbGaP) [7] and Genetic Testing Registry (GTR) [8] developed at NCBI, NIH. The genotypes are mapped to a reference genome assembly and typically include two major categories of variations: (1) single nucleotide polymorphisms (SNP) deposited in the dbSNP database [9] and (2) larger structural genetic variations in genomes, including long indels, inversions, and copy

Table 3
A summary of online resources for predicting the phenotypic effects of mutations

Name	Features and methods	URL	Reference
SIFT	Sequence homology and physical properties of amino acids	http://sift.bii.a-star.edu.sg	[62]
PolyPhen-2	Bayesian models based on substitution scores, phylogenetic conservation, and structural features	http://genetics.bwh.harvard.edu/pph2/	[31]
SNPs3D	Sequence conservation and changes in physical properties of amino acids affecting protein stability	http://snps3d.org	[63]
PMut	Neural network classifier	http://mmb2.pcb.ub.es/PMut/	[64]
SNAP	Neural network classifier based on protein structural properties predicted from protein sequence	https://www.rostlab.org/services/snap/	[65]
Fathmm	Cancer-specific and other disease-specific Hidden Markov Models for predicting the functional consequences of coding and noncoding variants	http://fathmm.biocompute.org.uk	[66]
MutationAssessor	Predicts the effects of mutations on subfamily specific sites	http://mutationassessor.org/	[67]
CHASM	Random Forest classifier for cancer somatic mutations	http://karchinlab.org/apps/appChasm.html/	[68]
MutPred	Uses sequence conservation and structural features, posttranslational modifications	http://mutpred.mutdb.org/	[18]
SNPs&GO	SVM classifier of disease-related variations based on protein functional annotation (GO)	http://snps.biofold.org/snps-and-go/	[69]
PROVEAN	Predicts the effects of amino acid substitutions, insertions, and deletions on protein function, allows scanning multiple mutations	http://provean.jcvi.org/	[30]

(continued)

Table 3 (continued)

Name	Features and methods	URL	Reference
FunSAV	Random Forest-based classifier, uses structural features and network properties of mutated proteins	http://sunflower.kuicr.kyoto-u.ac.jp/sjn/FunSA/	[70]
nsSNPAnalyzer	Random Forest-based classifier, uses structural and evolutionary information	http://snpanalyzer.utmem.edu/	[71]
PANTHER	Uses phylogenetic and evolutionary information	http://www.pantherdb.org/tools/csnpScoreForm.jsp/	[72]
PhD-SNP	SVM classifier, uses sequence profiles	http://gpcr.biocomp.unibo.it/cgi/predictors/PhD-SNP/PhD-SNP.cgi/	[73]
SAAP	Precalculated database of predicted effects of known variants, considers structural properties and clashes resulting from amino acid substitutions	http://www.bioinf.org.uk/saap/	[74]
SusPect	Incorporates sequence conservation and network-based features	www.sbg.bio.ic.ac.uk/suspect/	[75]
KinDriver	Annotates driver mutations in protein kinases with experimental evidence demonstrating their functional role	http://kin-driver.leloir.org.ar/	[76]
ProKinO	Unified resource for mining the cancer kinome	http://vulcan.cs.uga.edu/prokino	[77]

number variants (CNVs), deposited in the dbVar database [10]. The phenotypes in these databases are mainly organized by disease names or clinical conditions, and diseases are classified according to the Disease Ontology (DO) [11] and Medical Subject Headings (MeSH) terms. Additionally, the NCBI phenotype–genotype integrator (PheGenI) [12] merges genetic variations identified by genome-wide association studies (GWAS) with dbVar, OMIM, GTR, and dbSNP databases.

Table 4
A summary of online software resources for predicting the effects of mutations on protein stability. The second column indicates the kind of data a method requires as an input: protein sequence, structure, or any of the two. Here "$\Delta\Delta G$" refers to $\Delta\Delta G_{fold}$.

Resource	Input	Type of output, method, and energy function	URL	Reference
FoldX	Structure	ΔΔG using empirical force fields	http://foldxsuite.crg.eu/	[78]
PoPMuSiC-2.0	Structure	ΔΔG using a combination of statistical potentials and neural networks	http://dezyme.com/	[79]
ERIS	Structure	ΔΔG using physical force fields with atomic modeling	http://dokhlab.unc.edu/tools/eris/	[80]
CUPSAT	Structure	ΔΔG using mean force atom pair and torsion angle potentials	http://cupsat.tu-bs.de/	[81]
Hunter	Structure	ΔΔG using knowledge-based potentials	http://bioinfo41.weizmann.ac.il/hunter/	[82]
MultiMutate	Structure	ΔΔG using four-body scoring functions	http://www.math.wsu.edu/math/faculty/bkrishna/DT/Mutate/	[83]
AUTO-MUTE	Structure	ΔΔG using knowledge-based potentials	http://proteins.gmu.edu/automute	[84]

(continued)

Table 4 (continued)

Resource	Input	Type of output, method, and energy function	URL	Reference
NeEMO	Structure	$\Delta\Delta G$ using residue interaction networks	http://protein.bio.unipd.it/neemo/	[85]
DUET	Structure	$\Delta\Delta G$ using an integrated computational approach of mCSM and SDM	http://bleoberis.bioc.cam.ac.uk/duet/stability	[86]
MAESTRO	Structure	$\Delta\Delta G$ using multi agent stability prediction	http://biwww.che.sbg.ac.at/MAESTRO	[87]
I-Mutant3.0	Structure/Sequence	$\Delta\Delta G$ using SVMs	http://gpcr2.biocomp.unibo.it/cgi/predictors/I-Mutant3.0/I-Mutant3.0.cgi	[88]
MUPro	Structure/Sequence	Predicts qualitative decrease/increase of stability using SVM	http://mupro.proteomics.ics.uci.edu/	[89]
iStable	Structure/Sequence	$\Delta\Delta G$ using SVM	http://predictor.nchu.edu.tw/iStable	[90]
MuStab	Sequence	Predicts qualitative decrease/increase of stability using SVM	http://bioinfo.ggc.org/mustab/	[91]
iPTREE-STAB	Sequence	$\Delta\Delta G$ using decision tree methods	http://bioinformatics.myweb.hinet.net/iptree.htm	[92]

Table 5
A summary of online and software resources for predicting the effects of mutations on protein–protein binding affinity. Here "ΔΔG" refers to $\Delta\Delta G_{bind.}$ All resources need structure as an input

Resource	Type of output, method, and energy function	URL	Reference
MutaBind	ΔΔG using molecular mechanics force fields, statistical potentials, and fast side-chain optimization algorithms. Produces a model of mutant.	https://www.ncbi.nlm.nih.gov/projects/mutabind/	[46]
BeAtMuSiC	ΔΔG using a set of statistical potentials, does not produce a model of mutant	http://babylone.ulb.ac.be/beatmusic	[93]
ELASPIC	ΔΔG for mutations located on interface using a combination of semi-empirical energy terms and molecular features; does not produce a model of mutant	http://www.kimlab.org/software/elaspic	[94]
DrugScore^{PPI}	ΔΔG for alanine-scanning mutations located on interface using knowledge-based scoring functions; does not produce a model of mutant	http://cpclab.uni-duesseldorf.de/dsppi/	[95]
SNP-IN	classifies effects of mutations on protein–protein interactions using supervised and semi-supervised learning; does not produce a model of mutant	http://andromeda.rnet.missouri.edu/snpintool/	[96]
FoldX	ΔΔG using empirical force field, Produces a model of mutant.	http://foldxsuite.crg.eu/	[78]

The clinical implications of genetic variations are recorded in several other databases. The human gene mutation (HGMD) [13] and Online Mendelian inheritance in Man (OMIM) [14] are the main databases that integrate the information on genetic Mendelian disorders with genes and mutations reported to be causative. ClinVar [15] is another archive, which collects the data on genetic variations from dbSNP and dbVar and integrates them with the available clinical evidence of these variants obtained from multiple studies. In ClinVar each variant is assigned a score which shows the consistency of the reported clinical effect in different studies.

ClinVar annotates variants as: benign, likely benign, likely pathogenic, pathogenic, of uncertain significance, or variants with conflicting interpretations. Importantly, it also annotates the variants associated with individual drug effects.

Germline mutations constitute the bulk of mutations in OMIM, ClinVar, and HGMD, while somatic missense mutations are found predominantly in cancer cells and are not inherited. The information on location, tissue type, and frequency of somatic mutations in tumor samples together with the cancer type can be obtained from COSMIC[16] and TCGA databases [17], while other resources, for instance, MutDB [18] and SwissVar [19], integrate the data on germline and somatic mutations for specific genes and diseases.

Figure 1 describes a computational pipeline for exploring mutations and assessing their effects on protein structure, function, and interactions. Different sections of this chapter follow this pipeline and suggest the protocols to solve each specific problem.

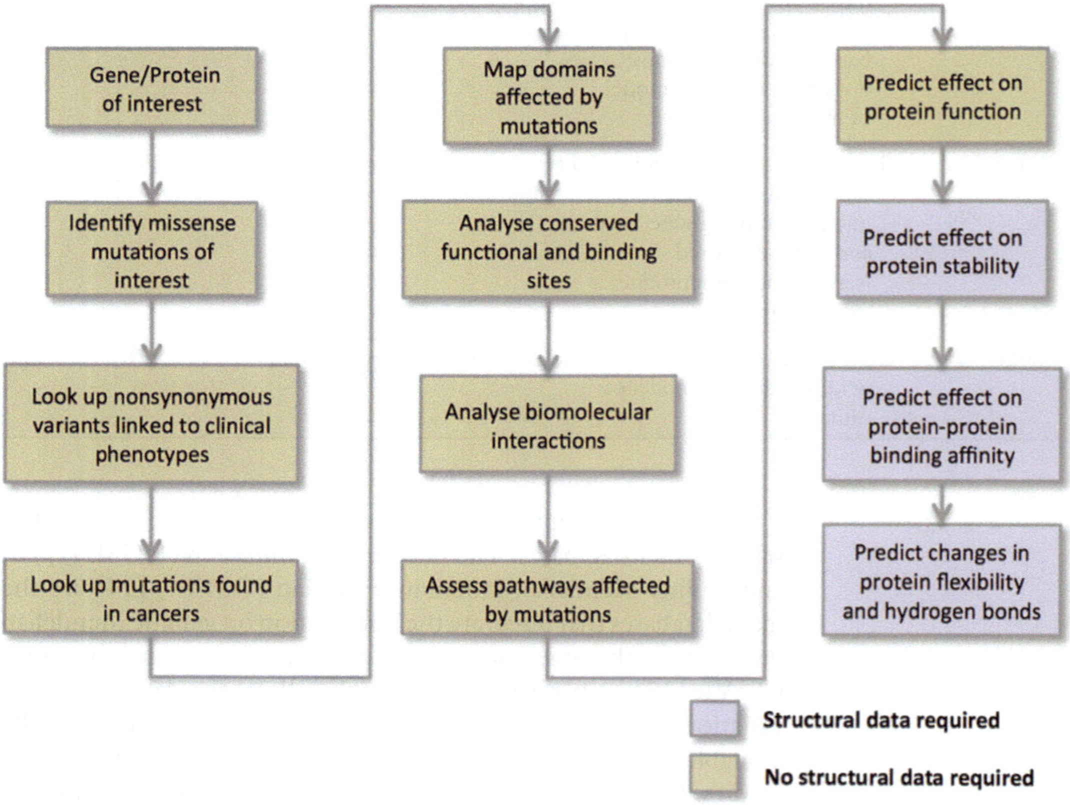

Fig. 1 Mutation assessment workflow

3.1 Collecting and Integrating the Data on Human Polymorphisms and Mutations

First, we will describe a protocol for extracting clinically relevant mutations from ClinVar, COSMIC, and TCGA databases to further analyze them with respect to their clinical and functional impacts. Here we use human monomeric Casitas B-lineage lymphoma c-CBL protein (CBL) as an example [20]. The links and references to web resources used in this protocol are listed in Table 1.

1. NCBI variation viewer (http://www.ncbi.nlm.nih.gov/variation/view/) [21] can be searched by gene name (CBL), Refseq accession (NP_005179), or Uniprot ID (P22681.2). A snapshot of variation viewer shows the genetic variation from ClinVar in the locus of CBL gene (Fig. 2a). Different variants in this viewer are depicted as separate tracks below the CBL gene locus. The ClinVar track displays multiple variants as boxes with the number of variants listed within each box. Variants with benign effect are shown using green color, whereas the purple boxes show pathogenic variants.
2. Additionally, the user can download all mutations from the COSMIC ftp server in VCF format (CosmicCodingMuts.vcf.gz) and display these mutations in a separate track in the variation viewer (choose menu "your data" and select "add track" option). Note that it is necessary to select the same genome assembly in both variation viewer and in COSMIC (e.g., GRCh37).
3. Each variant with a valid ClinVar annotation is linked to a corresponding dbVar or dbSNP record. Here we will focus on a single nucleotide variant of CBL gene with the dbSNP accession rs267606706. As shown in Fig. 2b, it is a missense mutation where nucleotide substitution T->C results in p.Y371H amino acid substitution. There is clinical evidence that this variant can cause a Noonan syndrome-like disorder and/or juvenile myelomonocytic leukemia. The GTR studies cited in ClinVar show [22] that p.Y371H is a heterozygous germline substitution.
4. The dbSNP page for accession rs267606706 contains a link to "3D structure mapping" (found under the "NCBI resources" section) which points to the SNP3D page where several synonymous and missense variants are mapped onto the CBL protein structure. By default, this variant is selected, but it is possible to select other variants for display by clicking on the "Cn3D selected" button. Additionally, the SNP3D page contains a link "CD" that shows conserved domains from the CDD database mapped onto this protein [23]. Figure 2c depicts the structure of CBL using Cn3D with mutated Tyr residue side chain colored in yellow.
5. As was shown previously using ClinVar, the germline p.Y371H mutation may be linked to leukemia, however, many cancer

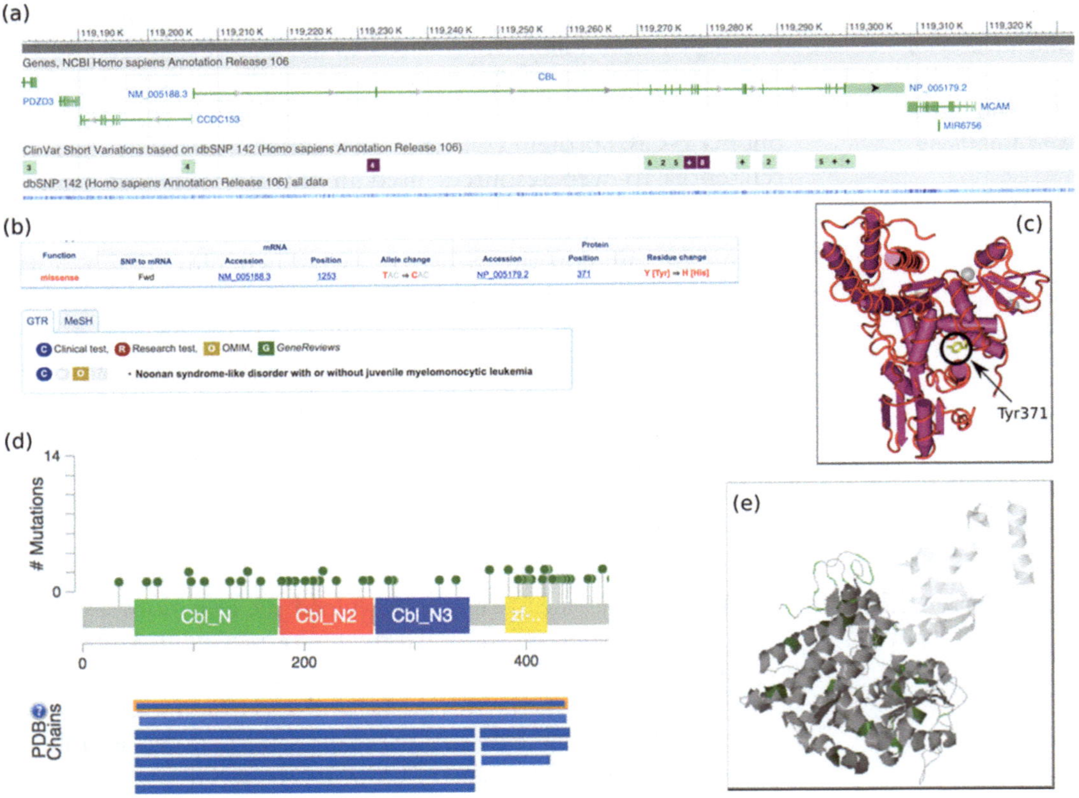

Fig. 2 Identification of clinically relevant mutations in ClinVar, COSMIC, and TCGA. (**a**) NCBI Variation Viewer showing the CBL gene locus on chromosome 11; ClinVar and dbSNP data shown as tracks below the gene, pathogenic mutations are presented as *purple squares* and closely-located mutations are grouped together in ClinVar track; (**b**) one of the pathogenic mutations p.Y371H shown in dbSNP with the corresponding disease annotation in GTR; (**c**) a representative structure of CBL protein visualized in SNP3D with Cn3D software, mutated p.Y371H is shown with the *yellow side chain*; (**d**) cBioPortal view of CBL protein with missense mutations mapped onto the corresponding domains; (**e**) a representative structure visualized with JMol directly in cBioPortal with all mapped missense mutations shown in *green*

mutations are somatic and therefore are not present in ClinVar. In order to explore somatic mutations we switch to cBioPortal, which allows exploring mutations from the TCGA and COSMIC databases. Open cBioPortal web page and submit a query "CBL" as the user-defined gene set. A summary for different types of cancers will be shown for the CBL gene. Open the second tab "Mutations", which display mutations on CBL protein sequence. Figure 2d depicts missense CBL mutations mapped by the cBioPortal onto the corresponding CDD domain context. Additionally, the cBioPortal provides locations of mutations on protein structures. The blue footprints in Fig. 2d show that several structures cover CBL protein sequence and could be explored interactively. All missense mutations from COSMIC and TCGA databases are mapped on the representative structure (in green) and are displayed in the web browser (Fig. 2e).

6. Each mutation in cBioPortal "Mutations" tab is shown as a pin indicating its position in CBL protein sequence. The height of the pin corresponds to the number of known mutations. Place the mouse cursor over the Znf domain (zinc finger, shown in yellow) and over the first pin from the left in Znf domain. As a result, a window with a list of mutations and cancer types will pop up. For zinc finger domain the first two missense mutations, C384R and C384W, are associated with glioblastoma and melanoma, respectively. By searching for Cys384 residue in the table below and pressing the "3D" button, the structural location of this mutation is displayed. We will explain how this mutation could be interpreted in the context of molecular interactions in the next section.

3.2 Finding Protein Domains and Functional Sites Affected by Mutations

Now we consider the domain context of CBL mutations described in the previous section (p.Y371H and p.C384R) and annotate their impacts on protein interactions and signaling pathways. The web resources used in this protocol are listed in Table 2.

1. Evolutionarily conserved sites in a multiple sequence alignment usually correspond to functionally important sites and mutations in these sites can be harmful to protein function. If a protein of interest has a known PDB structure, conservation profiles can be downloaded from the PDBsum resource otherwise ConSurf server [24] can be used. In addition, the CDD server can offer functional annotations of sites in conserved protein domains, whereas IBIS server provides locations of binding sites for different types of binding partners (protein, small molecule, nucleic acid, ion, and peptides), and facilitates the mapping of a comprehensive biomolecular interaction network for a given protein query (with or without structure) [25–27]. Similar binding sites in IBIS are clustered together based on their sequence and structure conservation.
2. Open IBIS web page and search for 1FBV structure, chain A. Go to "protein–protein" tab and click on the balloon with the annotation "RING" domain to display interactions of the CBL RING domain with other domains/proteins (Fig. 3a). Binding sites are shown on CBL sequence as triangles and highly conserved binding sites are shown in red color. In the list of interaction partners below, the first conserved binding site cluster is formed between RING domain of CBL and ubiquitin conjugating enzyme from UBCc family. By clicking on the plus sign next to "UBCc", one can see the corresponding binding site, the alignment of similar binding sites found in different CBL-UBCc complexes. By opening the link to Cn3D viewer, one can explore the interfaces and binding sites in these protein complexes (Fig. 3c).

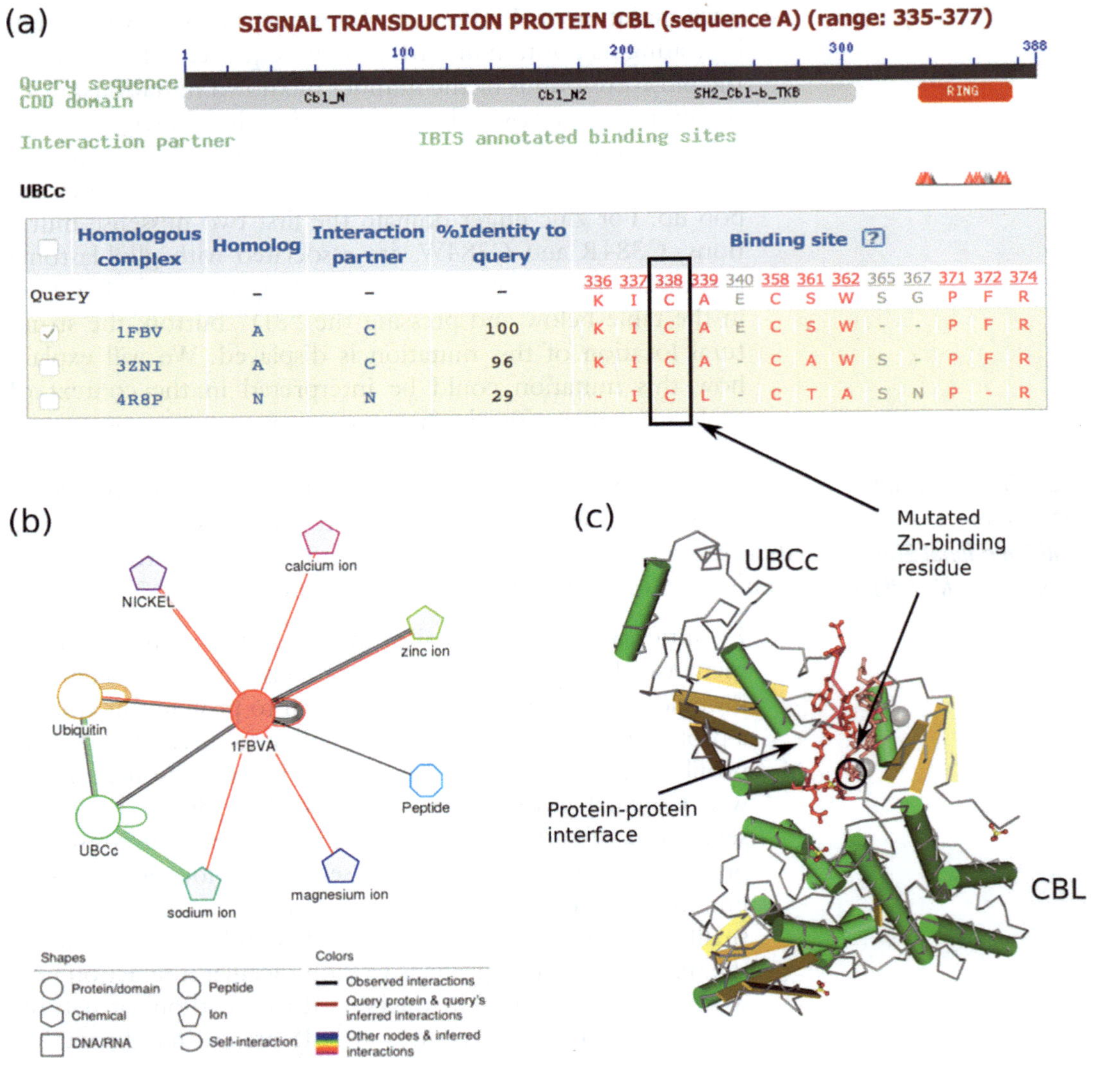

Fig. 3 Analysis of conserved functional and binding sites in mutated proteins using IBIS method. (**a**) conserved protein binding site in the RING domain of CBL; (**b**) interaction graph of CBL protein (represented by 1FBV PDB structure); (**c**) visualization of protein interface between CBL and UBCc in Cn3D software. Position 384 in 1FBV corresponds to position 338 in the full-length PDB sequence

3. The interaction graph in Fig. 3b shows the observed (black lines) and predicted interaction partners of CBL. Next, we will focus on interactions of CBL with zinc ions and UBCc ubiquitin ligase. Note that self-links indicate interactions between domains in CBL protein within or between CBL chains.

4. The structure in Fig. 3c shows that conserved cysteine residues in the binding site coordinate two zinc ions, apparently playing an important structural role. A substitution of cysteine by arginine disrupts the coordination of Zn, which affects the structure and stability of the zinc finger RING domain and may also affect CBL function.

5. Structural and biochemical analyses [28] show that CBL inactive state adopts an autoinhibited interaction. Substrate binding and Tyr371 phosphorylation activates CBL by producing a large conformational change in order to place the RING domain and UBCc in close proximity to the substrate necessary for effective catalysis. Importantly, mutation p.Y371H may prohibit activation of CBL by phosphorylation and may also affect the interaction with UBCc.
6. The impact of mutations on signaling pathways can be explored using recently developed PathiVar server [29]. Alternatively, it is possible to explore all pathways in which CBL interacts with UBCc using KEGG, Reactome, or NCBI Biosystems databases.
7. Search for 'CBL AND Y731 and "Homo sapiens"[Organism]' in the NCBI Biosystems portal. The first record will point to the "Regulation of signaling by CBL" pathway and the role of Y731 phosphorylation will be explained in the pathway description.

In addition to the NCBI IBIS server that allows analyzing the domain context and structural determinants of interactions, several other servers (DMDM, PolyDoms, and muPIT) provide the mapping of mutations on protein domains and protein structures.

3.3 Assessing If Mutations Have Damaging or Benign Effects on Proteins

Many methods have been proposed to predict the effects of missense mutations on proteins, classifying them as damaging or benign. These methods differ in terms of the properties of mutations or proteins used during the training procedure as well as in terms of the algorithms applied for decision-making. For example, machine learning algorithms train their models to distinguish known disease-associated from neutral mutations. Other methods do not explicitly train their models but almost all methods described in this section exploit the evolutionary conservation assuming that changes at conserved positions tend to be more deleterious. Besides sequence conservation various other sequence and structural features are used, which may include: changes in physicochemical properties between wild-type and substituted amino acid, structural features (mostly solvent accessibility), site mutability in DNA, and sequence context of the site.

An unbiased testing and comparison of machine learning methods is obviously an issue since they are usually trained on all available datasets of mutations and it is difficult to obtain a test set which would not overlap with training set. There are several experimental studies on variants in P53, LacI, and ABCA1 proteins which can be regarded as unbiased test cases. Comparisons of different methods on these experimental sets reported up to 70% TPR (True positive rate) at 10% FPR (False positive rate) [30]. Models trained to distinguish Mendelian variants with pronounced deleterious effects are more appropriate and accurate for predicting the effects of Mendelian mutations. The accuracy of these models is much higher than of those models that aim to assess the effects of mutations from complex diseases including cancer. This is evident from the evaluation of PolyPhen-2 performance which yields 0.70–0.77 TPR at 10% FPR when trained on the HumDiv dataset (Mendelian disease mutations) and drops to 0.50–0.52 TPR for HumVar (all human disease causing mutations) trained models [31]. It should be mentioned that there are several methods that are trained to distinguish cancer mutations from neutral polymorphisms (Table 3); however, no existing method can accurately identify driver and passenger mutations within the pool of cancer mutations.

One of the most comprehensive comparisons of different methods to predict phenotypic effects of mutations was performed recently [32]. To avoid any bias in evaluation of these methods, most of which were trained on all available sets of disease mutations and neutral polymorphisms, the authors of this study tested different methods on an independent set of experimental studies. They concluded that there was a variation between various methods in terms of their accuracy and applicability, with SNP&GO and MuPred being the most reliable predictors. Interestingly, despite the fact that different methods use similar sets of features, only half of their correctly predicted cases overlap [32]. Since this study was published, several new methods have been introduced (see Table 3). For example, in contrast to many methods, that assess amino acid frequency distribution in a given site of interest, a recently developed method PROVEAN accounts for the sequence context around the site of interest and poorly aligned regions/sites are assigned very low scores. Overall, the effect of alignment quality on the performance of all methods is largely undetermined but suspected to be very large. Therefore, a user-based construction of accurate alignments of homologous proteins would be very advantageous for accurate annotations of the effects of mutations.

3.4 Predicting the Impact of Mutations on Protein Stability

While methods, which provide a classification of damaging effects of mutations, are widely used by the genomics community, a new level of annotation is needed to offer an explanation of why and how these mutations damage proteins. Algorithms and servers described in the next several sections address these tasks. Proteins may evolve through the acquisition of new mutations, most of which are destabilizing but phenotypically neutral. Stability of a protein may be directly related to its functional activity and incorrect folding and decreased stability can be the major consequences of pathogenic missense mutations [33, 34]. However, protein stability is necessary but not sufficient for protein function, and proteins do not evolve to maximize their stability. Typically, the magnitude of effects of mutations on stability can be quantified by changes in unfolding free energy ($\Delta\Delta G_{\mathrm{fold}}$) (Fig. 4).

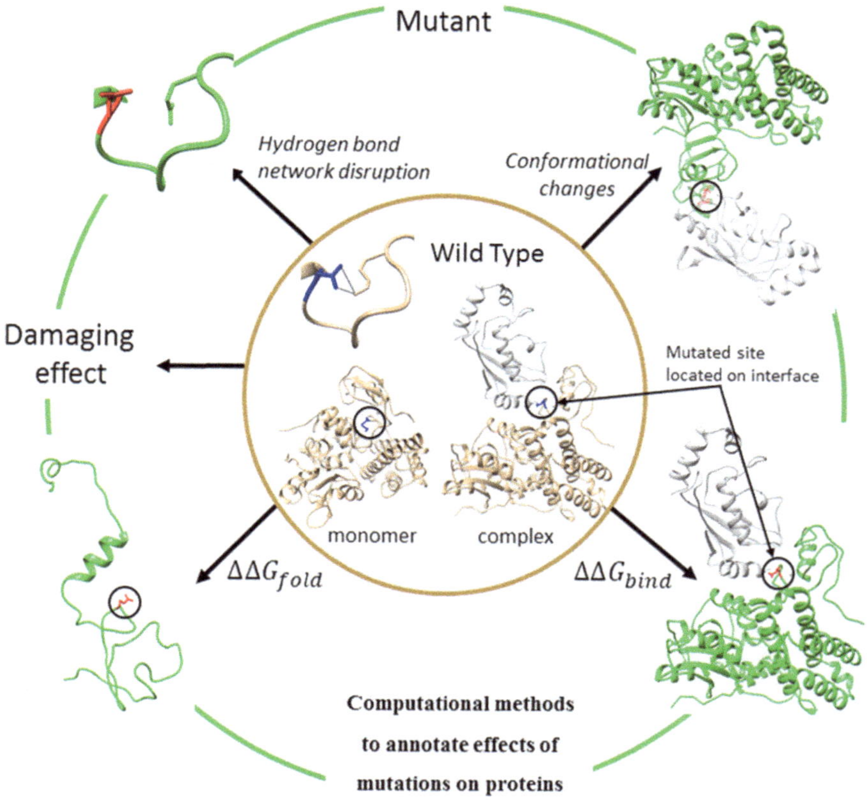

Fig. 4 Annotation of the effects of mutations on proteins with available structures

$$\Delta\Delta G_{\text{fold}} = \Delta G_{\text{fold}}^{\text{mut}} - \Delta G_{\text{fold}}^{\text{WT}} \quad (1)$$

Table 4 lists several state-of-the art methods for predicting the quantitative changes in unfolding free energy upon mutations and provides short descriptions and links to corresponding programs/servers. Methods described in this section differ in terms of energy functions, procedures used for optimization and sampling, and algorithms used for training, if applicable. Energy functions may vary from physics-based force fields, which describe fundamental physical forces between atoms, to knowledge-based potentials, which are based on statistical analyses of protein structures and residue properties. The majority of these methods require the coordinates of protein structures while methods like MuStab or iPTREE-STAB do not use structural data but their performance is also limited. The performance of different methods was evaluated in several studies [35–37]. In the first study [35] the following performance ranking was reported: EGAD > CC/PBSA > I-Mutant2.0 > FoldX > Hunter > Rosetta with correlation coefficients between experimental and predicted $\Delta\Delta G$ values in the range of 0.59 and 0.26 and standard deviation in the range of 0.95 and

2.32 kcal mol^{-1}. However, the servers of the top performing methods EGAD and CC/PBSA are no longer available. In the second study I-Mutant3.0, Dmutant, and FoldX were found to be the most reliable predictors [36].

There are several servers to assess the effects of mutations on stability that are straightforward and easy to use. Here we present a protocol on how to use FoldX software.

1. Run RepairPDB module of FoldX to correct errors in the structure produced during the refinement (nonstandard angles or distances) (Runfile can be obtained from http://foldxsuite.crg.eu/command/RepairPDB).
2. Run BuildModel module of FoldX to introduce a mutation on the optimized wild-type structure http://foldxsuite.crg.eu/command/BuildModel. BuildModel module optimizes the configurations of the side chains of amino acids in the vicinity of the mutated site and calculates the difference in unfolding free energy ($\Delta\Delta G_{\mathrm{fold}}$) between mutant and repaired wild-type structure. The total unfolding free energy and each energy term can be obtained from the "Dif_BuildModed_*.fxout" output file.

3.5 Predicting the Effects of Mutations on Protein–Protein Binding Affinity

Crucial prerequisite for proper biological function is a protein's ability to establish highly selective interactions with macromolecular partners. A missense mutation that affects protein interactions [38–40] may cause significant perturbations or complete abolishment of protein function, potentially leading to diseases. Typically, the change in binding free energy ($\Delta\Delta G_{\mathrm{bind}}$) is used to quantify the magnitude of mutational effects on protein–protein interactions (Fig. 4).

$$\Delta\Delta G_{\mathrm{bind}} = \Delta G_{\mathrm{bind}}^{\mathrm{mut}} - \Delta G_{\mathrm{bind}}^{\mathrm{WT}} \tag{2}$$

The binding energy is calculated as a difference between the free energies of a complex *AB* and unbound proteins *A* and *B*:

$$\Delta G_{\mathrm{bind}} = G^{\mathrm{AB}} - G^{\mathrm{A}} - G^{\mathrm{B}} \tag{3}$$

There are very few methods that estimate actual $\Delta\Delta G_{\mathrm{bind}}$ values and these methods require all-atom or at least protein backbone atom coordinates of a wild-type and/or mutated protein. Some of the methods use coarse-grained predictors based on statistical or empirical potentials, others apply molecular mechanics force fields with different solvation models. For example, the molecular mechanics Poisson–Boltzmann surface area (MM-PBSA) method and its derivatives have been shown to yield very good agreement between predicted and experimental values with correlation coefficients up to 0.69 [41]. For all methods, the right choices of minimization protocols, energy functions, and solvation models are crucial for achieving reasonable prediction accuracy. In addition, prediction accuracy strongly depends on the type of mutation and its location in a protein complex. For example, if a residue is located on the protein–protein interface, its mutation might have a larger effect on protein–protein interaction and binding affinity compared to a non-interfacial mutation [41]. A location of mutated sites can be mapped by SPPIDER (http://sppider.cchmc.org/) [42] or Meta-PPISP (http://pipe.scs.fsu.edu/meta-ppisp.html) [43] servers. These servers are recommended by two assessments of computational methods for predicting protein–protein interaction sites [44, 45]. Users can also analyze structures and locations of mutations using software Chimera or VMD.

Below is a step-by-step protocol reported in our previous paper [41] to predict the impact of mutations on binding affinity. This protocol combines molecular mechanics force fields with statistical (BeAtMuSiC) and empirical (FoldX) energy functions. All files are provided via the ftp site ftp://ftp.ncbi.nih.gov/pub/panch/Mutation_binding. The improved version can be available from our MutaBind server https://www.ncbi.nlm.nih.gov/projects/mutabind/.

1. Install software VMD, NAMD, and CHARMM.
2. Download a structure for your protein of interest from the Protein Data Bank (PDB).
3. Add hydrogen atoms, a rectangular box (10 Å) of water molecules, and Na^+ and Cl^- ions (ionic concentration of 150 mM) to the structure using VMD ("vmd.pgn").
4. Carry out 5000-step energy minimization with harmonic restraints (with the force constant of 5 kcal mol^{-1} Å^{-2}) applied on the backbone atoms of all residues ("minimization1.conf"), followed by a 35,000-step energy minimization on the whole system ("minimization2.conf") with NAMD program using the CHARMM27 force field.
5. Introduce a mutation using "mutator" plugin of VMD software on the final minimized model from **step 4**.
6. Run an additional 300-step minimization for the whole mutant structure ("minimization2.conf").
7. Run CHARMM program using the last frame from **step 4** (for wild-type structure) and **step 6** (for mutant structure) to obtain van der Waals interaction energy (ΔE_{vdw}), polar solvation energy of solute in water (ΔG_{solv}) for the wild-type and mutant, and interface area (ΔSA_{mut}) for mutant (Runfile is "binding_energy.str").
8. Submit your structure from **step 2** and a mutation to BeAtMuSiC webserver (Table 5) to obtain the binding affinity change $\Delta\Delta G_{BM}$.

9. Run AnalyseComplex module of FoldX to obtain the binding affinity change $\Delta\Delta G_{FD}$. (http://foldxsuite.crg.eu/command/AnalyseComplex).
10. Obtain the binding affinity changes using the following combination of energy terms from [41]:

$$\Delta\Delta G_{\text{bind}} = \alpha\Delta\Delta E_{\text{vdw}} + \beta\Delta\Delta G_{\text{solv}} + \gamma\Delta SA_{\text{mut}} + \varepsilon\Delta\Delta G_{BM} + \lambda\Delta\Delta G_{FD} + \delta$$

$$\begin{pmatrix} \alpha = 0.122, \beta = 0.101, \gamma = 0.043, \\ \varepsilon = 0.446, \lambda = 0.168 \, \textit{and} \, \delta = 1.326 \end{pmatrix}.$$

Recently a new computational method MutaBind [46] was developed to evaluate the effects of mutations on protein–protein interactions (http://www.ncbi.nlm.nih.gov/projects/mutabind/). The MutaBind method uses molecular mechanics force fields, statistical potentials, and fast side-chain optimization algorithms. It maps mutations on a protein complex structure, calculates the associated changes in binding affinity, determines the deleterious effect of a mutation, estimates the confidence of this prediction, and produces a mutant structural model for download.

3.6 Assessing the Changes in Protein Conformations and Hydrogen Bond Networks Induced by Mutations

Proteins may adopt different conformations along the pathway of a biochemical reaction and their intrinsic flexibility and ability to sample alternative conformations are crucial for protein function. Mutations might shift the equilibrium between different conformations (Fig. 4) and as a result, the most populated conformation of a mutated protein can be different in structure, stability, and functional activity from the wild-type conformation. It is extremely difficult to model structural changes in a protein backbone produced by mutations and large conformational shifts can be predicted correctly only in a few cases. In fact, most algorithms discussed in the previous sections do not account for the backbone flexibility. If several conformations of the same protein are available in the structural databank, all of them ideally should be used to provide a complete picture of dynamical and energetic effects of mutations [20].

Mutations can either change the global conformation of an entire molecule or have a localized effect in a small region. In a recent study of the NFAT5 transcription factor [47], different mutations from the same DNA-binding loop were analyzed and it was shown that effects of these mutations on protein dynamics and DNA binding were drastically different although they were located very close to each other in sequence and structure. Protein dynamics can be studied through performing molecular dynamics (MD) simulations using NAMD [2], CHARMM [3] and Amber [48] MD packages. NAMD, for example, is fast and easy to use; it can be applied with CHARMM or Amber force fields, whereas VMD or CHARMM packages can be used to analyze the MD trajectories produced by NAMD.

Changes in structure may also be assessed through the analyses of hydrogen bond networks and their differences between mutant and wild-type proteins since hydrogen bonds are important in determining protein stability. A mutation disrupting hydrogen bonds might have a significant impact on protein conformation, stability, and dynamics (reviewed in [49], Fig. 4). Hydrogen bonds can be calculated using HBOND (http://caps.ncbs.res.in/iws/hbond.html) [50] or PIC servers (http://pic.mbu.iisc.ernet.in/) [51] and visualized by Chimera.

Below is a step-by-step protocol to assess the conformational changes induced by mutations:

1. Download the structure and introduce a mutation using VMD (refer to **steps 2** and **5** from Subheading 3.5).
2. Build the model systems with VMD (Refer to **step 3** from Subheading 3.5).
3. Carry out the energy minimizations (Refer to **step 4** from Subheading 3.5). The number of minimization steps can be chosen based on the size of system.
4. Heat the system to 300 K over 300 ps with harmonic constraints applied to protein backbone atoms using NAMD ("heat.conf").
5. Perform unconstrained MD simulation on the system with NAMD ("md.conf").
6. Load MD trajectories into the VMD software to monitor the conformational changes and calculate the root-mean-square deviation (RMSD) between the wild-type and mutant structures.

   Assessing the effects of mutations on hydrogen bond networks using Chimera.

7. Load your protein structure of interest into Chimera.
8. Select residues of interest and input "findhbond selRestrict both distSlop 0.35 angleSlop 60.0 saveFile filename" in the command line. The hydrogen bonds will be shown on the screen and details will be saved in the file "filename". One can adjust the distance (distSlop) and angle (angleSlop) parameters in the definition of hydrogen bonds.
9. (Optional) Go to Tools ->Structural Analysis ->FindHBond to find the hydrogen bonds.

Acknowledgements

This work was supported by the Intramural Research Program of the National Library of Medicine.

References

1. Abecasis GR, Altshuler D, Auton A, Brooks LD, Durbin RM, Gibbs RA, Hurles ME, McVean GA, Genomes Project C (2010) A map of human genome variation from population-scale sequencing. Nature 467(7319):1061–1073
2. Phillips JC, Braun R, Wang W, Gumbart J, Tajkhorshid E, Villa E, Chipot C, Skeel RD, Kale L, Schulten K (2005) Scalable molecular dynamics with NAMD. J Comput Chem 26(16):1781–1802
3. Brooks BR, Bruccoleri RE, Olafson BD, States DJ, Swaminathan S, Karplus M (1983) Charmm – a program for macromolecular energy, minimization, and dynamics calculations. J Comput Chem 4(2):187–217
4. Humphrey W, Dalke A, Schulten K (1996) VMD: visual molecular dynamics. J Mol Graph 14(1):33–38, 27-38
5. Pettersen EF, Goddard TD, Huang CC, Couch GS, Greenblatt DM, Meng EC, Ferrin TE (2004) UCSF Chimera—a visualization system for exploratory research and analysis. J Comput Chem 25(13):1605–1612
6. Wang Y, Geer LY, Chappey C, Kans JA, Bryant SH (2000) Cn3D: sequence and structure views for Entrez. Trends Biochem Sci 25(6):300–302
7. Mailman MD, Feolo M, Jin Y, Kimura M, Tryka K, Bagoutdinov R, Hao L, Kiang A, Paschall J, Phan L, Popova N, Pretel S, Ziyabari L, Lee M, Shao Y, Wang ZY, Sirotkin K, Ward M, Kholodov M, Zbicz K, Beck J, Kimelman M, Shevelev S, Preuss D, Yaschenko E, Graeff A, Ostell J, Sherry ST (2007) The NCBI dbGaP database of genotypes and phenotypes. Nat Genet 39(10):1181–1186
8. Rubinstein WS, Maglott DR, Lee JM, Kattman BL, Malheiro AJ, Ovetsky M, Hem V, Gorelenkov V, Song G, Wallin C, Husain N, Chitipiralla S, Katz KS, Hoffman D, Jang W, Johnson M, Karmanov F, Ukrainchik A, Denisenko M, Fomous C, Hudson K, Ostell JM (2013) The NIH genetic testing registry: a new, centralized database of genetic tests to enable access to comprehensive information and improve transparency. Nucleic Acids Res 41(Database issue):D925–D935
9. Sherry ST, Ward M, Sirotkin K (1999) dbSNP-database for single nucleotide polymorphisms and other classes of minor genetic variation. Genome Res 9(8):677–679
10. Lappalainen I, Lopez J, Skipper L, Hefferon T, Spalding JD, Garner J, Chen C, Maguire M, Corbett M, Zhou G, Paschall J, Ananiev V, Flicek P, Church DM (2013) DbVar and DGVa: public archives for genomic structural variation. Nucleic Acids Res 41(Database issue):D936–D941
11. Kibbe WA, Arze C, Felix V, Mitraka E, Bolton E, Fu G, Mungall CJ, Binder JX, Malone J, Vasant D, Parkinson H, Schriml LM (2015) Disease Ontology 2015 update: an expanded and updated database of human diseases for linking biomedical knowledge through disease data. Nucleic Acids Res 43(Database issue):D1071–D1078
12. Ramos EM, Hoffman D, Junkins HA, Maglott D, Phan L, Sherry ST, Feolo M, Hindorff LA (2014) Phenotype-Genotype Integrator (PheGenI): synthesizing genome-wide association study (GWAS) data with existing genomic resources. Eur J Hum Genet 22(1):144–147
13. Stenson PD, Ball EV, Mort M, Phillips AD, Shiel JA, Thomas NS, Abeysinghe S, Krawczak M, Cooper DN (2003) Human Gene Mutation Database (HGMD): 2003 update. Hum Mutat 21(6):577–581
14. Amberger JS, Bocchini CA, Schiettecatte F, Scott AF, Hamosh A (2015) OMIM.org: Online Mendelian Inheritance in Man (OMIM(R)), an online catalog of human genes and genetic disorders. Nucleic Acids Res 43(Database issue):D789–D798
15. Landrum MJ, Lee JM, Riley GR, Jang W, Rubinstein WS, Church DM, Maglott DR (2014) ClinVar: public archive of relationships among sequence variation and human phenotype. Nucleic Acids Res 42(Database issue):D980–D985
16. Forbes SA, Bindal N, Bamford S, Cole C, Kok CY, Beare D, Jia M, Shepherd R, Leung K, Menzies A, Teague JW, Campbell PJ, Stratton MR, Futreal PA (2011) COSMIC: mining complete cancer genomes in the Catalogue of Somatic Mutations in Cancer. Nucleic Acids Res 39(suppl 1):D945–D950
17. Weinstein JN, Collisson EA, Mills GB, Shaw KR, Ozenberger BA, Ellrott K, Shmulevich I, Sander C, Stuart JM, Cancer Genome Atlas Research N (2013) The Cancer Genome Atlas Pan-Cancer analysis project. Nat Genet 45(10):1113–1120
18. Singh A, Olowoyeye A, Baenziger PH, Dantzer J, Kann MG, Radivojac P, Heiland R, Mooney SD (2008) MutDB: update on development of tools for the biochemical analysis of genetic variation. Nucleic Acids Res 36(Database issue):D815–D819
19. Mottaz A, David FP, Veuthey AL, Yip YL (2010) Easy retrieval of single amino-acid polymorphisms and phenotype information using SwissVar. Bioinformatics 26(6):851–852

20. Li M, Kales SC, Ma K, Shoemaker BA, Crespo-Barreto J, Cangelosi AL, Lipkowitz S, Panchenko AR (2015) Balancing protein stability and activity in cancer: a new approach for identifying driver mutations affecting CBL ubiquitin ligase activation. Cancer Res 76(3):561–571
21. NR Coordinators (2014) Database resources of the National Center for Biotechnology Information. Nucleic Acids Res 42(Database issue):D7–D17
22. Perez B, Mechinaud F, Galambrun C, Ben Romdhane N, Isidor B, Philip N, Derain-Court J, Cassinat B, Lachenaud J, Kaltenbach S, Salmon A, Desiree C, Pereira S, Menot ML, Royer N, Fenneteau O, Baruchel A, Chomienne C, Verloes A, Cave H (2010) Germline mutations of the CBL gene define a new genetic syndrome with predisposition to juvenile myelomonocytic leukaemia. J Med Genet 47(10):686–691
23. Marchler-Bauer A, Derbyshire MK, Gonzales NR, Lu S, Chitsaz F, Geer LY, Geer RC, He J, Gwadz M, Hurwitz DI, Lanczycki CJ, Lu F, Marchler GH, Song JS, Thanki N, Wang Z, Yamashita RA, Zhang D, Zheng C, Bryant SH (2015) CDD: NCBI's conserved domain database. Nucleic Acids Res 43(D1):D222–D226
24. Ashkenazy H, Erez E, Martz E, Pupko T, Ben-Tal N (2010) ConSurf 2010: calculating evolutionary conservation in sequence and structure of proteins and nucleic acids. Nucleic Acids Res 38(suppl 2):W529–W533
25. Shoemaker BA, Zhang D, Thangudu RR, Tyagi M, Fong JH, Marchler-Bauer A, Bryant SH, Madej T, Panchenko AR (2010) Inferred Biomolecular Interaction Server – a web server to analyze and predict protein interacting partners and binding sites. Nucleic Acids Res 38(Database issue):D518–D524
26. Goncearenco A, Shaytan AK, Shoemaker BA, Panchenko AR (2015) Structural perspectives on the evolutionary expansion of unique protein-protein binding sites. Biophys J 109(6):1295–1306
27. Shoemaker BA, Zhang D, Tyagi M, Thangudu RR, Fong JH, Marchler-Bauer A, Bryant SH, Madej T, Panchenko AR (2012) IBIS (Inferred Biomolecular Interaction Server) reports, predicts and integrates multiple types of conserved interactions for proteins. Nucleic Acids Res 40(Database issue):D834–D840
28. Dou H, Buetow L, Hock A, Sibbet GJ, Vousden KH, Huang DT (2012) Structural basis for autoinhibition and phosphorylation-dependent activation of c-Cbl. Nat Struct Mol Biol 19(2):184–192
29. Hernansaiz-Ballesteros RD, Salavert F, Sebastian-Leon P, Aleman A, Medina I, Dopazo J (2015) Assessing the impact of mutations found in next generation sequencing data over human signaling pathways. Nucleic Acids Res 43(W1):W270–W275
30. Choi Y, Sims GE, Murphy S, Miller JR, Chan AP (2012) Predicting the functional effect of amino acid substitutions and indels. PLoS One 7(10):e46688
31. Adzhubei IA, Schmidt S, Peshkin L, Ramensky VE, Gerasimova A, Bork P, Kondrashov AS, Sunyaev SR (2010) A method and server for predicting damaging missense mutations. Nat Methods 7(4):248–249
32. Thusberg J, Olatubosun A, Vihinen M (2011) Performance of mutation pathogenicity prediction methods on missense variants. Hum Mutat 32(4):358–368
33. Hashimoto K, Rogozin IB, Panchenko AR (2012) Oncogenic potential is related to activating effect of cancer single and double somatic mutations in receptor tyrosine kinases. Hum Mutat 33(11):1566–1575
34. Schlebach JP, Narayan M, Alford C, Mittendorf KF, Carter BD, Li J, Sanders CR (2015) Conformational stability and pathogenic misfolding of the integral membrane protein PMP22. J Am Chem Soc 137(27):8758–8768
35. Potapov V, Cohen M, Schreiber G (2009) Assessing computational methods for predicting protein stability upon mutation: good on average but not in the details. Protein Eng Des Sel 22(9):553–560
36. Khan S, Vihinen M (2010) Performance of protein stability predictors. Hum Mutat 31(6):675–684
37. Zhang Z, Wang L, Gao Y, Zhang J, Zhenirovskyy M, Alexov E (2012) Predicting folding free energy changes upon single point mutations. Bioinformatics 28(5):664–671
38. Nishi H, Tyagi M, Teng S, Shoemaker BA, Hashimoto K, Alexov E, Wuchty S, Panchenko AR (2013) Cancer missense mutations alter binding properties of proteins and their interaction networks. PLoS One 8(6):e66273
39. Teng S, Madej T, Panchenko A, Alexov E (2009) Modeling effects of human single nucleotide polymorphisms on protein-protein interactions. Biophys J 96(6):2178–2188
40. Ghersi D, Singh M (2014) Interaction-based discovery of functionally important genes in cancers. Nucleic Acids Res 42(3):e18
41. Li M, Petukh M, Alexov E, Panchenko AR (2014) Predicting the impact of missense

mutations on protein-protein binding affinity. J Chem Theory Comput 10(4):1770–1780
42. Porollo A, Meller J (2007) Prediction-based fingerprints of protein–protein interactions. Proteins 66(3):630–645
43. Qin S, Zhou H-X (2007) meta-PPISP: a meta web server for protein-protein interaction site prediction. Bioinformatics 23(24):3386–3387
44. Zhou H-X, Qin S (2007) Interaction-site prediction for protein complexes: a critical assessment. Bioinformatics 23(17):2203–2209
45. Porollo A, Meller J (2012) Computational methods for prediction of protein-protein interaction sites. Protein-Protein Interactions – Computational and Experimental Tools 472:3–26
46. Li M, Simonetti FL, Goncearenco A, Panchenko AR (2016) MutaBind estimates and interprets the effects of sequence variants on protein-protein interactions. Nucleic Acids Res. Jul 8;44(W1):W494-501.
47. Li M, Shoemaker BA, Thangudu RR, Ferraris JD, Burg MB, Panchenko AR (2013) Mutations in DNA-binding loop of NFAT5 transcription factor produce unique outcomes on protein-DNA binding and dynamics. J Phys Chem B 117(42):13226–13234
48. Case DA, Cheatham TE, Darden T, Gohlke H, Luo R, Merz KM, Onufriev A, Simmerling C, Wang B, Woods RJ (2005) The Amber biomolecular simulation programs. J Comput Chem 26(16):1668–1688
49. Stefl S, Nishi H, Petukh M, Panchenko AR, Alexov E (2013) Molecular mechanisms of disease-causing missense mutations. J Mol Biol 425(21):3919–3936
50. Mizuguchi K, Deane CM, Blundell TL, Johnson MS, Overington JP (1998) JOY: protein sequence-structure representation and analysis. Bioinformatics 14(7):617–623
51. Tina KG, Bhadra R, Srinivasan N (2007) PIC: protein interactions calculator. Nucleic Acids Res 35(suppl 2):W473–W476
52. Stenson P, Mort M, Ball E, Shaw K, Phillips A, Cooper D (2014) The Human Gene Mutation Database: building a comprehensive mutation repository for clinical and molecular genetics, diagnostic testing and personalized genomic medicine. Hum Genet 133(1):1–9
53. Thorn CF, Klein TE, Altman RB (2010) Pharmacogenomics and bioinformatics: PharmGKB. Pharmacogenomics 11(4):501–505
54. Cerami E, Gao J, Dogrusoz U, Gross BE, Sumer SO, Aksoy BA, Jacobsen A, Byrne CJ, Heuer ML, Larsson E, Antipin Y, Reva B, Goldberg AP, Sander C, Schultz N (2012) The cBio cancer genomics portal: an open platform for exploring multidimensional cancer genomics data. Cancer Discov 2(5):401–404
55. Peri S, Navarro JD, Amanchy R, Kristiansen TZ, Jonnalagadda CK, Surendranath V, Niranjan V, Muthusamy B, Gandhi TK, Gronborg M, Ibarrola N, Deshpande N, Shanker K, Shivashankar HN, Rashmi BP, Ramya MA, Zhao Z, Chandrika KN, Padma N, Harsha HC, Yatish AJ, Kavitha MP, Menezes M, Choudhury DR, Suresh S, Ghosh N, Saravana R, Chandran S, Krishna S, Joy M, Anand SK, Madavan V, Joseph A, Wong GW, Schiemann WP, Constantinescu SN, Huang L, Khosravi-Far R, Steen H, Tewari M, Ghaffari S, Blobe GC, Dang CV, Garcia JG, Pevsner J, Jensen ON, Roepstorff P, Deshpande KS, Chinnaiyan AM, Hamosh A, Chakravarti A, Pandey A (2003) Development of human protein reference database as an initial platform for approaching systems biology in humans. Genome Res 13(10):2363–2371
56. Kerrien S, Aranda B, Breuza L, Bridge A, Broackes-Carter F, Chen C, Duesbury M, Dumousseau M, Feuermann M, Hinz U, Jandrasits C, Jimenez RC, Khadake J, Mahadevan U, Masson P, Pedruzzi I, Pfeiffenberger E, Porras P, Raghunath A, Roechert B, Orchard S, Hermjakob H (2012) The IntAct molecular interaction database in 2012. Nucleic Acids Res 40(Database issue):D841–D846
57. Joshi-Tope G, Gillespie M, Vastrik I, D'Eustachio P, Schmidt E, de Bono B, Jassal B, Gopinath GR, Wu GR, Matthews L, Lewis S, Birney E, Stein L (2005) Reactome: a knowledgebase of biological pathways. Nucleic Acids Res 33(Database issue):D428–D432
58. Kanehisa M, Goto S, Hattori M, Aoki-Kinoshita KF, Itoh M, Kawashima S, Katayama T, Araki M, Hirakawa M (2006) From genomics to chemical genomics: new developments in KEGG. Nucleic Acids Res 34(Database issue):D354–D357
59. Niknafs N, Kim D, Kim R, Diekhans M, Ryan M, Stenson PD, Cooper DN, Karchin R (2013) MuPIT interactive: webserver for mapping variant positions to annotated, interactive 3D structures. Hum Genet 132(11):1235–1243
60. Peterson TA, Adadey A, Santana-Cruz I, Sun Y, Winder A, Kann MG (2010) DMDM: domain mapping of disease mutations. Bioinformatics 26(19):2458–2459
61. Jegga AG, Gowrisankar S, Chen J, Aronow BJ (2007) PolyDoms: a whole genome database for the identification of non-synonymous coding SNPs with the potential to impact disease.

Nucleic Acids Res 35(Database issue):D700–D706

62. Ng PC, Henikoff S (2003) SIFT: predicting amino acid changes that affect protein function. Nucleic Acids Res 31(13):3812–3814
63. Yue P, Melamud E, Moult J (2006) SNPs3D: candidate gene and SNP selection for association studies. BMC Bioinformatics 7:166
64. Ferrer-Costa C, Gelpi JL, Zamakola L, Parraga I, de la Cruz X, Orozco M (2005) PMUT: a web-based tool for the annotation of pathological mutations on proteins. Bioinformatics 21(14):3176–3178
65. Bromberg Y, Rost B (2007) SNAP: predict effect of non-synonymous polymorphisms on function. Nucleic Acids Res 35(11):3823–3835
66. Shihab HA, Gough J, Cooper DN, Stenson PD, Barker GL, Edwards KJ, Day IN, Gaunt TR (2013) Predicting the functional, molecular, and phenotypic consequences of amino acid substitutions using hidden Markov models. Hum Mutat 34(1):57–65
67. Reva B, Antipin Y, Sander C (2011) Predicting the functional impact of protein mutations: application to cancer genomics. Nucleic Acids Res 39(17):e118
68. Carter H, Chen S, Isik L, Tyekucheva S, Velculescu VE, Kinzler KW, Vogelstein B, Karchin R (2009) Cancer-specific high-throughput annotation of somatic mutations: computational prediction of driver missense mutations. Cancer Res 69(16):6660–6667
69. Capriotti E, Calabrese R, Fariselli P, Martelli PL, Altman RB, Casadio R (2013) WS-SNPs&GO: a web server for predicting the deleterious effect of human protein variants using functional annotation. BMC Genomics 14(Suppl 3):S6
70. Wang M, Zhao XM, Takemoto K, Xu H, Li Y, Akutsu T, Song J (2012) FunSAV: predicting the functional effect of single amino acid variants using a two-stage random forest model. PLoS One 7(8):e43847
71. Bao L, Zhou M, Cui Y (2005) nsSNPAnalyzer: identifying disease-associated nonsynonymous single nucleotide polymorphisms. Nucleic Acids Res 33(Web Server Issue):W480–482
72. Mi H, Muruganujan A, Thomas PD (2013) PANTHER in 2013: modeling the evolution of gene function, and other gene attributes, in the context of phylogenetic trees. Nucleic Acids Res 41(Database issue):D377–D386
73. Capriotti E, Calabrese R, Casadio R (2006) Predicting the insurgence of human genetic diseases associated to single point protein mutations with support vector machines and evolutionary information. Bioinformatics 22(22):2729–2734
74. Al-Numair NS, Martin AC (2013) The SAAP pipeline and database: tools to analyze the impact and predict the pathogenicity of mutations. BMC Genomics 14(Suppl 3):S4
75. Yates CM, Filippis I, Kelley LA, Sternberg MJ (2014) SuSPect: enhanced prediction of single amino acid variant (SAV) phenotype using network features. J Mol Biol 426(14):2692–2701
76. Simonetti FL, Tornador C, Nabau-Moreto N, Molina-Vila MA, Marino-Buslje C (2014) Kin-Driver: a database of driver mutations in protein kinases. Database 2014:bau104.
77. McSkimming DI, Dastgheib S, Talevich E, Narayanan A, Katiyar S, Taylor SS, Kochut K, Kannan N (2015) ProKinO: a unified resource for mining the cancer kinome. Hum Mutat 36(2):175–186
78. Guerois R, Nielsen JE, Serrano L (2002) Predicting changes in the stability of proteins and protein complexes: a study of more than 1000 mutations. J Mol Biol 320(2):369–387
79. Dehouck Y, Grosfils A, Folch B, Gilis D, Bogaerts P, Rooman M (2009) Fast and accurate predictions of protein stability changes upon mutations using statistical potentials and neural networks: PoPMuSiC-2.0. Bioinformatics 25(19):2537–2543
80. Yin S, Ding F, Dokholyan NV (2007) Eris: an automated estimator of protein stability. Nat Methods 4(6):466–467
81. Parthiban V, Gromiha MM, Schomburg D (2006) CUPSAT: prediction of protein stability upon point mutations. Nucleic Acids Res 34(Web Server Issue):W239–242
82. Potapov V, Cohen M, Inbar Y, Schreiber G (2010) Protein structure modelling and evaluation based on a 4-distance description of side-chain interactions. BMC Bioinformatics 11:374–374
83. Deutsch C, Krishnamoorthy B (2007) Four-body scoring function for mutagenesis. Bioinformatics 23(22):3009–3015
84. Willard L, Ranjan A, Zhang H, Monzavi H, Boyko RF, Sykes BD, Wishart DS (2003) VADAR: a web server for quantitative evaluation of protein structure quality. Nucleic Acids Res 31(13):3316–3319
85. Giollo M, Martin AJ, Walsh I, Ferrari C, Tosatto SC (2014) NeEMO: a method using residue interaction networks to improve prediction of protein stability upon mutation. BMC Genomics 15(Suppl 4):S7
86. Pires DE, Ascher DB, Blundell TL (2014) DUET: a server for predicting effects of mutations on protein stability using an integrated

computational approach. Nucleic Acids Res 42(Web Server Issue):W314–319

87. Laimer J, Hofer H, Fritz M, Wegenkittl S, Lackner P (2015) MAESTRO – multi agent stability prediction upon point mutations. BMC Bioinformatics 16(1):116
88. Capriotti E, Fariselli P, Rossi I, Casadio R (2008) A three-state prediction of single point mutations on protein stability changes. BMC Bioinformatics 9(Suppl 2):S6
89. Cheng J, Randall A, Baldi P (2006) Prediction of protein stability changes for single-site mutations using support vector machines. Proteins 62(4):1125–1132
90. Chen CW, Lin J, Chu YW (2013) iStable: off-the-shelf predictor integration for predicting protein stability changes. BMC Bioinformatics 14(Suppl 2):S5
91. Teng S, Srivastava A, Wang L (2010) Sequence feature-based prediction of protein stability changes upon amino acid substitutions. BMC Genomics 11(Suppl 2):1–8
92. Huang L-T, Gromiha MM, Ho S-Y (2007) iPTREE-STAB: interpretable decision tree based method for predicting protein stability changes upon mutations. Bioinformatics 23(10):1292–1293
93. Dehouck Y, Kwasigroch JM, Rooman M, Gilis D (2013) BeAtMuSiC: prediction of changes in protein–protein binding affinity on mutations. Nucleic Acids Res 41(W1):W333–W339
94. Berliner N, Teyra J, Çolak R, Garcia Lopez S, Kim PM (2014) Combining structural modeling with ensemble machine learning to accurately predict protein fold stability and binding affinity effects upon mutation. PLoS One 9(9):e107353
95. Kruger DM, Gohlke H (2010) DrugScorePPI webserver: fast and accurate in silico alanine scanning for scoring protein-protein interactions. Nucleic Acids Res 38(Web Server Issue):W480–486
96. Zhao N, Han JG, Shyu CR, Korkin D (2014) Determining effects of non-synonymous SNPs on protein-protein interactions using supervised and semi-supervised learning. PLoS Comput Biol 10(5):e1003592

Chapter 18

Protein Micropatterning Assay: Quantitative Analysis of Protein–Protein Interactions

Gerhard J. Schütz, Julian Weghuber, Peter Lanzerstorfer, and Eva Sevcsik

Abstract

Characterization, especially quantification, of protein interactions in live cells is usually not an easy endeavor. Here, we describe a straightforward method to identify and quantify the interaction of a membrane protein ("bait") and a fluorescently labeled interaction partner ("prey") (membrane-bound or cytosolic) in live cells using Total Internal Reflection Fluorescence microscopy. The bait protein is immobilized within patterns in the plasma membrane (e.g., via an antibody); the bait–prey interaction strength can be quantified by determining the prey bulk fluorescence intensity with respect to the bait patterns. This method is particularly suitable also for the analysis of weak, transient interactions that are not easily accessible with other methods.

Key words Micropatterning, Protein–protein interactions, Soft lithography, TIRF microscopy, Quantitive analysis, Membrane proteins

1 Introduction

Although there are many methods to analyze protein–protein interactions, quantitative analysis of protein interactions in live cells is still less than straightforward. Most approaches rely on immunoprecipitation, affinity purification or chemical crosslinking and, thus, analysis of cell lysates [1, 2]. In live cells, assays are rather challenging, laborious, suffer from detection of false positives or negatives, do not allow for easy quantification, and/or are not readily accessible for many labs (e.g., bimolecular fluorescence complementation [3], yeast two-hybrid screen [4], fluorescence resonance energy transfer [5], or single-molecule methods [6]).

Protein micropatterning is a technique that circumvents many of these problems: it is simple, inexpensive, does not need elaborate equipment, can also capture transient interactions, and is performed in live cells, and data analysis is uncomplicated. The method is based on the work of several groups who forced membrane proteins into

Lucio Comai et al. (eds.), *Proteomics: Methods and Protocols*, Methods in Molecular Biology, vol. 1550, DOI 10.1007/978-1-4939-6747-6_18, © Springer Science+Business Media LLC 2017 specific patterns within the plasma membrane of living cells [7, 8]. We have extended this approach to use it as a tool for characterization and quantification of protein interactions: One interaction partner (bait) is restricted to specific regions (typically regular micropatterns) in the live cell plasma membrane and the lateral distribution of a fluorescently labeled interaction partner (prey) is monitored. In case of an interaction, prey molecules will follow the bait pattern; homogeneous distribution of prey protein in the plasma membrane indicates the absence of an interaction (Fig. 1). Quantification can be achieved by comparing the prey signal intensity within and outside the bait regions: the signal contrast between these regions provides a measure of the interaction strength.

While patterned surfaces can be generated by different methods (e.g., photolithography [9] or dip-pen nanolithography [10]), soft lithography [11] is probably the most convenient: it is fast, simple, and lends itself to high throughput routines. In this protocol, the patterned cell substrate is produced by printing streptavidin patterns on a glass coverslip, to which a bait-specific biotinylated antibody is then attached. We have first used this approach to

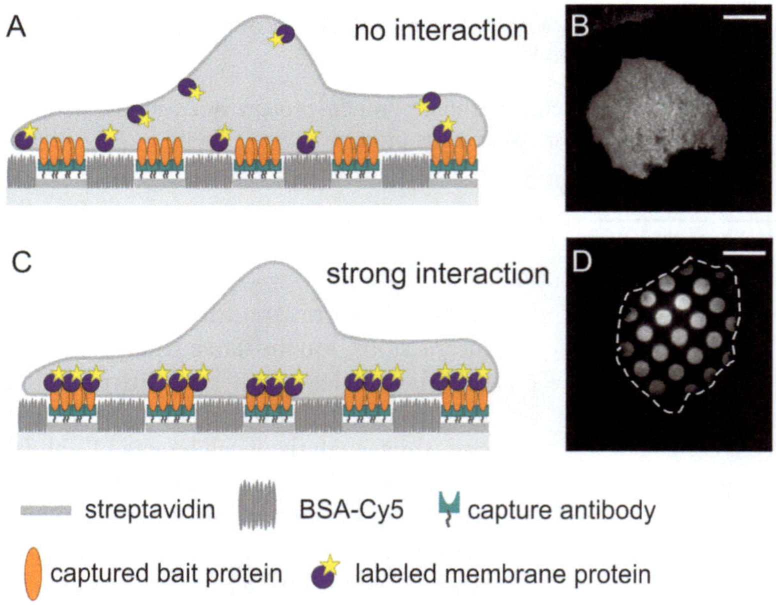

Fig. 1 Principle of protein micropatterning in the plasma membrane. (**a**) Sketch and (**b**) TIRF image of a cell grown on a micropatterned substrate. Bait antibody is arranged in a regular pattern of 3 μm sized dots with 3 μm interspaces. The bait protein (unlabeled) reorganizes according to the antibody patterns, but the fluorescently labeled prey protein is distributed homogeneously in the plasma membrane, indicating no interaction between bait and prey protein. Scale bar is 7 μm. (**c**, **d**) As in (**a**, **b**), but here the prey protein interacts strongly with the bait protein and localizes according to the bait patterns. The cell outline is indicated by a *dashed white contour line*

characterize the interaction of two proteins involved in immunosignaling: CD4, a transmembrane protein, and the tyrosine kinase Lck, a palmitoylated protein that is transiently associated with the plasma membrane [12]. Since then, it has been applied to characterize various protein–protein interactions in several different cell types [10, 13–17] and has been used to determine protein binding curves [18] and dissociation constants [18]. Recently, we have also used Protein Micropatterning to interrogate lipid-mediated protein interactions [19]. Versions of the Protein Micropatterning assay have been reviewed in [20, 21].

2 Materials

Prepare all work solutions fresh each time. Store epoxy-coated coverslips in the desiccator after opening. This protocol is optimized for PDMS stamps; if a different material is used, conditions may need to be adjusted for optimal printing results.

1. Polydimethylsiloxane (PDMS) stamps (*see* **Note 1**).
2. Epoxy-coated coverslips: NEXTERION® slide E (Schott, Germany).
3. Streptavidin stock solution: dissolve 0.5 mg/ml streptavidin (Sigma, USA) in phosphate buffered saline (PBS) pH 7.4. Store aliquots at −20 °C. Do not freeze and thaw.
4. Streptavidin work solution: dilute streptavidin stock solution to 50 μg/ml in PBS pH 7.4.
5. Secure Seal™ Hybridization chambers (Grace Biolabs, USA).
6. BSA-Cy5 stock solution (*see* **Note 2**): dissolve Cy5-labeled bovine serum albumin (BSA-Cy5; Nanocs, USA) to 1 mg/ml in PBS pH 7.4. Store aliquots at −20 °C. Do not freeze and thaw.
7. BSA-Cy5 work solution: dilute BSA-Cy5 stock solution to 100 μg/ml in PBS pH 7.4.
8. Antibody work solution: dilute biotinylated antibody to 10 μg/ml in PBS pH 7.4 containing 1 % BSA.
9. Imaging buffer: Hank's Balanced Salt Solution (HBSS) with Ca^{2+} and Mg^{2+} and 2 % fetal calf serum (FCS) (*see* **Note 3**).
10. Cells expressing bait proteins and fluorescent prey proteins (*see* **Note 4**), Accutase (Sigma, USA) (*see* **Note 5**).

3 Methods

Carry out all procedures at room temperature unless otherwise specified.

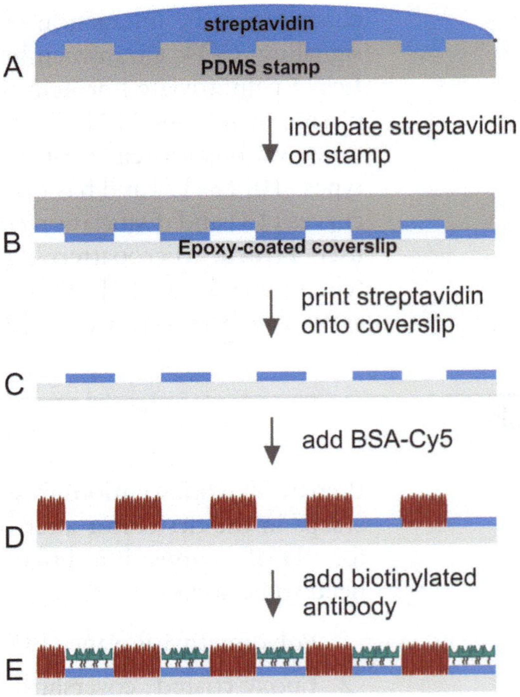

Fig. 2 Soft lithography and functionalization. (**a**) Streptavidin work solution is incubated on a PDMS stamp. (**b**) After washing and drying of the stamp, streptavidin is printed onto an epoxy-coated coverslip. (**c**, **d**) The stamp is removed; BSA-Cy5 is added to fill the interspaces. (**e**) When biotinylated antibody is added, it binds specifically to the streptavidin patterns

3.1 Soft Lithography and Functionalization

The workflow of "3.1 Soft lithography and functionalization" is sketched in Fig. 2.

1. Wash PDMS stamp by rinsing with ethanol (p.a.) and ultrapure water. Dry the PDMS stamp under a stream of a dry inert gas such as nitrogen or argon.
2. Place ~50 μL of streptavidin work solution (50 μg/ml) on the PDMS stamp (the whole pattern area should be covered). Let protein adsorb to stamp for 15 min at room temperature (*see* **Note 6**).
3. Wash the PDMS stamp by rinsing carefully with water and dry under a stream of nitrogen or argon.
4. Place the PDMS stamp face-down under its own weight onto an epoxy-coated coverslip and incubate for 30 min at room temperature or overnight at 4 °C in a humidified atmosphere (e.g., a petri dish with a wet tissue) (*see* **Note 7**).
5. Mark the position of the patterned area on the back of the coverslip with a water-resistant marker and separate the stamp from the slide using tweezers (*see* **Note 8**).

6. Stick a Secure Seal™ Hybridization chamber over the marked area.
7. Add BSA-Cy5 work solution (100 μg/ml) to the hybridization chamber and incubate for 15 min at room temperature (*see* **Note 9**).
8. Wash with 500 μl PBS by adding the buffer into one port of the hybridization chamber and removing it at the second port.
9. Add antibody work solution (10 μg/ml) to the hybridization chamber and incubate for 15 min at room temperature.
10. Wash with 500 μl PBS.
11. Store the micropatterned surfaces with PBS in the dark at room temperature until seeding of cells (*see* **Note 10**).

3.2 Seeding Cells

1. Grow adherent cells expressing bait and prey proteins of interest to 70% confluency in a 10 cm tissue culture dish.
2. Detach cells with Accutase® solution and centrifuge 4 min at 300×*g*. This protocol has been tested for T24, HeLa and CHO cells (*see* **Note 11**).
3. Pellet cells by spinning for 5 min at ~300×*g*.
4. Discard the supernatant and resuspend the cell pellet in 1 ml of the appropriate growth medium. Then, dilute this ~1:10 in growth medium (*see* **Note 12**).
5. Remove the PBS from the hybridization chamber on the micropatterned coverslip and seed cell suspension.
6. Check cell density on a light microscope. Cells should be single but not too sparse.
7. Put coverslips in a petri dish humidity chamber to prevent the sample from running dry and incubate for 1.5–2 h at 37 °C in a 5% CO_2 atm.
8. Before analyzing the cells on the microscope, replace the medium with imaging buffer.

3.3 Total Internal Reflection Fluorescence (TIRF) Microscopy

1. Place the coverslip on a TIRF microscope in a suitable mount (*see* **Note 13**).
2. The BSA-Cy5 grid needed for quantitative analysis is recorded at 647 nm.
3. Distribution of fluorescent prey protein (tagged with e.g., GFP) is recorded (at e.g., 488 nm (*see* **Note 14**)).

3.4 Contrast Quantitation

1. Export microscopy images as 8-bit TIF image. For contrast quantitation it is necessary to export images of the fluorescent prey/bait protein (Fig. 3a) and the respective image with the BSA-Cy5 grid (Fig. 3b). Figure 3c shows the overlaid images.

2. 8-bit TIF images are imported in the semi-automated micropatterning analysis software ("Spotty", *see* **Note 15**).
3. An automatic gridding algorithm is used to calculate the grid-size and the rotation angle ϕ of the used image. The algorithm automatically determines the grid parameters that correctly fit the micropatterned structure (*see* **Note 16**). Cells to be analyzed are detected automatically or can also be selected manually (Fig. 3d).
4. Based on the correct identification of the grid position with respect to fluorescent patterns, the fluorescence contrast can be calculated for each pattern in the image as $C = (F^{+} - F^{-})/(F^{+} - FBG)$, where F^{+} denotes the average intensity of the inner pixels of the micropatterns, F^{-} the average intensity of the pixels surrounding the micropatterns, and FBG the intensity of the global background (*see* **Note 17**) (Fig. 3e).

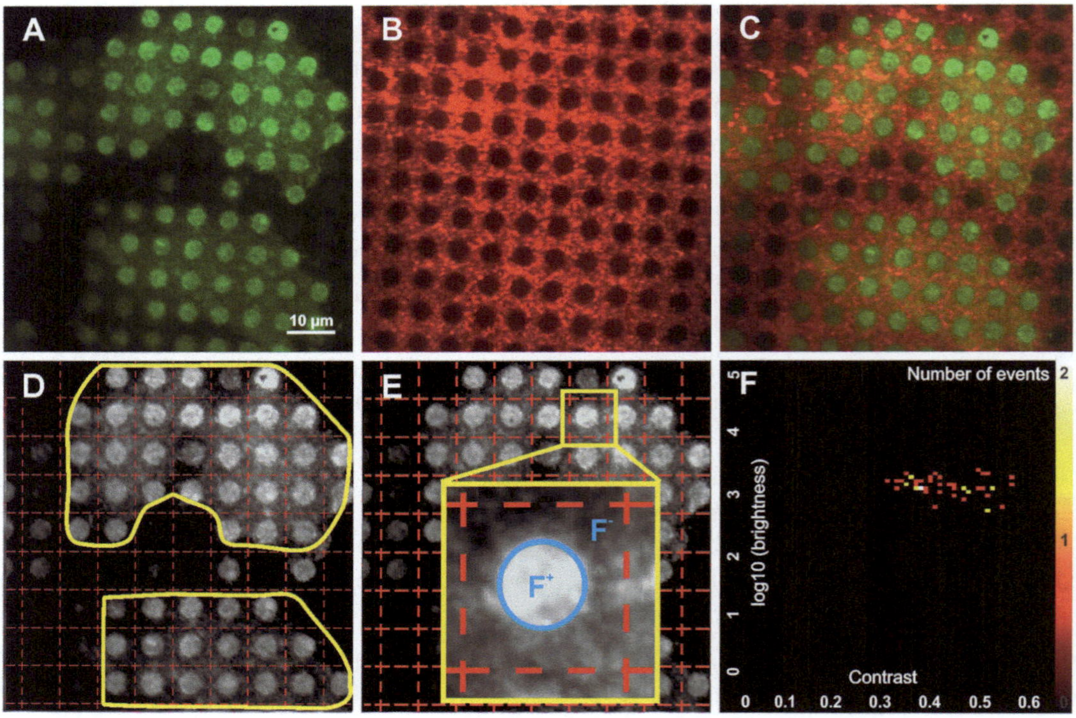

Fig. 3 Quantitation of protein interactions using "Spotty". Image recorded of the fluorescently labeled prey protein (**a**) and the corresponding BSA-Cy5 grid (**b**). (**c**) Overlay. (**d**) An automatic gridding algorithm automatically optimizes the grid parameters and produces a grid that correctly fits the micropatterns. *Yellow lines* denote the cell areas to be used for analysis. (**e**) The grid subdivides the total image into adjacent squares, each of which is quantified according to the average signal within a central circle comprising the micropattern spot (F^{+}) and the signal outside this circle (F^{-}). (**f**) Statistical analysis of multiple cells is shown in a 2D histogram of the fluorescence brightness and contrast. The color scale corresponds to the number of events (i.e., individual analyzed spots)

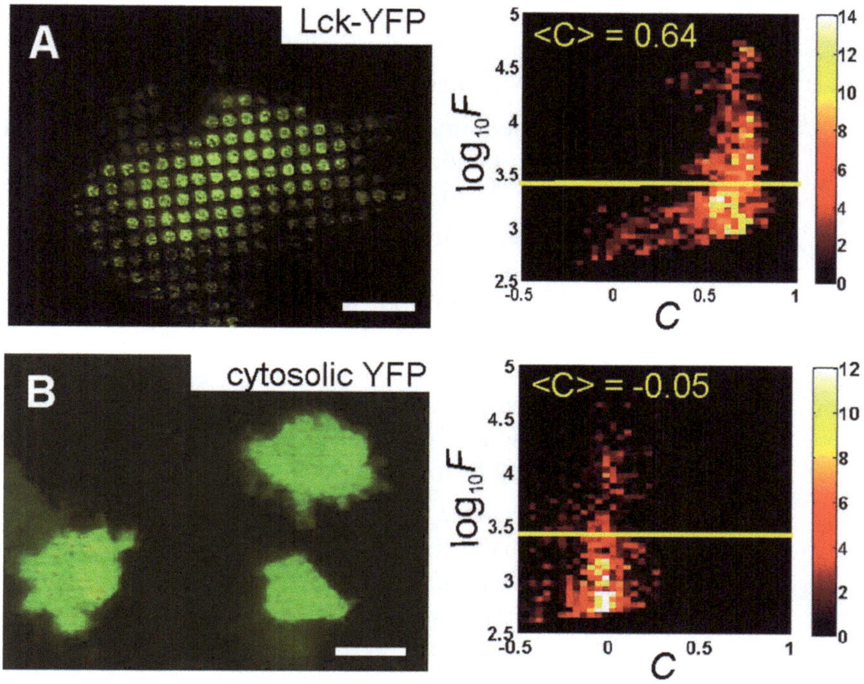

Fig. 4 Examples of generated 2D histograms (**a**) T24 cell transiently expressing CD4 and Lck-YFP grown on CD4 antibody patterns. Lck-YFP interacts strongly with the patterned CD4, which is also reflected in the high contrast values shown in the 2D histogram on the *right*. The low contrast values at lower fluorescence intensities are probably a result of low CD4 (and Lck-YFP) expression levels of a cell subpopulation. For calculating the mean contrast <C>, we only consider data points above a certain intensity threshold (indicated by the *yellow line*). (**b**) T24 cell transiently expressing CD4 and cytosolic YFP grown on CD4 antibody patterns. No copatterning of YFP with CD4 can be observed, the contrast values fluctuate around zero. Scale bars are 10 μm. The color scale corresponds to the number of events (i.e., individual analyzed spots). Figure modified from [12]

5. Several fluorescence parameters (e.g., mean brightness, background fluorescence, contrast,…) as well as graphical descriptions can be extracted from the software for further processing. For statistical analysis of multiple cells, we find it useful to present the data in two-dimensional histograms, with the fluorescence brightness $F = F^{+} - F_{BG}$ on the ordinate against the signal contrast C on the abscissa (Fig. 3f) (*see* **Note 18**). To facilitate comparison of two-dimensional histograms, we use the mean contrast <C>. Figure 4 shows examples of 2D histograms in the presence and absence of protein–protein interaction, yielding high contrast and low contrast values in the 2D histograms, respectively.

4 Notes

1. PDMS is an often-used and reliable material for soft lithography, but it is rather soft. Stamp feature sizes need to be above 1 μm. We prefer stamps featuring regularly spaced dots (3 μm in size, with 3 μm interspaces).

2. BSA can also be labeled with a different fluorophore. Its fluorescence should be spectrally separated from the fluorescence of the prey protein.
3. Growth medium is exchanged for imaging buffer (a) to reduce background fluorescence (if Phenol Red-containing medium is used) and (b) to keep cells at pH 7.4 during measurements.
4. For initial tests, it is convenient to use cells expressing a fluorescent bait protein. This way, successful immobilization of bait protein at the antibody patterns can be evaluated. Alternatively, this can also be verified by staining patterned bait protein with a fluorescently labeled antibody targeting a different epitope than the biotinylated capture antibody. It may be useful, however, to use antibody Fab-fragments, since full antibodies may be excluded from very densely populated patterns.
5. We use Accutase to detach cells because it is gentler than trypsin but equally efficient for most cell types. We found that e.g., loss of glycosylphosphatidylinositol-anchored proteins from the cell surface was significantly reduced when using Accutase instead of trypsin.
6. You can use the pipet tip to spread the streptavidin drop. Do not touch the stamp surface.
7. Water is needed for the streptavidin binding covalently to the epoxy-coated coverslips. In their protocol for protein printing onto Nexterion E coverslips, the manufacturer suggests a humidity of 75 % during printing. We found that using a wet tissue in a petri dish gives satisfactory results.
8. Be careful to lift the stamp without dragging it across the surface.
9. BSA-Cy5 serves two purposes: (a) blocking areas of the coverslips not covered with streptavidin (interspaces) and (b) providing the grid necessary for quantitative analysis.
10. We have found that micropatterned surfaces with the stamps still attached can be stored at 4 °C for at least 2 days without losing imprint quality.
11. Other adherent cell types may be suited for micropatterning as well. For some cell types it may be beneficial to replace BSA-Cy5 (completely or partially) with fibronectin or polylysine to promote cell adhesion in the interspaces between streptavidin regions.
12. Best results will be obtained when cells are plated to ~30–50 % confluency. We use growth medium without Phenol Red to reduce background fluorescence.
13. TIRF microscopy is used to ensure that only membrane-bound prey protein is detected. Otherwise, detection of cytosolic prey protein can lead to an apparently reduced contrast.

14. When using this assay for the first time, we recommend using a fluorescently labeled bait protein as described in **Note 4**. If the fluorescence signals of bait, prey and analysis grid are sufficiently spectrally separated, labeled bait protein can be used for all measurements.
15. "Spotty" can be obtained from www.protein-interaction-lab.at upon request.
16. Evolutionary computation strategies are used for optimized grid identification and detection of micropatterns in biological samples.
17. A relevant factor for the success of contrast evaluation is the size of the F^+ region. It has to be adjusted to fit the actual size of the printed patterns (as shown in Fig. 3e).
18. Taking into account the fluorescence brightness is especially useful when dealing with a heterogeneous cell population with very different expression levels of bait and prey protein (*see* also Fig. 4). It may be advantageous to analyze cell subpopulations of different expression levels separately, or to apply an intensity threshold as shown in Fig. 4.

Acknowledgments

This work was funded by the Austrian Science Fund (FWF projects P 26337 and P 25730), the Austrian Research Promotion Agency (FFG project 842379), the program 'Regionale Wettbewerbsfähigkeit OÖ 2007–2013' with the financial support of the European Fund for Regional Development, as well as the Federal State of Upper Austria.

References

1. Barrios-Rodiles M, Brown KR, Ozdamar B, Bose R, Liu Z, Donovan RS, Shinjo F, Liu Y, Dembowy J, Taylor IW, Luga V, Przulj N, Robinson M, Suzuki H, Hayashizaki Y, Jurisica I, Wrana JL (2005) High-throughput mapping of a dynamic signaling network in mammalian cells. Science 307:1621–1625
2. Rigaut G, Shevchenko A, Rutz B, Wilm M, Mann M, Seraphin B (1999) A generic protein purification method for protein complex characterization and proteome exploration. Nat Biotechnol 17:1030–1032
3. Kerppola TK (2006) Design and implementation of bimolecular fluorescence complementation (BiFC) assays for the visualization of protein interactions in living cells. Nat Protoc 1:1278–1286
4. Young KH (1998) Yeast two-hybrid: so many interactions, (in) so little time. Biol Reprod 58:302–311
5. Maurel D, Comps-Agrar L, Brock C, Rives ML, Bourrier E, Ayoub MA, Bazin H, Tinel N, Durroux T, Prezeau L, Trinquet E, Pin JP (2008) Cell-surface protein-protein interaction analysis with time-resolved FRET and snap-tag technologies: application to GPCR oligomerization. Nat Methods 5:561–567
6. Suzuki KG, Fujiwara TK, Sanematsu F, Iino R, Edidin M, Kusumi A (2007) GPI-anchored receptor clusters transiently recruit Lyn and Ga for temporary cluster immobilization and Lyn activation: single-molecule tracking study 1. J Cell Biol 177:717–730

7. Orth RN, Wu M, Holowka D, Craighead HG, Baird B (2003) Mast cell activation on patterned lipid bilayers of subcellular dimensions. Langmuir 19:1599–1605
8. Mossman KD, Campi G, Groves JT, Dustin ML (2005) Altered TCR signaling from geometrically repatterned immunological synapses. Science 310:1191–1193
9. Waichman S, You C, Beutel O, Bhagawati M, Piehler J (2011) Maleimide photolithography for single-molecule protein-protein interaction analysis in micropatterns. Anal Chem 83(2):501–508
10. Gandor S, Reisewitz S, Venkatachalapathy M, Arrabito G, Reibner M, Schröder H, Ruf K, Niemeyer C, Bastiaens P, Dehmelt L (2013) A protein-interaction array inside a living cell. Angewandte Chemie 52:4790–4794
11. Kane RS, Takayama S, Ostuni E, Ingber DE, Whitesides GM (1999) Patterning proteins and cells using soft lithography. Biomaterials 20:2363–2376
12. Schwarzenbacher M, Kaltenbrunner M, Brameshuber M, Hesch C, Paster W, Weghuber J, Heise B, Sonnleitner A, Stockinger H, Schütz GJ (2008) Micropatterning for quantitative analysis of protein-protein interactions in living cells. Nat Methods 5:1053–1060
13. Weghuber J, Sunzenauer S, Plochberger B, Brameshuber M, Haselgrubler T, Schutz GJ (2010) Temporal resolution of protein-protein interactions in the live-cell plasma membrane. Anal Bioanal Chem 397:3339–3347
14. Alexander RA, Prager GW, Mihaly-Bison J, Uhrin P, Sunzenauer S, Binder BR, Schutz GJ, Freissmuth M, Breuss JM (2012) VEGF-induced endothelial cell migration requires urokinase receptor (uPAR)-dependent integrin redistribution. Cardiovasc Res 94:125–135
15. Lanzerstorfer P, Borgmann D, Schutz G, Winkler SM, Hoglinger O, Weghuber J (2014) Quantification and kinetic analysis of Grb2-EGFR interaction on micro-patterned surfaces for the characterization of EGFR-modulating substances. PLoS One 9:e92151
16. Lanzerstorfer P, Yoneyama Y, Hakuno F, Muller U, Hoglinger O, Takahashi S, Weghuber J (2015) Analysis of insulin receptor substrate signaling dynamics on microstructured surfaces. FEBS J 282:987–1005
17. Bashour KT, Gondarenko A, Chen H, Shen K, Liu X, Huse M, Hone JC, Kam LC (2014) CD28 and CD3 have complementary roles in T-cell traction forces. Proc Natl Acad Sci U S A 111:2241–2246
18. Sunzenauer S, Zojer V, Brameshuber M, Trols A, Weghuber J, Stockinger H, Schutz GJ (2013) Determination of binding curves via protein micropatterning in vitro and in living cells. Cytometry A 83:847–854
19. Sevcsik E, Brameshuber M, Folser M, Weghuber J, Honigmann A, Schutz GJ (2015) GPI-anchored proteins do not reside in ordered domains in the live cell plasma membrane. Nat Commun 6:6969
20. Weghuber J, Brameshuber M, Sunzenauer S, Lehner M, Paar C, Haselgrubler T, Schwarzenbacher M, Kaltenbrunner M, Hesch C, Paster W, Heise B, Sonnleitner A, Stockinger H, Schutz GJ (2010) Detection of protein-protein interactions in the live cell plasma membrane by quantifying prey redistribution upon bait micropatterning. Methods Enzymol 472:133–151
21. Weghuber J, Sunzenauer S, Brameshuber M, Plochberger B, Hesch C, Schutz G. J (2010) In-vivo detection of protein-protein interactions on micro-patterned surfaces. J Vis Exp 37

Chapter 19

Designing Successful Proteomics Experiments

Daniel Ruderman

Abstract

Because proteomics experiments are so complex they can readily fail, and do so without clear cause. Using standard experimental design techniques and incorporating quality control can greatly increase the chances of success. This chapter introduces the relevant concepts and provides examples specific to proteomic workflows. Applying these notions to design successful proteomics experiments is straightforward. It can help identify failure causes and greatly increase the likelihood of inter-laboratory reproducibility.

Key words Design of experiments, Randomization, Bias, Variance

1 Introduction

This chapter's goal is to help researchers design proteomics experiments that succeed. I will present the concepts of experimental design in the context of proteomics workflows. The particular combination of experimental factors that impact proteomic offers specific challenges for their design. Since each experiment is unique, some extrapolation to particular circumstances will be needed. I hope to provide the tools to do so. Additionally, there are many good experimental design texts (e.g., [1–3]) and online resources (e.g., [4]), which I encourage the reader to explore.

1.1 Why Proteomics Experiments Benefit from Design

Proteomics aims to quantify thousands of proteins in complex biological backgrounds, such as tissue and plasma. It is typically done using sensitive instruments like mass spectrometers following multiple preparation steps, including protein extraction, enrichment, fractionation, and digestion. Because these complex laboratory processes can lack stability across a full study, they can negatively impact the outcome unless the experiment is designed to take this into account.

In their 2005 paper [5], Hu et al. describe three proteomics studies that failed due to poor experimental design. In one study, cancer sample data showed strong dissimilarity between run dates.

Lucio Comai et al. (eds.), *Proteomics: Methods and Protocols*, Methods in Molecular Biology, vol. 1550,
DOI 10.1007/978-1-4939-6747-6_19, © Springer Science+Business Media LLC 2017

Because one cancer subtype's samples were all run on a single day, any biological signal was confounded with any technical changes that day. In another study, samples were found to group in similarity not by cancer subtype, as expected, but instead by the date the collection protocol was changed. Their third example was the victim of erroneous mass spectrometer calibration and sample degradation. These studies would likely have succeeded with proper experimental design and inclusion of quality controls. Poor experimental design is cited as a major impediment to translational proteomics [6].

Science is ultimately held to the standard of reproducibility. So it is surprising that attempts to reproduce the results of high-profile biomedical studies mostly failed [7, 8]. Poor experimental design has been highlighted as a cause reproducibility failure [9]. The US National Institutes of Health has recently emphasized reproducibility, stating that grant applications will be reviewed specifically for "rigorous experimental design for robust and unbiased results" [10]. Thus experimental design is not only important in executing research, but in securing the funds to do so in the first place.

2 Materials

As best you can, gather and enumerate the following:

1. The scientific question you are addressing with the study.
2. A list of available samples and their annotations (including experimental factors such as treatment group and "nuisance" factors like sample processing date and technician).
3. Response variables you will measure (e.g., peptide abundance, protein identity, retention time).
4. A list of experimental factors that will be varied to answer the question (e.g., disease vs. control, drug dose, time point).
5. A list of nuisance factors which need to vary across the experiment (but ideally would not) and may affect response variables (e.g., run date, reagent batches, LC columns).
6. Experimental protocols.
7. Ground truth analytes and quality control samples.
8. Experimental design and data analysis software (*see* **Note 1**).
9. Estimates of the effect size and biological/technical variability in the response variable (typically through a pilot experiment).
10. Statistician consultant/collaborator.

3 Methods: Seven Steps to Executing a Well-Designed Proteomics Experiment

A successful experiment is one that can be replicated by others. This means that all the factors specific to one laboratory—like equipment, reagents and technique—must not influence its outcome. The experimental result must accurately reflect a sample weighing micrograms, and yet be unaffected by any variability in the tons of lab equipment used to measure it.

The techniques of experimental design have been recognized as an important component of proteomics research for over a decade [11–16]. An in-depth treatment can be found in the seminal work of Vitek and colleagues [11, 12]. The basic steps outlined below should help researchers achieve unbiased, efficient proteomics experiments. A visual outline of the approach is shown in Fig. 1.

3.1 Define the Research Goal and Relevant Response Variables

A clear research goal determines the experiment you'll want to run. It embodies questions like: "What effect will you measure?" and "Which sample groups will you compare?" These in turn identify the response variables you will measure and the experimental factors you will change to determine their impact.

Response variables are the experimental outputs (numbers) used to answer the research question. In quantitative proteomics they may be the abundances of peptides or proteins. In proteome surveys they may instead be the number of distinct proteins found across samples. Core to designing an experiment is choosing the experimental methodology. Other chapters in this volume can aid in the decision, for example, between MRM (Multiple Reaction Monitoring), DDA (Data-Dependent Acquisition), DIA (Data-Independent Acquisition), or other methods [17]. The best choice will optimize the ability to accurately measure the response variables, possibly at the expense of reduced performance by other metrics. For example, in setting a mass isolation window there is a trade-off between analyte selectivity and quantitation sensitivity [18].

It is important to have an estimate of the effect size, as it will help determine how many samples will need to be run (*see* below). For early stage biomarker screening any significant difference may be of interest, no matter how small. In other cases, such as toxicology studies, only physiologically meaningful differences are of value.

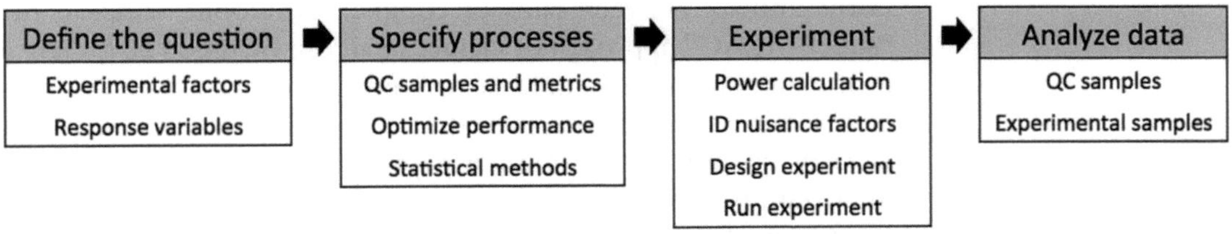

Fig. 1 Four major activities and their component steps to designing an unbiased and efficient experiment

3.2 List Relevant Factors

Factors are those things that potentially affect response variables. Ideally the response would only be impacted by the physical property the experimental system is designed to measure. But this is almost never the case. For example, the measured peptide peak area depends not just on that peptide's abundance but also on electrospray stability, chromatography column carryover, and the efficiency of proteolytic digestion. Digestion efficiency may in turn depend on the trypsin reagent lot number and which lab tech performs the digestion.

Those factors we change as part of the experimental investigation are called *experimental factors*. For example, when comparing the proteome between subcellular fractions, the proteomic fraction (e.g., cytoplasm or nucleus) is such a factor. If we are instead testing the effect of chromatography on peak height, then flow rate might be an experimental factor. Once the experimental factors have been chosen to answer the research question, all those remaining in the list are called *nuisance factors*. These are conditions that can affect response variables but do not remain fixed for all samples or runs. Examples include reagent batches, intra-day LC-MS run order, and personnel. The goal of identifying nuisance factors is to minimize their effect on response variables while allowing experimental factors to systematically alter those responses.

3.3 Develop Control Samples and Quality Control Procedures

Two kinds of control samples are important for verifying that the experimental platform actually works. *Experimental controls* ensure accuracy: Quantitation should reflect what is in the sample; actual differences should be detected; lack of change between sample groups should not. *Quality control samples* (QCs) ensure that processes and equipment perform within defined specifications [19, 20].

Experimental controls for accuracy are used to ensure that quantities like retention time, *m/z*, and protein/peptide abundance are well measured. They are typically created by spiking compounds into the experimental samples. Controls can also be run through the mass spectrometer at different times than the experimental samples (e.g., lock masses during alternate MS scans, separate QC samples). Krokhin and Spicer [21] describe a set of spike-in peptides for normalizing reverse phase HPLC retention times. The peptides enable both retention time normalization of data between runs and hydrophobicity index prediction of sample peptides [22] for improved identification. For protein abundance normalization between runs of complex proteomic samples, the "super-SILAC" approach of Geiger et al. can be used [23]. Here a proteomic standard sample is derived from species-specific cells cultured in SILAC media, providing mass shifts to tryptic peptides so they do not overlap sample peptides. This approach has two important benefits. First the proteins are matched to the sample, so most peptides will have a corresponding quantitation standard. Second, because the standard can be spiked into the sample prior to fractionation and digestion, the ratio of sample to standard peptide abundance is unaffected by any variability from those upstream processes.

Positive and negative experimental controls can be used to nest an "orthogonal" experiment within the main experiment. In these controls protein abundances differ by a known amount, which enables testing of the experiment's ability to detect them. These control samples should closely resemble the experimental samples. For example, if studying the plasma proteome, the control samples are ideally based on a single plasma sample that is available in sufficient quantity to run periodically during the entire experiment (*see* **Note 2**). A positive control for detecting protein differences between two samples might contrast a human plasma sample with and without spiked-in proteins [24]. The corresponding negative control would be multiple runs of the same sample to ensure that few proteins are found that differ (false discoveries). To instead control an experiment quantifying peak area differences between sample groups, proteins can be spiked into control samples at two different concentrations (which vary from protein to protein, e.g., UPS2 from Sigma-Aldrich). The concentration differences might cover a range from multiple fold-changes down to a few percent in order to detect the assay's lower limit of change detection. Running both control samples multiple times during the course of the experiment will mimic an experiment on two separate sample groups (though without biological variability within each group).

The purpose of quality control samples is to indicate when processes or equipment have failed, or, better, reassure us that they have not. Measurements on quality control samples should indicate quantitatively how well processes and equipment perform. Multiple samples can be used to QC different experimental steps. For example, a protein fractionation workflow can be periodically assessed using a cell lysate or other complex mixture, with quantitative readout given by UV absorption across fractions. LC-MS performance can be measured using a simple mixture of peptides, quantifying the peak areas across peptides and charge states as well as other peaks reflecting contamination or carry-over. Alternatively, a QC sample derived from biological samples of interest (or a vendor-provided reference sample [25]) can be periodically run through the entire workflow. Sets of performance metrics and analysis software are available from the NIST [26], the *QuaMeter* tool [27, 28], and SProCoP [29]. Metrics specific to isobaric labeling can be found in [30]. Finally, QCs should be implemented to detect any previously identified failure modes.

Accurate peptide/protein quantitation is a general goal of proteomics. In unlabeled experiments, this hinges on chromatographic reproducibility between runs, for two reasons. First, quantitation itself is impacted by variations in elution peak width. Second, the ability of software to reproducibly identify the same peak across runs is degraded by retention time shifts, which can lead to missing values and artificially increased unique peptide counts. Use of performance metrics can assist in optimizing chromatography [26, 27, 29]. For isobarically labeled experiments (e.g., iTRAQ, TMT), peptide/protein abundances are compared to a known standard that is labeled and present in all runs. To reduce variability, it is best to label the standard once as a large batch. Although including a standard in this way reduces experimental efficiency, it both enables LC-MS quality control and data normalization between runs. For some experiments there are alternative computational normalization methods that do not require a control standard [31].

3.4 Optimize Process

Protocols such as found in this volume and in vendor application notes give detailed roadmaps for key parts of the experiment. But there remain many parameters to adjust in making the experimental system work well as a whole. For example, Agilent's Jet Stream electrospray ionization source has six adjustments, such as drying gas temperature, nebulizer voltage, and capillary voltage [32]. Additionally, MS data analysis packages have their own settings. The OMSSA search engine [33], for example, has 9 of them, not including PTM selection. An experiment's success may hinge on getting them right.

Fortunately, there are systematic methods for optimizing these settings. The statistics sub-field of *Response Surface Methodology* (RSM) [34] is used to design and analyze an experiment to estimate the settings that maximize a performance metric. Although process optimization is not strictly part of experimental design (it need not be done for each individual experiment), it is equally important and often overlooked. The basic idea is to perform a designed experiment across a set of parameter adjustments. RSM then fits the performance metric to a (usually) quadratic surface in the space of parameters, and estimates the parameter settings that optimize performance. To optimize data analysis parameters no additional experiments need be performed. Instead, the analysis software can be run on a single representative data set under different parameter settings. A number of examples demonstrating proteomics performance gains through process optimization have been published [35–42].

It is not always obvious how to choose the performance metric. For example, when trying to discover biomarkers, is it best to maximize the total number of spectral "features" detected or instead to maximize the number of features with high signal-to-noise ratio? Is the number of identified proteins also important? A standard approach in such situations is to mathematically combine a number of relevant metrics using Derringer's *desirability* metric [34] (*see* **Note 1**). It tends to compromise well between competing criteria. One such tradeoff is found during label-free MS1 quantitation in tandem mass spectrometry, where higher accuracy can be achieved by dedicating more scans per eluted peak to MS1 while sacrificing MS2-based identification scans. The desirability metric can be used to find a principled compromise between quantitation accuracy and depth of protein identification.

3.5 Estimate the Number of Samples Needed

Underpowered studies do not have sufficient independent samples or technical replicates to answer research questions with high probability. They can lead to either false positive or false negative conclusions [43]. The probability of detecting an effect when it is actually present is called the *power*. Researchers typically aim for 80–90 % power. To create a study that is likely to find the effect we are looking for, we must estimate how many samples will be needed and design the experiment accordingly. This involves knowing two things: how big an effect we are looking for (the signal) and how much the measured response varies between samples in the same group (the noise).

Measuring the noise level through pilot experiments is a critical step in the design. I recommend running at least ten independent biological samples from within the same group (e.g., disease, control) to assess "all-in" variability. This includes noise both from the technical experimental aspects and those from biological variation between replicate samples. Additionally, one biological sample should be processed and measured at least five times to assess the technical variability magnitude. While these experiments may seem to be a costly investment, the knowledge they provide is key to running a successful experiment the first time around.

In designing experiments, we quantify noise as the statistical variance in a measurement when the "signal" (in this case, the sample group or the sample itself) does not change. In proteomics there is often more than one signal of interest and thus many such variances to measure [44]. For MS1 data there may be more than 10^5 individual peaks quantified, each with its signal-to-noise ratio (*see* **Note 3**). In such cases a good shortcut is estimate a single typical noise variance (say, the 75th percentile across all peaks) and use it to compute the number of samples needed to detect a range of biologically relevant peak area differences (*see* below). Note that for abundance measurements noise processes are usually multiplicative, and logarithmic transforms of the data are typically used instead of raw abundance values [11].

In general, a study's power calculation method depends on the actual statistical analysis to be performed. However, a simple and useful method for estimating sample sizes uses the Gaussian noise approximation. This assumes the response variable is either continuous (e.g., a peptide's abundance) or even a large integer (e.g., the number of proteins identified), and that noise due to both intra-group and technical variability is approximately Gaussian and additive. In this case the well-known sample size estimate

$$N = \left[\left(z_{1-\alpha/2} + z_{1-\beta}\right)\sigma / \delta\right]^2$$

holds, where N is the minimum number of samples needed per group, zq is the qth quantile of the normal distribution, α is the desired significance level (typically 0.05), 1-β is the power, σ is the estimated total noise standard deviation (biological and technical), and δ is the expected effect size. At a significance level of 0.05 and power of 80%, $z_{1-\alpha/2} = 1.96$ and $z_{1-\beta} = 0.84$, so $N = 7.84(\sigma / \delta)^2$. Thus for a signal-to-noise ratio (σ/δ) of 1, at least 8 samples per group are needed to power such a study. For a more complete discussion on choosing the number of technical and biological replicates, *see* Ref. 11. As noted above, running pilot experiments will provide the biological and technical noise levels needed to make these important decisions. Software such as *JMP* and *Stat-Ease* provide methods for determining sample size (*see* **Note 1**).

Often the study goal is to estimate a number, like a peptide's average abundance across samples, to some precision. Assuming independent Gaussian variability (combined biological and technical) in each sample's measurement with standard deviation σ, the number of samples to achieve a precision of τ is approximately $N = (\sigma / \tau)^2$.

Although outside the scope of this chapter, determining the sample size while controlling the false discovery rate instead of the false positive rate (significance) can also be achieved [45].

3.6 Configure the Experiment

An experiment's design maps its idealized set of measurements onto laboratory reality. Which samples should be analyzed? In what order should they be run? How many QC samples should be included? A good experimental design will provide enough statistical power to answer the research question and do so without biasing the result. In what follows I will assume there are two sample groups being compared. While many of the same ideas apply to more complex (e.g., multi-factorial) investigations, I direct the reader to standard texts for more complete treatments [1, 2].

Modern design of experiments goes back to R.A. Fisher's work, first published in 1926 [46]. With the joint goals of efficiency and validity, he identified three cornerstones of experimental design: *blocking*, *randomization*, and *replication*. Blocking and randomization serve to eliminate biased answers. Replication increases the signal-to-noise ratio to improve precision of the experiment's result. **The concepts of blocking, randomization and replication are the most important takeaways of this chapter**. I present them and other design considerations below.

3.6.1 Blocking

Many factors change discretely over the course of an experiment. Some examples are reagent batches, personnel, date, and run order. As one of these factors changes it can alter the response of the experimental system. Lab members may use different techniques. If two sample groups are isolated by levels of such a factor (e.g., all responding patients' sample run on one day and all non-responders on another), then a significant technical change may be mistaken for a biological effect.

The solution is to use blocking. For each level of a nuisance factor (e.g., date, personnel) a small balanced experiment is run. As an example, if only eight samples per day can be run, then four from each of two sample groups would be run per day. Any day-to-day variation in the response variable would then affect both groups nearly equally, preventing bias which might lead to a false discovery. When there are multiple such factors, it can be challenging to solve the design in a balanced way, particularly if there are interactions between those factors. When a balanced design is not possible (e.g., sample counts are not multiples of the block size), randomization should be used instead (see below). George Box offered the sound advice: "Block what you can, randomize what you cannot."

If a nuisance factor is known to have little or no effect on the response variable, then it is safe to forego blocking it. One such factor might be pipette tip lot numbers; it is simply very unlikely that it makes any difference. The magnitude of a nuisance factor's impact can be assessed by running a control sample across that factor's levels, followed by appropriate data analysis (e.g., linear fixed-effects or mixed-effects modeling). It is a good idea to run QC samples periodically to monitor longitudinal performance and detect any such effects. Furthermore, once a balanced experiment has been run its data can be analyzed to determine the magnitude of variation and temporal trends due to nuisance factors. This can be a good way to detect process issues even though they may not directly impact the experiment's results.

3.6.2 Randomization

There may be nuisance factors with many levels where it is not obvious how to perform blocking, for example the assignment of samples to wells in a 96-well plate. Although it is known that plate-based experiments can have systematic changes toward plate edges, it is very effective simply to randomize sample placement. Similarly, the order in which a day's samples are processed can be randomized in case of changing equipment or personnel attentiveness. Temporal randomization (both within and across days) in proteomics is particularly important since there are many potential sources of drift, including LC column degradation, electrospray instability, transfer tube build-up, and tuning loss.

Randomization has the added advantage of counteracting bias due to unrecognized influences. For example if a contiguous subset of vials gets contaminated, and they were randomized relative to sample groups, then the experimental result will show increased variability but without a biased effect on the conclusion (although with increased confidence intervals). Randomization is particularly

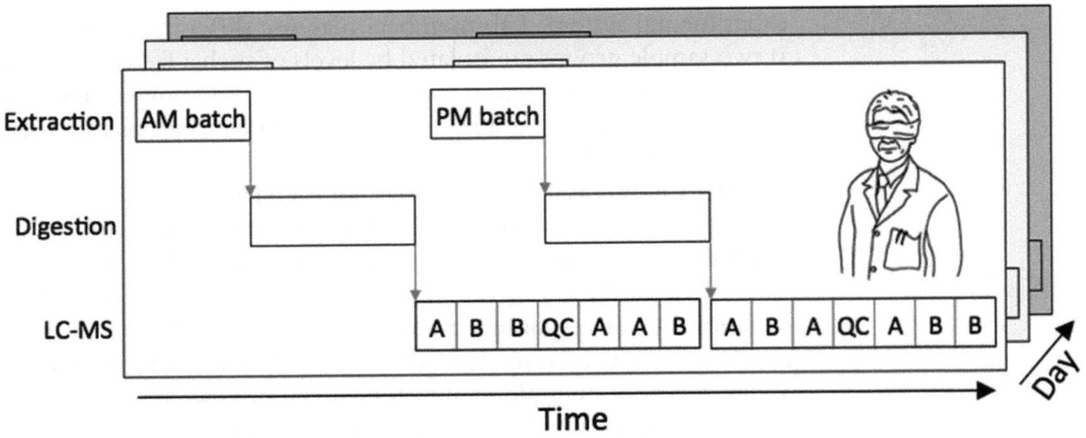

Fig. 2 A randomized block design for a proteomics LC-MS workflow. Two sample groups (*A* and *B*) are being measured for proteomic differences. The experimental day is a blocking factor, and morning (AM) versus afternoon (PM) is a blocking factor nested within the day. Each such block contains six samples run (balanced between three samples of each group). The 6-sample run order is randomized to avoid confounding temporal drift with the A/B proteomic difference, with laboratory personnel blinded to the ordering. At the middle of each LC-MS run a QC sample is analyzed to monitor process. QC and sample data should be analyzed to ensure no systematic AM versus PM differences or large day-to-day variation impact the signals of interest

important in the case of longitudinally collected samples. They should be processed in random order to ensure that any procedural changes are not mistaken for biological trends. Many experiments randomize samples within blocks, which is called a *block-randomized design*. An example is shown in Fig. 2.

Multi-run isobaric labeling (e.g., iTRAQ, TMT) experiments [47] are cases where blocking and randomization should be used. An example using 4-plex iTRAQ to contrast two sample groups of eight samples each in a completely balanced design is shown in Fig. 3a. Here each of the four labels is applied twice per sample group (Fig. 3b). Also, each pooled MS run contains two samples from each group (Fig. 3c). Thus bias due to either a particular label or run will have limited impact on the inter-group comparison. When sample counts are unbalanced between groups or not divisible by the number of labels, assignment of sample to label and/or run should be made randomly. Although many have reported only minimal label bias (e.g., [30, 48]), it is best to start with label randomization and discontinue only if statistical testing demonstrates no bias. **When using randomization, there is a trade-off between reducing bias and increasing both experimental effort and the risk of introducing procedural error.**

3.6.3 Replication

Unlike blocking and randomization, which seek to remove systematic bias that could be mistaken for or mask a real effect, the goal of replication is to reduce the impact of random measurement variation. This leads to more precise estimates of the effects that answer

Run/ pool	114	115	116	117
1	A1	A2	B1	B2
2	A3	B3	A4	B4
3	B5	A5	B6	A6
4	B7	B8	A7	A8

(a)

Label	As	Bs
114	2	2
115	2	2
116	2	2
117	2	2

(b)

Run/ pool	As	Bs
1	2	2
2	2	2
3	2	2
4	2	2

(c)

Fig. 3 (**a**) Blocked and balanced design for a 4-plex iTRAQ experiment contrasting two sample groups (*A* and *B*) with eight samples per group (A1 through A8, B1 through B8) in four runs/pools. (**b**) Frequency table for labels across sample groups showing balanced design. (**c**) Frequency table for runs across sample groups showing balanced design

the research question. Replication comes in two types. By repeatedly drawing independent random samples from a population we achieve more and more accurate estimates of that population's mean response value (e.g., a protein expression level). This is *biological replication*. Additionally, any given sample can be processed multiple times in the laboratory and the results averaged to reduce the impact of process variability. This is *technical replication*.

Biological replication serves to accurately reflect a population of interest using only a subset of samples. The population may, for example, be patients who responded to a therapy or cell cultures grown under specific conditions. Because there is inherent variability among patients and the cells that happen to grow best in culture, a single sample from those populations may deviate greatly from the population average of interest. The number of biological replicates to employ in a study depends in large part on the desired power to detect an estimated effect, as mentioned above. However, there are other considerations.

Some sample populations are particularly heterogeneous, having "long tails" that provide sample values far from the mean more frequently than expected from Gaussian statistics [49]. In my experience clinical samples often have this property. In such cases it is important to run enough samples to include some from the tails, otherwise replication will fail when they show up in validation sets. I thus recommend at least 10 samples from each group when heterogeneity is expected. In contrast, replicates of laboratory samples, like cell culture, are more likely to have Gaussian statistics. In this case the estimates from power calculations are likely to work well, even if fewer than 10 samples.

Technical replication reduces the impact of process noise. It is useful particularly when laboratory procedures introduce variability which is on the scale of or larger than the population variability. This may particularly be the case for low concentration analytes that are inherently difficult to measure precisely. Protein identification in complex samples is often improved by replicating data-dependent LC-MS/MS runs and accumulating the list of proteins identified across them. Liu et al. [50] found that 10 such runs are needed to identify 95 % of the proteins in yeast lysate. To replicate a process the sample must be split and processed multiple times, meaning sufficient sample must be available to rerun the protocol in question. To determine which steps in a workflow cause the most variability, an *analysis of variance* can be performed [51]. Highly variable steps can be remedied through process improvement or technical replication [13].

3.6.4 Blinding

Modern clinical trials—arguably some of the most carefully designed experiments anywhere—are blinded for good reason. When patients, clinicians, or data analysts know which subjects belong to which group, results become biased [52]. There is no reason to expect scientific studies to be any different. Indeed, many biological results were found not to be reproducible by the original investigators once they are blinded [53]. Data analysts also need to be blinded [54], which suggests that confirmation bias has strong impact on biomedical research.

3.6.5 Control Samples

An experiment's design should include control samples to ensure process. This is most crucial in complex experiments involving many samples measured across multiple days or instruments. However, even a simple experiment on just a few samples will benefit, particularly when a negative result is buoyed by knowledge that the measurements are reliable. The ~15 % additional cost in time and resources is well worth the payoff.

Quality control samples measure whether technology is functioning consistently and within specifications. Since most proteomic workflows involve multiple steps (e.g., capture, fractionation, LC-MS) it is best to insert the QC samples as early in the workflow as possible to test them all. A daily series of sample preparations can be "bookended" by QC samples at either end to measure performance changes. If intra-day changes are unlikely then a single QC sample could instead be placed in the middle of a day's experimental samples. In either case, if a QC sample demonstrates the equipment has failed, those runs can be rejected and the samples possibly rerun. Additionally, the response measures on the QC samples can be used to normalize experimental data for possible drift.

Figure 4 shows a design exemplifying blocking, randomization, and QCs. This design was motivated by a set of blood samples collected from eight subjects in pairs, one prior to and one after a seizure. The research goal was to determine whether there are detectable changes in the plasma proteome following seizure. The design specifies how samples are run on LC-MS on a system with two LC columns (LC-1 and LC-2) alternately driving a single

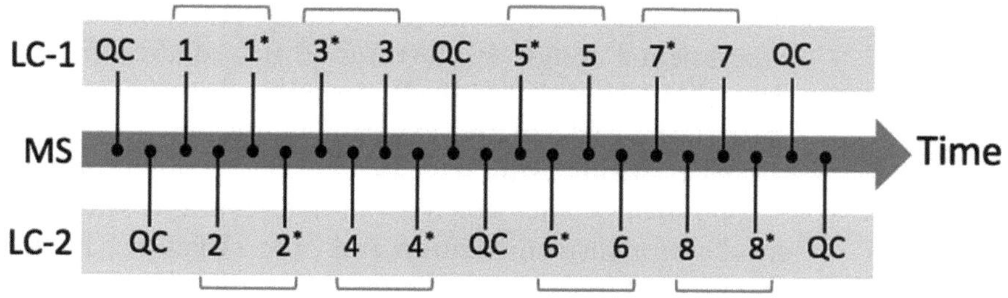

Fig. 4 Example of blocking by LC system and inclusion of QCs in a paired sample (pre- [*no asterisk*] and post-seizure [*asterisk*]) workflow. *See* Subheading 3.6.5 for details

MS. It includes technical QC samples on every fifth elution from each column to monitor performance. Because samples are evaluated pairwise for differences, it is important that they are run in a manner that minimizes any technical changes. Thus each LC forms a block that contains all samples from four patients so that no patient's samples are subject to inter-column differences. Furthermore, to mitigate temporal drift, both samples from a patient are run as close in time as possible. Finally, to ensure there is no systematic effect of temporal drift on the pre- vs post-seizure effect, half of the patients have the pre-seizure sample (no asterisk) run first and half had the post-seizure sample (asterisk) run first.

As previously mentioned, positive and negative experimental control samples enable researchers to demonstrate the capability of the experiment to correctly discover positive results while ignoring negatives. More complex controls may instead be designed to quantify the sensitivity to detect effects of various sizes against a fixed specificity (e.g., $p < 0.05$). To best mimic the measurement conditions of experimental samples, these controls should be randomized into the general sample population and their data analyzed separately.

3.7 Choose the Statistical Analysis

In addition to vendor-specific software, there are now many freely available analysis platforms for proteomics data [55–58]. Analysis steps include protein identification, quantification, and statistical testing. The most common methods are t-tests and ANOVA. Since these tests make certain assumptions about the data statistics (e.g., Gaussian noise, homogeneity of variance, independence across samples) one should be sure these preconditions are met (particularly for small data sets) [3]; if not, other methods such as nonparametric tests (e.g., Mann–Whitney U) can be used. Quantitation through spectral counting, for example, can often have non-Gaussian noise statistics. With blocked designs one must ensure blocking is accounted for in the analysis (so that the possibly large variation between blocks is ignored when estimating the noise level). This can typically be accomplished using a blocking factor in ANOVA [59] or mixed effects models [60]. As a general rule the statistical analysis must match both the research question and the experimental design. It must not be altered after the collection of data ("HARKing" [61]). **I strongly recommend engaging a statistician early in the experiment design process to address these questions**. After data are collected is often too late.

Control samples also must be analyzed. QC samples should be checked for instrumentation and process failure. These QCs can also be used to detect performance variation between blocks (e.g., across days) to characterize changes in equipment performance. Analysis of positive and negative control samples should validate that the entire experimental workflow functions properly.

4 Conclusions

Oberg and Vitek state that "...all quantitative investigations fail to deliver reproducible and accurate results if proper attention is not devoted to the experimental design" [11]. A well-designed experiment does not take much more effort to execute than a poorly designed one. Yet, it can be crucial to the experiment's success or understanding why it failed. R. A. Fisher's basic prescription was for blocking, randomization, and replication. Additionally, by instituting appropriate QCs, not only can outlier samples caused by technical failures be recognized and removed, but long-term changes in process performance can also be detected before they negatively impact experiments. The goal is results that others can consistently reproduce. Finally, I encourage the sharing of proteomic data via public repositories (e.g., *proteomexchange.org*). It not only reduces unneeded replication of research, but also enables the comparison of similar data sets to evaluate quality and accuracy. Although standards for LC-MS quality control runs do not currently exist, sharing this information together the experimental data could one day greatly ease the process of debugging and optimizing proteomics experiments. This would help make efficient, reproducible proteomics accessible to a far wider group of researchers.

5 Notes

1. For select tasks, the following software are useful:

   The *MSStats R* package for statistical analysis of quantitative mass spectrometry data [60].

   Using *Bioconductor* in *R* to analyze mass spectrometry data [55].

   The *R* package *desirability* [62] aids in formulating and optimizing multiple experiment performance read-outs (e.g., mass accuracy and quantitation noise), using the methods of Derringer and Suich [63].

JMP (www.jmp.com) and *Stat-Ease* (www.statease.com) offer tools for designing and powering experiments.

2. To ensure sufficient quantity, a pool of multiple plasma samples could also be used. But its characteristics may differ in unknown but important ways from an actual plasma sample. Thus I (weakly) prefer using a single patient's sample.
3. Many of these peaks may be redundant since they correspond to different peptides from the same protein.

Acknowledgments

I thank Dr. Parag Mallick for introducing me to the field of experimental design and Dr. Nicholas Graham for comments on the manuscript.

References

1. Montgomery DC (2013) Design and analysis of experiments, 8th edn. John Wiley, Hoboken, NJ
2. Box GEP, Hunter JS, Hunter WG (2005) Statistics for experimenters : design, innovation, and discovery. Wiley series in probability and statistics, 2nd edn. Wiley-Interscience, Hoboken, NJ
3. Quinn GP, Keough MJ (2002) Experimental design and data analysis for biologists. Cambridge University Press, Cambridge, UK
4. NIST/SEMATECH e-Handbook of statistical methods. http://www.itl.nist.gov/div898/handbook/index.htm
5. Hu J, Coombes KR, Morris JS (2005) The importance of experimental design in proteomic mass spectrometry experiments: some cautionary tales. Brief Funct Genomic Proteomic 3(4):322–331
6. Maes E, Cho WC, Baggerman G (2015) Translating clinical proteomics: the importance of study design. Expert Rev Proteomics 12(3):217–219
7. Prinz F, Schlange T, Asadullah K (2011) Believe it or not: how much can we rely on published data on potential drug targets? Nature: pp 1–2. doi: papers3://publication/doi/10.1038/nrd3439-c1
8. Begley CG, Ellis LM (2012) Drug development: raise standards for preclinical cancer research. Nature 483 (7391).[Online]. doi: papers3://publication/uuid/0C526949-4171-407F-834A-A82A37275234
9. Begley CG, Ioannidis JPA (2015) Reproducibility in science: improving the standard for basic and preclinical research. Circ Res 116(1):116–126
10. The National Institutes of Health (2015) Enhancing Reproducibility through Rigor and Transparency. http://grants.nih.gov/grants/guide/notice-files/NOT-OD-15-103.html
11. Oberg AL, Vitek O (2009) Statistical design of quantitative mass spectrometry-based proteomic experiments. J Proteome Res 8(5):2144–2156
12. Riter LS, Vitek O, Gooding KM, Hodge BD, Julian RK (2005) Statistical design of experiments as a tool in mass spectrometry. J Mass Spectrom 40(5):565–579
13. Karp NA, Lilley KS (2007) Design and analysis issues in quantitative proteomics studies. Proteomics 7(Suppl 1):42–50
14. Leek JT, Scharpf RB, Bravo HC, Simcha D, Langmead B, Johnson WE, Geman D, Baggerly K, Irizarry RA (2010) Tackling the widespread and critical impact of batch effects in high-throughput data. Nat Rev Genet 11(10):733–739
15. Cairns DA (2011) Statistical issues in quality control of proteomic analyses: good experimental design and planning. Proteomics 11(6):1037–1048

16. Cairns DA (2014) Statistical issues in the design and planning of proteomic profiling experiments. Clin Proteomics 18:223–236
17. Leitner A, Aebersold R (2013) SnapShot: mass spectrometry for protein and proteome analyses. Cell 154(1):252–252e251. doi:10.1016/j.cell.2013.06.025
18. Gallien S, Bourmaud A, Kim SY, Domon B (2014) Technical considerations for large-scale parallel reaction monitoring analysis. J Proteomics 100:147–159. doi:10.1016/j.jprot.2013.10.029
19. Montgomery DC (2013) Introduction to statistical quality control, 7th edn. Wiley, Hoboken, NJ
20. Bramwell D (2013) An introduction to statistical process control in research proteomics. J Proteomics 95(C):3–21, doi: papers3://publication/doi/10.1016/j.jprot.2013.06.010
21. Krokhin OV, Spicer V (2009) Peptide retention standards and hydrophobicity indexes in reversed-phase high-performance liquid chromatography of peptides. Anal Chem 81(22):9522–9530, doi: papers3://publication/doi/10.1021/ac9016693
22. Krokhin OV (2006) Sequence-specific retention calculator. algorithm for peptide retention prediction in ion-pair RP-HPLC: application to 300- and 100-Å pore size C18 sorbents. Anal Chem 78(22):7785–7795. doi:10.1021/ac060777w
23. Geiger T, Wisniewski JR, Cox J, Zanivan S, Kruger M, Ishihama Y, Mann M (2011) Use of stable isotope labeling by amino acids in cell culture as a spike-in standard in quantitative proteomics. Nat Protoc 6(2):147–157, doi: papers3://publication/doi/10.1038/nprot.2010.192
24. Levin Y, Hradetzky E, Bahn S (2011) Quantification of proteins using data-independent analysis (MSE) in simple and-complex samples: a systematic evaluation. Proteomics 11(16):3273–3287
25. Ivanov AR, Colangelo CM, Dufresne CP, Friedman DB, Lilley KS, Mechtler K, Phinney BS, Rose KL, Rudnick PA, Searle BC, Shaffer SA, Weintraub ST (2013) Interlaboratory studies and initiatives developing standards for proteomics. Proteomics 13(6):904–909, doi: papers3://publication/doi/10.1002/pmic.201200532
26. Rudnick PA, Clauser KR, Kilpatrick LE, Tchekhovskoi DV, Neta P, Blonder N, Billheimer DD, Blackman RK, Bunk DM, Cardasis HL, Ham A-JL, Jaffe JD, Kinsinger CR, Mesri M, Neubert TA, Schilling B, Tabb DL, Tegeler TJ, Vega-Montoto L, Variyath AM, Wang M, Wang P, Whiteaker JR, Zimmerman LJ, Carr SA, Fisher SJ, Gibson BW, Paulovich AG, Regnier FE, Rodriguez H, Spiegelman C, Tempst P, Liebler DC, Stein SE (2010) Performance metrics for liquid chromatography-tandem mass spectrometry systems in proteomics analyses. Mol Cell Proteomics 9(2):225–241
27. Ma Z-Q, Polzin KO, Dasari S, Chambers MC, Schilling B, Gibson BW, Tran BQ, Vega-Montoto L, Liebler DC, Tabb DL (2012) QuaMeter: multivendor performance metrics for LC–MS/MS proteomics instrumentation. Anal Chem 84(14):5845–5850
28. Wang X, Chambers MC, Vega-Montoto LJ, Bunk DM, Stein SE, Tabb DL (2014) QC metrics from CPTAC raw LC-MS/MS data interpreted through multivariate statistics. Anal Chem 86(5):2497–2509
29. Bereman MS, Johnson R, Bollinger J, Boss Y, Shulman N, MacLean B, Hoofnagle AN, MacCoss MJ (2014) Implementation of statistical process control for proteomic experiments via LC MS/MS. J Am Soc Mass Spectrom 25(4):581–587. doi:10.1007/s13361-013-0824-5
30. Burkhart JM, Vaudel M, Zahedi RP, Martens L, Sickmann A (2011) iTRAQ protein quantification: a quality-controlled workflow. Proteomics 11(6):1125–1134, doi: papers3://publication/doi/10.1002/pmic.201000711
31. Herbrich SM, Cole RN, West KP Jr, Schulze K, Yager JD, Groopman JD, Christian P, Wu L, O'Meally RN, May DH, McIntosh MW, Ruczinski I (2013) Statistical inference from multiple iTRAQ experiments without using common reference standards. J Proteome Res 12(2):594–604. doi:10.1021/pr300624g
32. Greco G, Boltner A, Letzel T (2014) Optimization of jet stream ESI parameters when coupling agilent 1260 infinity analytical SFC system with agilent 6230 TOF LC/MS., Agilent Technologies, https://www.agilent.com/cs/library/applications/5991-4510EN.pdf
33. Geer LY, Markey SP, Kowalak JA, Wagner L, Xu M, Maynard DM, Yang X, Shi W, Bryant SH (2004) Open mass spectrometry search algorithm. J Proteome Res 3(5):958–964. doi:10.1021/pr0499491
34. Bezerra MA, Santelli RE, Oliveira EP, Villar LS, Escaleira LA (2008) Response surface methodology (RSM) as a tool for optimization in analytical chemistry. Talanta 76(5):965–977
35. Coscollà C, Navarro-Olivares S, Martí P, Yusà V (2014) Application of the experimental design of experiments (DoE) for the determination of organotin compounds in water samples

using HS-SPME and GC–MS/MS. Talanta 119:544–552

36. Eliasson M, Rännar S, Madsen R, Donten MA, Marsden-Edwards E, Moritz T, Shockcor JP, Johansson E, Trygg J (2012) Strategy for optimizing LC-MS data processing in metabolomics: a design of experiments approach. Anal Chem 84(15):6869–76
37. Székely G, Henriques B, Gil M, Ramos A, Alvarez C (2012) Design of experiments as a tool for LC–MS/MS method development for the trace analysis of the potentially genotoxic 4-dimethylaminopyridine impurity in glucocorticoids. J Pharmaceut Biomed Anal 70:251–8
38. Maes K, Van Liefferinge J, Viaene J, Van Schoors J, Van Wanseele Y, Béchade G, Chambers EE, Morren H, Michotte Y, Vander Heyden Y, Claereboudt J, Smolders I, Van Eeckhaut A (2014) Improved sensitivity of the nano ultra-high performance liquid chromatography-tandem mass spectrometric analysis of low-concentrated neuropeptides by reducing aspecific adsorption and optimizing the injection solvent. J Chromatogr A 1360:217–228
39. Passeport E, Guenne A, Culhaoglu T, Moreau S, Bouyé J-M, Tournebize J (2010) Design of experiments and detailed uncertainty analysis to develop and validate a solid-phase microextraction/gas chromatography–mass spectrometry method for the simultaneous analysis of 16 pesticides in water. J Chromatograph A 1217(33):5317–5327
40. Raji MA, Schug KA (2009) Chemometric study of the influence of instrumental parameters on ESI-MS analyte response using full factorial design. Int J Mass Spectrom 279:100–106
41. Zhou Y, Song J-Z, Choi FF-K, Wu H-F, Qiao C-F, Ding L-S, Gesang S-L, Xu H-X (2009) An experimental design approach using response surface techniques to obtain optimal liquid chromatography and mass spectrometry conditions to determine the alkaloids in Meconopsi species. J Chromatogr A 1216(42):7013–7023
42. Switzar L, Giera M, Lingeman H, Irth H, Niessen WMA (2011) Protein digestion optimization for characterization of drug–protein adducts using response surface modeling. J Chromatogr A 1218(13):1715–1723, doi: papers3://publication/doi/10.1016/j.chroma.2010.12.043
43. Mischak H, Allmaier G, Apweiler R, Attwood T, Baumann M, Benigni A, Bennett SE, Bischoff R, Bongcam-Rudloff E, Capasso G, Coon JJ, D'Haese P, Dominiczak AF, Dakna M, Dihazi H, Ehrich JH, Fernandez-Llama P, Fliser D, Frokiaer J, Garin J, Girolami M, Hancock WS, Haubitz M, Hochstrasser D, Holman RR, Ioannidis JPA, Jankowski J, Julian BA, Klein JB, Kolch W, Luider T, Massy Z, Mattes WB, Molina F, Monsarrat B, Novak J, Peter K, Rossing P, Sánchez-Carbayo M, Schanstra JP, Semmes OJ, Spasovski G, Theodorescu D, Thongboonkerd V, Vanholder R, Veenstra TD, Weissinger E, Yamamoto T, Vlahou A (2010) Recommendations for biomarker identification and qualification in clinical proteomics. Science Transl Med 2 (46): 46ps42
44. Anderle M, Roy S, Lin H, Becker C, Joho K (2004) Quantifying reproducibility for differential proteomics: noise analysis for protein liquid chromatography-mass spectrometry of human serum. Bioinformatics 20(18):3575–3582
45. Liu P, Hwang JTG (2007) Quick calculation for sample size while controlling false discovery rate with application to microarray analysis. Bioinformatics 23(6):739–746
46. Fisher RA, Others (1926) The arrangement of field experiments. doi: citeulike-article-id:10709753
47. Westbrook JA, Noirel J, Brown JE, Wright PC, Evans CA (2015) Quantitation with chemical tagging reagents in biomarker studies. Proteomics Clin Appl 9(3-4):295–300. doi:10.1002/prca.201400120
48. Oberg AL, Mahoney DW, Eckel-Passow JE, Malone CJ, Wolfinger RD, Hill EG, Cooper LT, Onuma OK, Spiro C, Therneau TM, Bergen Iii HR (2008) Statistical analysis of relative labeled mass spectrometry data from complex samples using ANOVA. J Proteome Res 7(1):225–233, doi: papers3://publication/doi/10.1021/pr700734f
49. Niepel M, Spencer SL, Sorger PK (2009) Non-genetic cell-to-cell variability and the consequences for pharmacology. Curr Opin Chem Biol 13(5–6):556–561, doi: http://dx.doi.org/10.1016/j.cbpa.2009.09.015
50. Liu H, Sadygov RG, Yates JR 3rd (2004) A model for random sampling and estimation of relative protein abundance in shotgun proteomics. Anal Chem 76(14):4193–4201. doi:10.1021/ac0498563
51. Mercier C, Truntzer C, Pecqueur D, Gimeno J-P, Belz G, Roy P (2009) Mixed-model of ANOVA for measurement reproducibility in proteomics. J Proteomics 72(6):974–981, doi: http://dx.doi.org/10.1016/j.jprot.2009.05.005
52. Day SJ, Altman DG (2000) Blinding in clinical trials and other studies. BMJ 321(7259):504
53. Begley CG (2013) Six red flags for suspect work. Nature 497(7450):433–4
54. MacCoun R, Perlmutter S (2015) Blind analysis: hide results to seek the truth. Nature 526(7572):187–189

55. Gatto L, Christoforou A (2014) Using R and Bioconductor for proteomics data analysis. Biochim Biophys Acta 1844(1 Pt A):42–51
56. Colangelo CM, Chung L, Bruce C, Cheung K-H (2013) Review of software tools for design and analysis of large scale MRM proteomic datasets. Methods 61(3):287–298
57. Schwacke JH, Hill EG, Krug EL, Comte-Walters S, Schey KL (2009) iQuantitator: a tool for protein expression inference using iTRAQ. BMC Bioinformatics 10:342. doi:10.1186/1471-2105-10-342
58. Gonzalez-Galarza FF, Lawless C, Hubbard SJ, Fan J, Bessant C, Hermjakob H, Jones AR (2012) A critical appraisal of techniques, software packages, and standards for quantitative proteomic analysis. OMICS 16(9):431–442, doi: papers3://publication/doi/10.1089/omi.2012.0022
59. Kleinbaum DG, Kleinbaum DG (1998) Applied regression analysis and other multivariable methods, 3rd edn. Duxbury, Pacific Grove
60. Choi M, Chang CY, Clough T, Broudy D, Killeen T, MacLean B, Vitek O (2014) MSstats: an R package for statistical analysis of quantitative mass spectrometry-based proteomic experiments. Bioinformatics 30(17):2524–2526
61. Kerr NL (1998) HARKing: hypothesizing after the results are known. Pers Soc Psychol Rev 2(3):196–217
62. Kuhn M. Desirability: desirability function optimization and ranking. https://github.com/topepo/desirability
63. Derringer G, Suich R (1980) Simultaneous optimization of several response variables. J Qual Technol 12:214–219

Chapter 20

Automated SWATH Data Analysis Using Targeted Extraction of Ion Chromatograms

Hannes L. Röst, Ruedi Aebersold, and Olga T. Schubert

Abstract

Targeted mass spectrometry comprises a set of methods able to quantify protein analytes in complex mixtures with high accuracy and sensitivity. These methods, e.g., Selected Reaction Monitoring (SRM) and SWATH MS, use specific mass spectrometric coordinates (assays) for reproducible detection and quantification of proteins. In this protocol, we describe how to analyze, in a targeted manner, data from a SWATH MS experiment aimed at monitoring thousands of proteins reproducibly over many samples. We present a standard SWATH MS analysis workflow, including manual data analysis for quality control (based on Skyline) as well as automated data analysis with appropriate control of error rates (based on the OpenSWATH workflow). We also discuss considerations to ensure maximal coverage, reproducibility, and quantitative accuracy.

Key words Targeted proteomics, SWATH, SWATH MS, SWATH acquisition, Data-independent acquisition, DIA, OpenSWATH, pyProphet, TRIC aligner, Skyline

1 Introduction

Over the past decade, protein analysis strategies based on mass spectrometry (MS) have steadily gained popularity. Untargeted shotgun proteomics is currently the most widely used technique for qualitative and quantitative measurements of proteins on a large scale. It allows the detection of hundreds to several thousands of proteins in a single run. However, its focus on high proteome coverage leads to some curtailments in reproducibility, quantitative accuracy, and sample throughput [1]. The targeted proteomic technique selected/multiple reaction monitoring (SRM) alleviates these limitations by focusing the mass spectrometer on a defined set of proteins of interest [2]. SRM excels at consistent and accurate quantification of proteins over large sample sets with coefficients of variation (CV) below 15 % and offers the largest dynamic range of all MS-based techniques available today [3]. However, SRM measurements are limited to a few dozens of target proteins per sample injection. Novel MS-based proteomic methods, such as

Lucio Comai et al. (eds.), *Proteomics: Methods and Protocols*, Methods in Molecular Biology, vol. 1550,
DOI 10.1007/978-1-4939-6747-6_20, © Springer Science+Business Media LLC 2017

SWATH MS, utilize untargeted, data-independent acquisition (DIA) with targeted data extraction to improve the throughput of targeted proteomics by massively multiplexing the targeted detection of peptides. They thereby combine the comprehensiveness of the shotgun method with the quantitative accuracy and reproducibility of SRM. In terms of quantification reproducibility, SWATH MS performs similarly to SRM and offers a dynamic range of at least three orders of magnitude, positioning it as an ideal technology for large-scale and high-quality proteome measurements.

1.1 SWATH MS

In the SWATH MS workflow, proteins are enzymatically cleaved to produce a mixture of homogeneous peptides and then separated by online liquid chromatography (LC) coupled to tandem mass spectrometry (MS/MS). Similar to other DIA methods [4, 5], the mass spectrometer recursively cycles through a large *m/z* range and co-fragments all peptide ions in relatively large isolation windows (or "swathes") of several *m/z* (Fig. 1a) [6]. For each isolation window, the resulting fragment ions are recorded in a high-resolution composite mass spectrum. Thus, the instrument does not explicitly target single precursors as in shotgun or SRM but rather fragments all precursor ions falling into one isolation window simultaneously. Window size and accumulation time per window are chosen such that the instrument can cycle through all windows relatively quickly, allowing every peptide to be fragmented 8–10 times during its chromatographic elution (Fig. 1b). In SWATH acquisition, all ionized species of the sample are thus fragmented and recorded in a systematic, unbiased fashion, independently of their abundance. Compared to shotgun approaches, which record fragment ion spectra based on precursor ion intensity, SWATH acquisition systematically records fragment ion spectra every few seconds, thereby allowing reconstruction of the LC elution profile of each fragment ion (Fig. 1b). Similar to SRM, the deterministic acquisition strategy makes SWATH acquisition highly reproducible. The acquisition of full, high-resolution fragment ion data provides additional information compared to SRM, similar to parallel reaction monitoring [7, 8].

What distinguishes SWATH MS from other DIA methods is the way the data, that is, the highly multiplexed fragment ion spectra, are analyzed [6, 9]. In SWATH MS, intensities of specific fragment ions are extracted from the highly multiplexed fragment ion spectra in a targeted manner to produce extracted ion chromatograms (XICs) (Fig. 1b). These XICs are similar to SRM traces and can be analyzed analogously. This approach facilitates data analysis by reducing the complexity of the data significantly, but requires prior knowledge on the peptides and proteins to be analyzed. This prior knowledge consists of a set of MS coordinates uniquely describing the protein of interest (also called an "assay"). The complete set of coordinates describes (1) which peptides of a

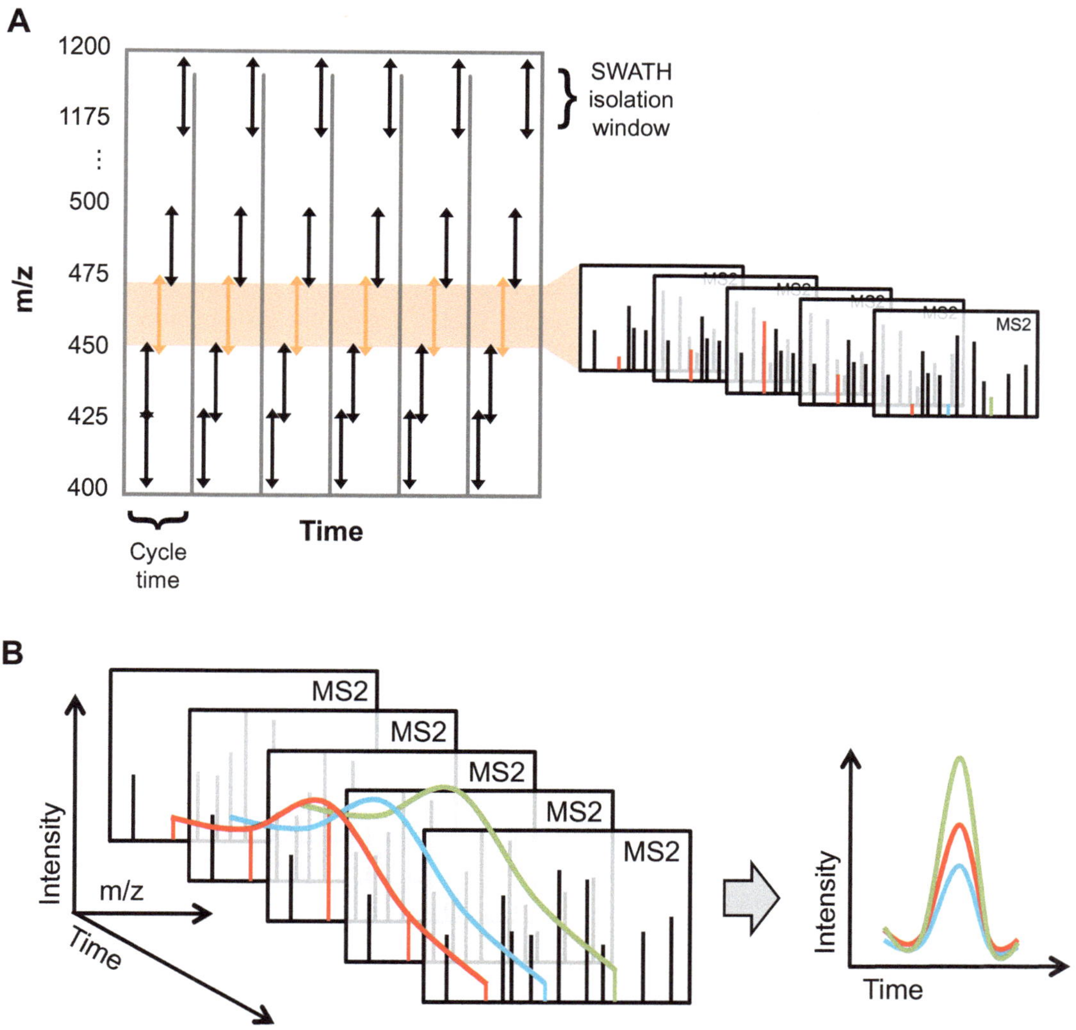

Fig. 1 SWATH MS. (**a**) While peptides elute from the LC, the mass spectrometer cycles through a large *m/z* range (here from 400 to 1200 *m/z*) and co-fragments all peptide ions in relatively large isolation windows (or "swathes") of several *m/z* (here 25 *m/z*) [6]. For each isolation window, the resulting fragment ions are recorded in a high-resolution composite mass spectrum (MS2). For every isolation window, the instrument accumulates fragment ions for 100 ms, for 32 isolation windows covering 400–1200 *m/z* this results in a cycle time of 3.2 s. At the beginning of each cycle a survey scan (MS1) is recorded (not shown). (**b**) In SWATH MS, intensities of specific fragment ions are extracted from the highly multiplexed fragment ion spectra in a targeted manner to produce extracted ion chromatogram peak groups

protein are most representative, i.e., unique and well detectable, (2) the elution time of these peptides from the LC, (3) their predominant charge state, (4) the most abundant fragment ions formed during fragmentation, and (5) their relative intensity. Comprehensive, high-quality assay libraries are a critical prerequisite for the success of a SWATH MS analysis [10].

1.2 Automated SWATH MS Analysis

While small-scale analysis of SWATH MS data can be performed manually (similarly to SRM data analysis), this strategy is not feasible any more for proteome-wide analysis. Automated software packages provide fast, unbiased and comprehensive data analysis solutions using advanced machine learning and data processing algorithms. These software packages are able to automatically import assay libraries, extract chromatographic traces from the raw data and identify peak groups resulting from co-eluting fragment ions (Fig. 2). These peak groups are then scored and statistically evaluated using a noise model (negative distribution of scores). This step allows the software to assign each peak group a quality score, which reflects the probability of making a correct identification. Researchers can then use this score to apply an experiment-wide false discovery rate (FDR) cutoff based on how many false quantifications they are willing to tolerate in the final dataset. If multiple runs are present, a cross-run alignment can be performed that integrates all available data to further increase the consistency of the final data matrix (Fig. 2).

In this chapter we provide a detailed workflow for automated analysis of SWATH MS data using the software suite OpenSWATH, including pyProphet and TRIC Aligner [9, 11]. Because data quality is key for a successful automated SWATH MS analysis, we start

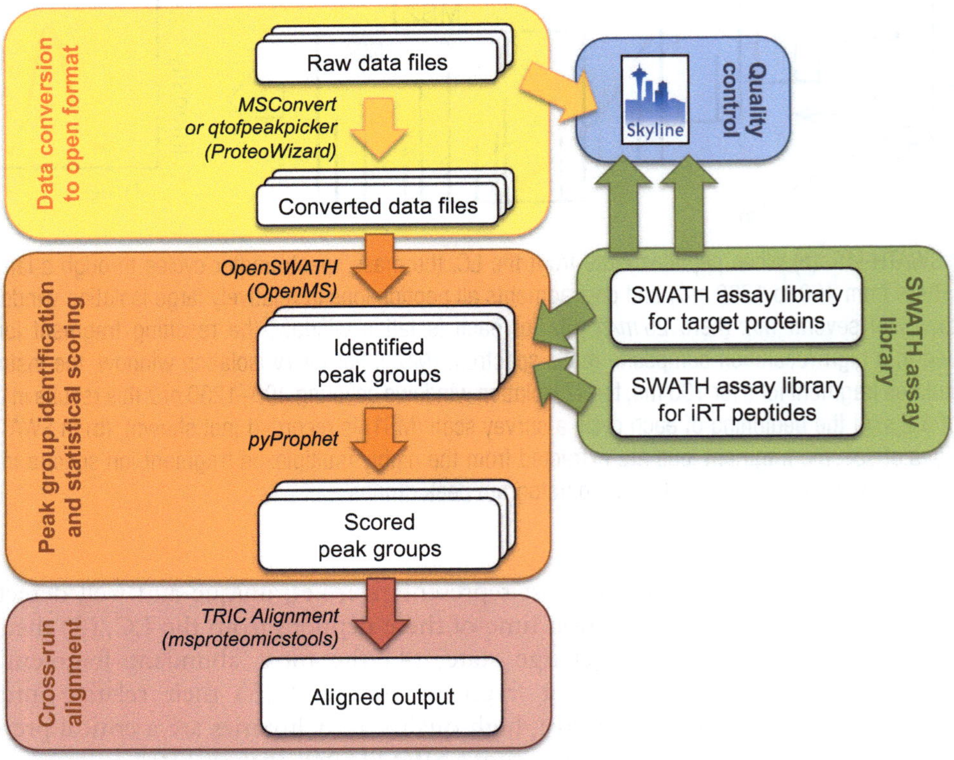

Fig. 2 Overview of the workflow for automated SWATH data analysis

with a brief description on how to load SWATH MS data into Skyline [12] to visually inspect a small number of peptides in every run for quality control.

2 Materials

Any files described in this chapter, including example files to test the protocol, can be downloaded from the following website: http://www.peptideatlas.org/PASS/PASS00779. Most of the example files are related to a recent study in *Mycobacterium tuberculosis* that used SWATH MS [13]. The installation of software tools is described for a 64-bit Microsoft Windows 7 system. However, many of the tools can be installed on other platforms, such as Mac OS X and Linux, as well (OpenMS, Python, pyProphet, TRIC aligner). Specific software versions used to develop this protocol are noted in the SoftwareVersions.txt file available at the above mentioned website. It is important to have enough free disk space available as the data files and the intermediary analysis files tend to be very large (dozens of GB).

1. SWATH data files. A detailed protocol of how to obtain high-quality SWATH MS data on a TripleTOF instrument is provided in Chapter 16 by Hunter et al., but other instrument types can be used as well (*see* **Note 1**). See **Note 2** for a discussion of parameters that are important during acquisition, such as LC gradient, SWATH window size, and acquisition time. Importantly, the samples have to contain retention time calibration (iRT) peptides at the time they are measured (*see* **Note 3**). In this protocol, we assume that the 11 synthetic iRT peptides provided in the iRT-kit by Biognosys have been spiked into each sample [14]. Please note that the AB SCIEX TripleTOF instrument produces two data files per SWATH run, a .wiff and a .wiff.scan file, which should always be stored together (see three .wiff and corresponding .wiff.scan example files).
2. File describing SWATH window setup. The selection of appropriate SWATH windows is an important part of a SWATH MS experiment (Chapter 16 by Hunter et al.). For the current protocol, we assume that a fixed window setup (32 windows of 26 *m/z* each, 1 *m/z* overlap: 400–425 *m/z*, 424–450 *m/z*, 449–475 *m/z*, ... 1149–1175 *m/z*, 1174–1200 *m/z*) has been used for data acquisition (see example file SWATHwindows_acquisition.tsv). In contrast to the SWATH windows used for data acquisition, the SWATH windows used for data analysis should not be overlapping. Therefore, the 1 *m/z* overlap that was used for acquisition among the neighboring SWATH windows is split, such that each window is only 25 *m/z* (first and last windows are slightly different): 400–424.5 *m/z*, 424.5–449.5 *m/z*, 449.5–474.5 *m/z*, ... 1149.5–1174.5 *m/z*, 1174.5–1200 *m/z* (see example file SWATHwindows_analysis.tsv).

3. SWATH assay library. In addition to the SWATH data files, the following workflow requires as an input a SWATH assay library, containing precomputed decoy transition groups as well as assays for the retention time peptides. These assays can be built from experimental data (*see* Ref. 10 for a detailed protocol) or downloaded from the SWATHAtlas database (*see* **Note 4**). As an example library for this protocol we use a comprehensive SWATH assay library of *M. tuberculosis* (see example file Mtb_TubercuList-R27_iRT_UPS_decoy.tsv).
4. iRT peptide SWATH assays. In order to perform targeted extraction, the retention time of each SWATH MS injection needs to be transformed into a normalized retention time space, which is also used by the assay library (*see* **Note 3**). The protocol we describe here assumes that the 11 synthetic iRT peptides from the iRT-Kit by Biognosys have been spiked into every sample. For this protocol we provide an iRT peptide SWATH assay library in TraML format for OpenSWATH analysis (*see* example file iRTassays.TraML) and a reduced table for import into Skyline (*see* example file iRTassays_Skyline.tsv).
5. Skyline. Skyline is a free, open-source software for targeted data analysis of various types of proteomics data [12] and provides great visualization options. The software can be downloaded from the website: http://skyline.maccosslab.org. It is only available for Microsoft Windows operating systems.
6. Proteowizard. To convert vendor raw data files into an open format, such as mzML or mzXML, we use the msconvert tool, which is part of the Proteowizard software suite (Chapter 23 by Mallick et al.). Download Proteowizard from http://proteowizard.sourceforge.net/downloads.shtml and install it on your machine. Due to vendor licensing constraints, the data conversion functionality of ProteoWizard is only available for Microsoft Windows operating systems.
7. OpenMS. OpenMS is a cross-platform and open-source analysis package for mass spectrometric data [15–18]. It is specifically suited for automated, large-scale analysis and can be run on any machine, from a Desktop computer to a high-performance Linux cluster [19, 20]. Download the appropriate version of OpenMS for your operating system from https://sourceforge.net/projects/open-ms/files/OpenMS/OpenMS-2.0/. Install OpenMS into a local folder, for example C:\Program Files\OpenMS-2.0. On newer Windows versions (Windows Vista, 7, and higher) some of the.NET dependencies will already be installed, which will be indicated during the install and removes the need to download and install them. (The warning during.NET installation "Turn Windows features on or off" can be ignored.)

8. Python. Several scripts used in this protocol require Python 2.7. On Windows, the easiest way to install Python is through Anaconda, which can be downloaded from https://store.continuum.io/cshop/anaconda/. Choose the "Graphical Installer" under the "PYTHON 2.7" tab appropriate for your system (64-bit or 32-bit). Install Anaconda as Administrator on your system (right-click, "Run as" and then select the Administrator user). Select the installation for all users and choose C:\Anaconda2\ as install path (should be the default). Unless you have another Python installation on your system, allow Anaconda to be added to your PATH and register it as the default Python 2.7 installation (these options become available during the installation process).
9. pyProphet. pyProphet is a tool to calculate a discriminating score separating target from decoy assays and to compute a false discovery rate (FDR) cutoff based on this score [21]. Open the "Anaconda Prompt" by opening the Start Menu (Windows icon in lower left corner) and type "Anaconda" in the search field. Right-click the "Anaconda Prompt" entry and select "Run as administrator". To install the pyProphet package, enter:
```
pip install scikit-learn==0.15.2
pip install pyprophet==0.13.3
```
If running a 32-bit system, there are no pre-built packages and a working C++ compiler is required on your machine. To get such a compiler, go to http://aka.ms/vcpython27 and download the file called VCForPython27.msi. After installing this file, use the command above to install pyProphet.
10. TRIC feature aligner. The TRansfer of Identification Confidence (TRIC) software tool integrates information from multiple OpenSWATH runs using cross-run alignment and retention time correction [11]. Open the "Anaconda Prompt" by opening the Start Menu (Windows icon in lower left corner) and type "Anaconda" in the search field. Right-click the "Anaconda Prompt" entry and select "Run as administrator". To install the TRIC aligner package, enter:
```
conda install biopython
pip install msproteomicstools==0.3.2
```

3 Methods

3.1 Visualization of Selected Peptides for Quality Control

Before proceeding with automated SWATH data analysis, it is crucial to ensure adequate quality of the raw data. Skyline offers great visualization options to assess SWATH data quality and we recommend loading all data files to be subjected to automated analysis by OpenSWATH first into Skyline to manually inspect a few quality control peptides, such as the spiked-in iRT peptides. Optimally, every SWATH data file is loaded into Skyline right after acquisition to monitor LC stability as well as performance of the mass spectrometer.

1. Open a blank Skyline document and go to Settings → Peptide Settings → "Filter" tab. Uncheck the option to "Auto-select all matching peptides" and click "OK".
2. Go to Settings → Transition Settings → "Filter" tab. Uncheck the option to "Auto-select all matching transitions" and click "OK".
3. Go to Settings → Transition Settings → "Full-Scan" tab. Set the following parameters: Acquisition method: DIA; Product mass analyzer: TOF; Isolation scheme: From the drop-down menu, select "Add..." and fill the "Edit Isolation Scheme" window as shown in Fig. 3a (for this protocol, the window boundaries can be copy-pasted from the file SWATHwindows_analysis.tsv). Resolving power: 15,000 (depends on the state of the instrument during SWATH data acquisition); Retention time filtering: Include all matching scans. Click "OK".
4. Go to Edit → Insert → Transition list. In the opening window, paste assays of peptides you would like to monitor. For this protocol, the assays for the iRT peptides can be copy-pasted from the file iRTassays_Skyline.tsv. Click "Insert".
5. Save the Skyline file.
6. Go to File → Import → Results and select "Add single-injection replicates in files" and click "OK".
7. Select the raw SWATH data .wiff files (the .wiff.scan files are not showing up here, but need to be located in the same folder) and click "Open". When asked about removing the common prefix of the file name, click "Remove". Now it will take a while to import all the data depending on file size and number of files.
8. After the data is loaded, arrange the Skyline file such that it is convenient to inspect the runs (Fig. 3b). First, get the peak area and retention time overview plots: View → Retention Times → Click "Replicate Comparison" and View → Peak Areas → Click "Replicate Comparison". Next, arrange the panels by drag-and-drop to the desired location. Then go to "Settings" and click "Integrate all". Finally, right-click on a chromatogram plot → Auto-zoom X-axis → Best Peak.
9. Inspect the SWATH peak groups to judge performance of LC (peak shape, retention time stability) and mass spectrometer (signal intensity and signal-to-noise ratio for low abundant peaks) over time (Fig. 3b). Note that Skyline uses a scoring system to determine the correct peak group but it can still happen that it does not pick the correct one. In these cases, peak boundaries can be changed manually by click-and-drag in the retention time axis of the chromatogram plot. A confidently

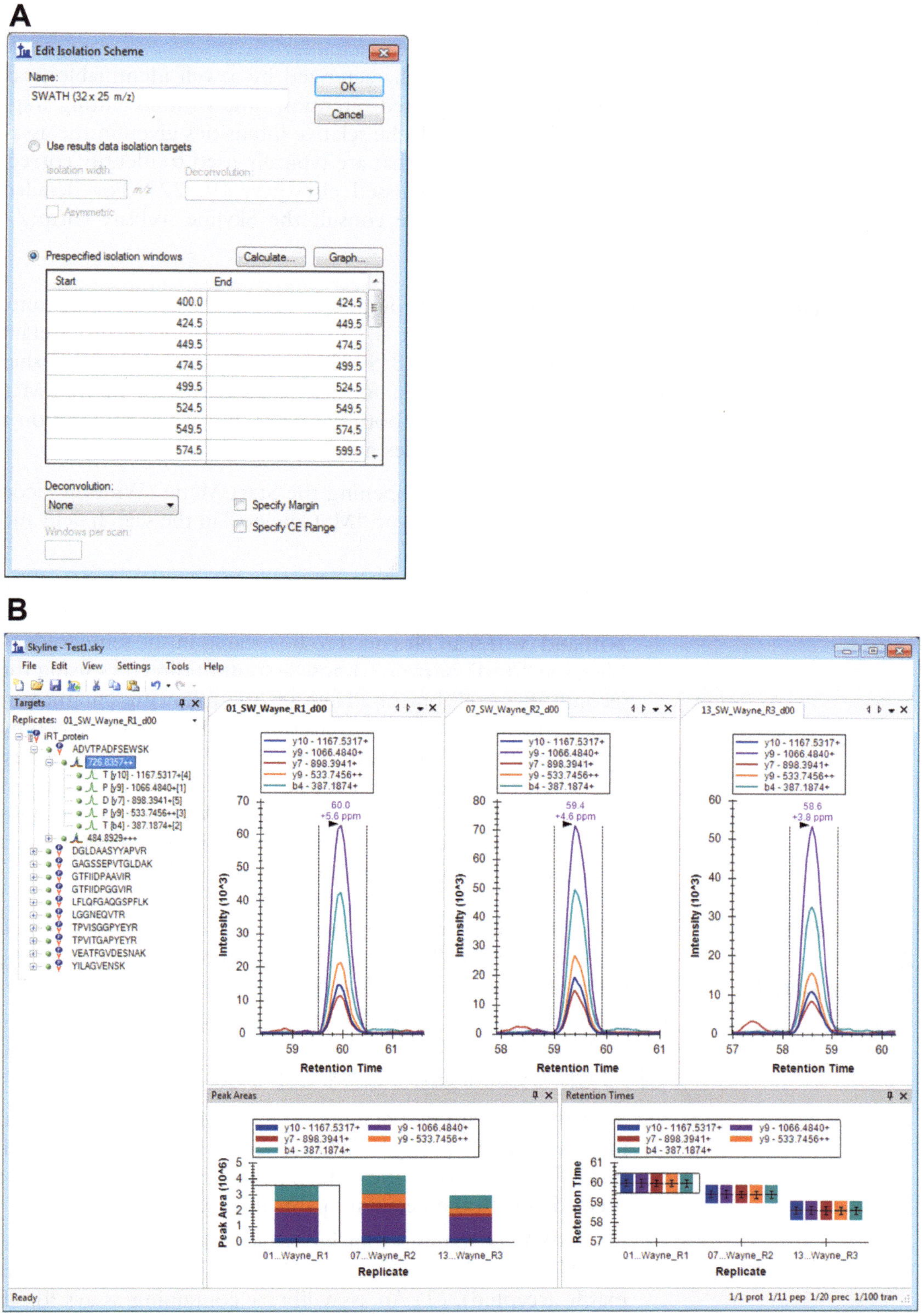

Fig. 3 SWATH data visualization in Skyline. (**a**) Screen shot showing how SWATH windows are defined in Skyline. This is essential for the software to know from which SWATH window the specified fragment ions should be extracted. (**b**) Example of how to organize the Skyline window for fast and efficient monitoring of large numbers of SWATH runs. The example here shows just three runs, but Skyline can easily handle dozens of runs

identified peptide is characterized by a well identifiable peak group near the expected retention time with co-eluting fragment ions that match the relative intensities given in the assay library. More criteria that are typically used to identify correct peak groups are discussed elsewhere [9, 22]. For detailed Skyline tutorials please consult the Skyline website (http://skyline.maccosslab.org).

3.2 Raw Data Conversion Into mzML

MSConvert, provided through the ProteoWizard software suite (Chapter 23 by Mallick et al.), enables conversion of proprietary raw data files (.wiff for AB SCIEX, .raw files for Thermo Fisher instruments) into an open format, such as mzML or mzXML. OpenSWATH can handle both file types, but when running on a Windows PC, the input files need to be in mzML format.

1. Start the software by opening the Start Menu (Windows icon in lower left corner), type "MSConvert" in the search field and click on "MSConvert".
2. In the MSConvert window, select "List of Files" and use the "Browse" button to locate the raw files to be converted (the .wiff and .wiff.scan files need to be located in the same folder). Click the "Add" button. Choose an output directory using the second "Browse" button. Then set the following parameters: Output format: mzML; Extension: mzML; Binary encoding precision: 64-bit; Write index: yes; Use zlib compression: yes. All other checkboxes should be unchecked and no filters should be set (*see* Fig. 4). While OpenSWATH produces best results on profile data, it is possible to run it on centroided data to reduce disk space and execution time (*see* **Note 5**).

3.3 Automated Analysis with OpenSWATH

OpenSWATH can be run as a single command, which executes all steps of the OpenSWATH data analysis pipeline automatically. The individual steps can be controlled through flags passed on the command line. Use the `--helphelp` command to see all options and to learn about particular speed and memory optimizations available. Table 1 summarizes the most important options.

1. Start the software by opening the Start Menu (Windows icon in lower left corner), type "TOPP" in the search field and click on "TOPP command line" which will start a command prompt preloaded with the necessary paths to execute OpenSWATH.
2. OpenSWATH requires four input files: (1) The actual data in mzXML or mzML format (on Windows PC only mzML format is accepted). (2) An assay library containing assays for all target peptides. OpenSWATH can take assay libraries in both tab-separated table (tsv) and TraML format. For more details *see* **Notes 4** and **6**. (3) An assay library containing assays for all iRT peptides (*see* **Note 3**) in TraML format. 4) A file with a

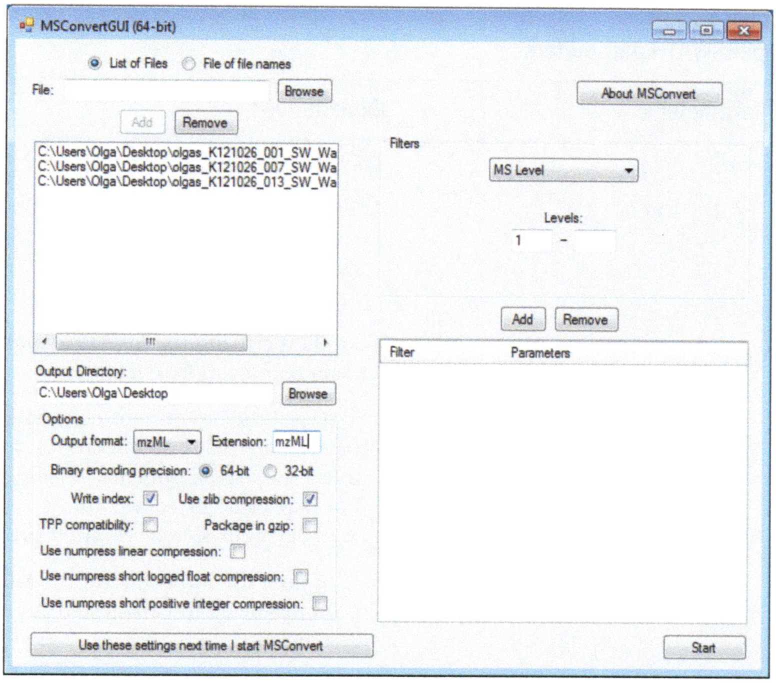

Fig. 4 SWATH data conversion with MSConvert. Conversion of raw data to mzML format using the MSConvert GUI, showing the individual options that should be selected during the conversion step for profile-mode conversion

small table specifying the SWATH windows out of which the fragment ion traces should be extracted. Note that the extraction windows are slightly different from the windows specified for data acquisition, i.e., they should not contain overlaps (see example files described in the Materials section). A full command for running OpenSWATH looks like this (enter entire command on a single line):

```
OpenSwathWorkflow.exe
- in data.mzML
- tr library.tsv
- sort_swath_maps
- readOptions cache
- tempDirectory C:\Temp
- batchSize 1000
- tr_irt iRT_assays.TraML
- swath_windows_file SWATHwindows_analysis.tsv
- out_tsv osw_output.tsv
```

Table 1
OpenSWATH parameters

Parameter	Explanation
`--helphelp`	To see all options
`-out_chrom`	Additionally also output the generated XICs. This option is required if the OpenSWATH results are to be inspected manually with TAPIR [23] or for other downstream processing not discussed here
`-swath_windows_file`	Text file specifying the SWATH windows (*see* example file and main text for more details)
`-rt_extraction_window`	The size of the retention time extraction window in seconds (centered around the theoretical retention time calculated using the iRT alignment), usually 300–600 s are reasonable, but may be adjusted according to the chromatography and the accuracy of expected retention times in the assay library
`-mz_extraction_window`	Size of the extraction window in Dalton. The default value of 0.05 may need to be adjusted depending on the instrument resolution
`-ppm`	Indicates that `-mz_extraction_window` is given in ppm instead of Dalton
`-Scoring:TransitionGroupPicker:min_peak_width`	Minimal peak width in seconds preventing spurious peaks from appearing in the results. Defaults to 14 s
`-TransitionGroupPicker:background_subtraction`	Indicates whether background subtraction should be performed during quantification. This option may increase the accuracy of the quantification
`-Scoring:TransitionGroupPicker:PeakPicker MRM:sgolay_frame_length`	Smoothing parameter for Savitzky-Golay smoothing (expressed in number of scans). Increasing this parameter leads to stronger smoothing. Alternatively, Gaussian smoothing may be used by `-Scoring:TransitionGroupPicker:PeakPickerMRM:use_gauss`
`-Scoring:stop_report_after_feature`	Determines the number of peak groups per assay that are reported in the result. A smaller number produces less output; generally 5–10 is appropriate
`-batchSize`	Determines the size of the batch of chromatograms that are loaded into memory and analyzed at once. The smaller this number, the less memory is required
`-sort_swath_maps`	This parameter ensures that the SWATH maps in the input are sorted by `m/z` (this assumes that the swath window file given in `-swath_windows_file` also contains windows sorted by *m/z*)
`-readOptions`	Setting this option to `cache` instructs the software to not load all data into memory at once but to create a local cache of the data. Make sure to specify a directory for the temporary files using the `-tempDirectory` parameter
`-tempDirectory`	Defines the directory to store temporary files (required if `-readOptions cache` is used)

The output of the OpenSWATH command is a large table containing one scored peak group per row (usually more than one peak group is scored per assay). The properties of each peak group are given in columns (i.e., the assay used to generate the peak group, the retention time, the individual scores etc.).

3.4 Error Rate Estimation with pyProphet

After running OpenSWATH, q-values corresponding to the FDR of peak identification can be estimated with the pyProphet software tool. The only input for pyProphet is the OpenSWATH result table file (osw_output.tsv).

1. Use the same command line window as above and enter (on a single line):
```
C:\Anaconda2\Scripts\pyprophet.exe
--ignore.invalid_score_columns
--d_score.cutoff=0.5
osw_output.tsv
```
   This command will output a set of files, of which osw_output_with_dscore_filtered.csv will be used in the next step.
2. pyProphet will also output a table containing the number of true and false positives for different q-value cutoffs (_summary_stat.csv). A q-value cutoff (or FDR) between 1 and 10% is appropriate for most studies. For further information, the pdf report file should be consulted, which contains plots of the fitted distributions as well as ROC curves. This step provides an opportunity to identify common errors, such as using an assay library without decoys or using an assay library unsuited for the measured sample (e.g., from another organism). In a successful run, target and decoy distributions should be clearly separated as shown in Fig. 5.

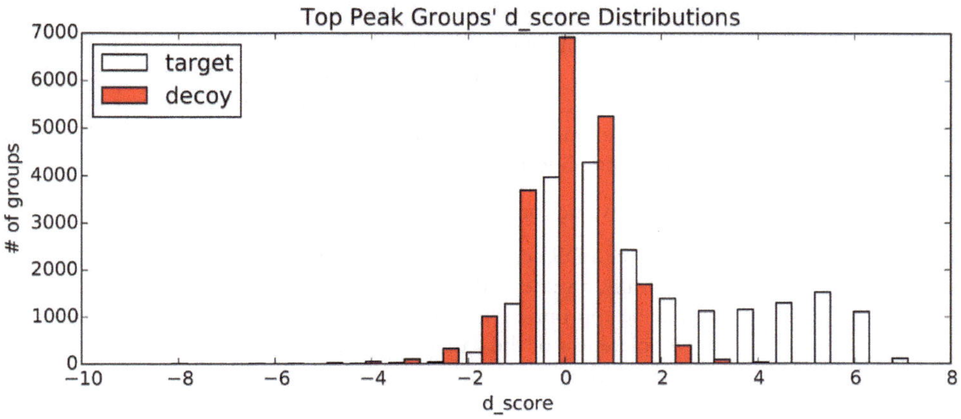

Fig. 5 Peak group scoring with pyProphet. To ensure that the FDR estimation step using pyProphet worked properly, the output should be manually inspected to identify potential problems. The plotted distributions should be bimodal and the decoy distribution should be close in shape to the false positive distribution (left part of the bimodal distribution)

3.5 Cross-Run Alignment with TRIC Aligner

If multiple SWATH MS runs were analyzed with OpenSWATH, the resulting quantitative data matrix may contain a substantial number of missing values. One way to reduce these missing values is to use the TRIC algorithm to perform cross-run alignment. This algorithm aligns peak groups across all runs using either a reference-based alignment or a tree-based alignment strategy and is implemented in the `feature_alignment.py` tool from the msproteomicstools package. Table 2 describes the individual parameters. For further information you may also consider the online documentation at https://github.com/msproteomicstools/msproteomicstools.

Table 2
TRIC alignment parameters

Parameter	Explanation
`--help`	To see all options
`--method`	Alignment method, either reference-based alignment (`best_overall` or `global_best_overall`) or reference-free, tree-based alignment (`LocalMST`). The method `best_overall` is most conservative as it does not change peak groups with a good q-value, thereby avoiding errors introduced during alignment. Alternatively, `global_best_overall` may be chosen, which, however, requires that the maximal shift in retention time is low, otherwise even good peak groups may be removed from the output if their retention time is too far off. The method `LocalMST` is recommended for large-scale SWATH MS analysis
`--realign_method`	Describes how the alignment between individual runs is performed and may be either linear (`linear` or `diRT`) or nonlinear (`lowess` or `CVSpline`). For small datasets, linear alignment methods typically perform sufficiently well and are faster than the nonlinear methods
`--max_rt_diff`	Maximal shift in retention time that is tolerated after alignment (in seconds). The software will automatically estimate a sensible value if it is set to `auto_3medianstdev`. Adjust this parameter to your chromatography. Nonlinear alignment methods typically allow for lower values than linear alignment methods
`--target_fdr`	Desired q-value (or FDR) on assay level (estimated using decoys)
`--max_fdr_quality`	q-value (or FDR) cutoff to still consider a feature for quantitation. Typically, a value of 0.01 (1%) is appropriate. Higher values (e.g., 0.05) may be appropriate for small to medium size datasets as they increase completeness of the data matrix, but may also introduce lower quality peaks groups

1. Use the same command line window as above and enter (on a single line):

```
C:\Anaconda2\python.exe
C:\Anaconda2\Scripts\feature_alignment.py
--in file1_with_dscore.csv file2_with_dscore.
csv file3_with_dscore.csv
--out aligned.tsv
--method best_overall
--realign_method diRT
--max_rt_diff 90
--target_fdr 0.01
--max_fdr_quality 0.01
```

   The command exemplified here will run alignment on three files using linear iRT alignment and pick an appropriate peak group in each run within the aligned window using a reference-based alignment. Only peptides that pass the identification threshold estimated by the TRIC algorithm will be reported in the final output (set to 1% FDR using the `--target_fdr` parameter in the example above). Note that recent results suggest substantial improvements when using reference-free, nonlinear retention time alignment, which can be enabled using the "`LocalMST`" tree-based alignment option and the "`lowess`" RT correction option.

4 Notes

1. While SWATH MS was originally developed on an AB SCIEX TripleTOF instrument, SWATH-like data-independent acquisition can also be obtained from other types of instruments. One of the major considerations is the acquisition speed, which needs to allow sufficient sampling during elution of an analyte from the LC column. The software tools described here, i.e., Skyline and OpenSWATH, can both analyze data from multiple vendors, including data acquired on Waters Synapt TOF instruments, Thermo Fisher Q Exactive instruments and AB SCIEX TripleTOF instruments [9]. Multiple groups have reported SWATH-like acquisition on Thermo Q Exactive instruments and successfully used OpenSWATH and other SWATH MS software for data analysis [24–26].
2. SWATH acquisition requires careful optimization of multiple parameters including chromatographic peak width, *m/z* window width, acquisition time per window and overall *m/z* coverage. In the initial implementation of SWATH MS, these parameters were set to ca. 30 s for the chromatographic peak width, 25 *m/z* for the window width, 100 ms for the acquisition time per window and a precursor mass range of 400–1200 *m/z* was chosen, resulting in 32 SWATH windows.

Since then, multiple improvements to the scheme have been proposed (Chapter 16 by Hunter et al.). It is important to remember that one of the main constraints is to retain sufficient sampling of each peptide precursor during its chromatographic elution. In the original implementation, no more than 100 ms acquisition time could be allowed in order to complete a 32-window cycle within 3.3 s. Assuming an average chromatographic peak width of 30 s, this would enable the sampling of on average 9 data points across the peak which is sufficient for the OpenSWATH algorithm to reconstruct a peak and perform fragment ion trace cross-correlation. If more powerful chromatographic separation were employed (as offered by UPLC, for example), the other parameters would have to be adjusted accordingly. For example, the acquisition time of each SWATH window could be reduced in order to retain sufficient sampling. Also, the precursor mass range could be decreased or the size of each window increased, both resulting in a smaller number of SWATH windows allowing the instrument to complete each cycle faster. Also, working with flexible window sizes has been shown to improve performance (Chapter 16 by Hunter et al.).

3. The presence of iRT retention time reference peptides, the so-called iRT peptides, is crucial to calibrate the retention time information present in the SWATH assay library to the retention times recorded in the SWATH data. In principle, any set of well-detectable peptides spanning a wide retention time range can be used [14]. Even endogenous peptides can serve as iRT peptides, making it unnecessary to purchase and spike synthetic peptides into every sample. A set of conserved endogenous peptides found in most eukaryotic samples has recently been described [27]. These Common internal Retention Time standards (CiRT) can be used in the same way as the iRT peptides without the need to spike in additional peptides. However, for some specialized applications, i.e., blood plasma analysis, the CiRT approach may be suboptimal and the use of the commercial iRT peptides is recommended.

4. Comprehensive, ready-to-use SWATH assay libraries have been published for a number of organisms, including human [28] and yeast [29], and can be downloaded from http://www.SWATHAtlas.org. Schubert et al. provide an extensive discussion on various aspects of SWATH assay libraries and provide a detailed protocol on how to build high-quality assay libraries from shotgun data [10].

5. The OpenSWATH software is designed and tested on profile data and produces optimal results on profile data. However, it is possible to run OpenSWATH on centroided data, reducing

the required disk space and execution time of the whole workflow substantially. As centroiding reduces the number of peaks in the data and sometimes may remove low-intensity peaks, the XICs generated by OpenSWATH become less smooth and peak detection becomes less sensitive. If centroiding is performed, it is crucial to compare the results of OpenSWATH to those obtained on profile data on the same dataset to obtain an accurate estimation of how centroiding affects the identification rate. Multiple centroiding algorithms are freely available, including the internal msconvert centroiding (enabled during conversion by the "Filter" settings) as well as the separate executable `qtofpeakpicker.exe` (found in the same folder as `msconvert.exe` after installation of ProteoWizard). The qtofpeakpicker has been specifically designed for data derived from QTOF instruments and we recommend to use it to centroid TripleTOF data [10]. The qtofpeakpicker is run directly on the .wiff files and we recommend using the same parameters as Schubert et al. to generate centroided mzML files [10]: `--resolution=20000 --area=1 --threshold=1 --smoothwidth=1.1`.

6. A high-quality SWATH assay library is an essential feature for any targeted analysis of DIA or SWATH MS data. The OpenSWATH software uses this information to extract XICs for the targeted peptide from the appropriate fragment ion spectra and to score the peaks according to how well they match the information in the assay library (expected retention time, fragment ion *m/z*, etc.). Alongside the assays for the proteins of interest (target assays), the library needs to contain a set of "decoy assays" which represent peptides not present in the sample (usually shuffled target peptide sequences). These assays are scored in the same fashion as the targeted assays and used by pyProphet to estimate the null distribution of different OpenSWATH scores and compute q-values (for FDR control). Having decoys in the assay library is essential for this step to work. Repositories such as http://www.SWATHAtlas.org provide assay libraries that already contain decoys. Please refer to Röst et al. and Schubert et al. for more information on the scoring algorithm and the importance of decoys and FDR control [9, 10]. OpenSWATH accepts assay libraries in both tab-separated table (tsv) and TraML format. For large assay libraries, the tsv format is more memory efficient than the TraML format. OpenMS comes with the tools `ConvertTraMLToTSV` and `ConvertTSVToTraML`, which enable conversion between the two formats (regardless of the names of these conversion tools, the file ending for the table format has to be.csv, while the actual data format is tab-separated).

References

1. Domon B (2012) Considerations on selected reaction monitoring experiments: implications for the selectivity and accuracy of measurements. Proteomics Clin Appl 6:609–614. doi:10.1002/prca.201200111
2. Picotti P, Aebersold R (2012) Selected reaction monitoring-based proteomics: workflows, potential, pitfalls and future directions. Nat Methods 9:555–566. doi:10.1038/nmeth.2015
3. Picotti P, Bodenmiller B, Mueller LN et al (2009) Full dynamic range proteome analysis of S. cerevisiae by targeted proteomics. Cell 138:795–806. doi:10.1016/j.cell.2009.05.051
4. Venable JD, Dong M-Q, Wohlschlegel J et al (2004) Automated approach for quantitative analysis of complex peptide mixtures from tandem mass spectra. Nat Methods 1:39–45. doi:10.1038/nmeth705
5. Chapman JD, Goodlett DR, Masselon CD (2013) Multiplexed and data-independent tandem mass spectrometry for global proteome profiling. Mass Spectrom Rev. doi:10.1002/mas.21400
6. Gillet LC, Navarro P, Tate S et al (2012) Targeted data extraction of the MS/MS spectra generated by data-independent acquisition: a new concept for consistent and accurate proteome analysis. Mol Cell Proteomics 11:O111.016717. doi:10.1074/mcp.O111.016717
7. Gallien S, Duriez E, Crone C et al (2012) Targeted proteomic quantification on quadrupole-orbitrap mass spectrometer. Mol Cell Proteomics 11:1709–1723. doi:10.1074/mcp.O112.019802
8. Peterson AC, Russell JD, Bailey DJ et al (2012) Parallel reaction monitoring for high resolution and high mass accuracy quantitative, targeted proteomics. Mol Cell Proteomics. doi:10.1074/mcp.O112.020131
9. Röst HL, Rosenberger G, Navarro P et al (2014) OpenSWATH enables automated, targeted analysis of data-independent acquisition MS data. Nat Biotechnol 32:219–223. doi:10.1038/nbt.2841
10. Schubert OT, Gillet LC, Collins BC et al (2015) Building high-quality assay libraries for targeted analysis of SWATH MS data. Nat Protoc 10:426–441. doi:10.1038/nprot.2015.015
11. Röst HL, Liu Y, D'Agostino G et al (2016) TRIC: an automated alignment strategy for reproducible protein quantification in targeted proteomics. Nat Methods 13:777–783. doi:10.1038/nmeth.3954
12. MacLean B, Tomazela DM, Shulman N et al (2010) Skyline: an open source document editor for creating and analyzing targeted proteomics experiments. Bioinformatics 26:966–968. doi:10.1093/bioinformatics/btq054
13. Schubert OT, Ludwig C, Kogadeeva M et al (2015) Absolute proteome composition and dynamics during dormancy and resuscitation of Mycobacterium tuberculosis. Cell Host Microbe. doi:10.1016/j.chom.2015.06.001
14. Escher C, Reiter L, MacLean B et al (2012) Using iRT, a normalized retention time for more targeted measurement of peptides. Proteomics 12:1111–1121. doi:10.1002/pmic.201100463
15. Kohlbacher O, Reinert K, Gröpl C et al (2007) TOPP--the OpenMS proteomics pipeline. Bioinformatics 23:e191–7. doi:10.1093/bioinformatics/btl299
16. Sturm M, Bertsch A, Gröpl C et al (2008) OpenMS – an open-source software framework for mass spectrometry. BMC Bioinformatics 9:163. doi:10.1186/1471-2105-9-163
17. Röst HL, Schmitt U, Aebersold R, Malmström L (2014) pyOpenMS: a Python-based interface to the OpenMS mass-spectrometry algorithm library. Proteomics 14:74–77. doi:10.1002/pmic.201300246
18. Röst HL, Schmitt U, Aebersold R, Malmström L (2015) Fast and efficient XML data access for next-generation mass spectrometry. PLoS One 10:e0125108. doi:10.1371/journal.pone.0125108
19. Junker J, Bielow C, Bertsch A et al (2012) TOPPAS: a graphical workflow editor for the analysis of high-throughput proteomics data. J Proteome Res 11:3914–3920. doi:10.1021/pr300187f
20. Aiche S, Sachsenberg T, Kenar E et al (2015) Workflows for automated downstream data analysis and visualization in large-scale computational mass spectrometry. Proteomics 15:1443–1447. doi:10.1002/pmic.201400391
21. Teleman J, Röst HL, Rosenberger G et al (2014) DIANA-algorithmic improvements for analysis of data-independent acquisition MS data. Bioinformatics. Oxford, England. doi:10.1093/bioinformatics/btu686
22. Reiter L, Rinner O, Picotti P et al (2011) mProphet: automated data processing and statistical validation for large-scale SRM experiments. Nat Methods 8:430–435. doi:10.1038/nmeth.1584
23. Röst HL, Rosenberger G, Aebersold R, Malmström L (2015) Efficient visualization of high-throughput targeted proteomics experiments: TAPIR. Bioinformatics. Oxford, England. doi:10.1093/bioinformatics/btv152

24. Malmström L, Bakochi A, Svensson G et al (2015) Quantitative proteogenomics of human pathogens using DIA-MS. Proteomics 129:98–107. doi:10.1016/j.jprot.2015.09.012
25. Bruderer R, Bernhardt OM, Gandhi T et al (2015) Extending the limits of quantitative proteome profiling with data-independent acquisition and application to acetaminophen treated 3D liver microtissues. Mol Cell Proteomics. doi:10.1074/mcp.M114.044305
26. Egertson JD, Kuehn A, Merrihew GE et al (2013) Multiplexed MS/MS for improved data-independent acquisition. Nat Methods 10:744–746. doi:10.1038/nmeth.2528
27. Parker SJ, Röst HL, Rosenberger G et al (2015) Identification of a set of conserved eukaryotic internal retention time standards for data-independent acquisition mass spectrometry. Mol Cell Proteomics 14:2800–2813. doi:10.1074/mcp.O114.042267
28. Rosenberger G, Koh CC, Guo T et al (2014) A repository of assays to quantify 10,000 human proteins by SWATH-MS. Sci Data 1:140031. doi:10.1038/sdata.2014.31
29. Selevsek N, Chang C-Y, Gillet LC et al (2015) Reproducible and consistent quantification of the Saccharomyces cerevisiae proteome by SWATH-MS. Mol Cell Proteomics 14:739–749. doi:10.1074/mcp.M113.035550

Chapter 21

Virtualization of Legacy Instrumentation Control Computers for Improved Reliability, Operational Life, and Management

Jonathan E. Katz

Abstract

Laboratories tend to be amenable environments for long-term reliable operation of scientific measurement equipment. Indeed, it is not uncommon to find equipment 5, 10, or even 20+ years old still being routinely used in labs. Unfortunately, the Achilles heel for many of these devices is the control/data acquisition computer. Often these computers run older operating systems (e.g., Windows XP) and, while they might only use standard network, USB or serial ports, they require proprietary software to be installed. Even if the original installation disks can be found, it is a burdensome process to reinstall and is fraught with "gotchas" that can derail the process—lost license keys, incompatible hardware, forgotten configuration settings, etc. If you have running legacy instrumentation, the computer is the ticking time bomb waiting to put a halt to your operation.

In this chapter, I describe how to virtualize your currently running control computer. This virtualized computer "image" is easy to maintain, easy to back up and easy to redeploy. I have used this multiple times in my own lab to greatly improve the robustness of my legacy devices.

After completing the steps in this chapter, you will have your original control computer as well as a virtual instance of that computer with all the software installed ready to control your hardware should your original computer ever be decommissioned.

Key words Legacy hardware, Virtual computers, System reliability, Mass Spectrometry, VirtualBox, Cloning, Systems management

1 Introduction

Scientific measurement devices with attached control computers often have a viable life that greatly exceeds the supported life of the operating system that hosts the vendor software. It is also often the case that the vendors have either gone out of business or have discontinued support for their older devices in this same time frame. Even for the vendors still in business, the process of, for example, reinstalling and reconfiguring a mass spectrometer, an LC system and its associated autosampler is not a fast nor simple process.

Lucio Comai et al. (eds.), *Proteomics: Methods and Protocols*, Methods in Molecular Biology, vol. 1550,
DOI 10.1007/978-1-4939-6747-6_21,

Many labs often have multiple deployments built on (now) legacy operating systems, such as Windows XP, for which there is no driver support under more current versions of Windows. In spite of being well past their amortization, these devices still provide invaluable daily service. A failed computer alone (all too common an occurrence) should not be deciding factor forcing procurement of replacement funds.

In this chapter, I show you how to take the required steps to virtualize a legacy control computer. A virtual computer is computer that is created in software that runs on a host computer. In its most common deployment, there is a definition file that describes the features of the computer (the memory, the processor type, what interface cards are available, etc.) and there is an image file that contains the contents of the hard drive. Modern computers are more than capable of storing, simultaneously, the definition and image files of multiple virtual computers. While there is a minor performance penalty for running the virtual equivalent of a real computer, this deficit is greatly outweighed by the administrative and logistic advantages. Effectively, a virtualized computer is entirely recapitulated by a single file; a file as easily copied, moved, or otherwise manipulated as any other. Physical computers are often different from one another, but, from the perspective of the operating system and the applications, every virtual computer with the same definition file looks the same.

When deploying a new physical control computer (say to replace one with a failed motherboard), it requires installation of an operating system, all the applications and then configuring all the components—hours of work. On the other hand, to deploy a duplicate of a configured virtual machine requires copying a file and starting the virtual machine—minutes of work.

Creating a new virtual computer instance is fairly straightforward and not unlike deploying a new physical computer. It requires installation of an operating system, installation of applications and then configuration of these components. However, for legacy control, this has the same problems as installing a new physical computer—copies of operating systems need to be found, original control software and drivers needs to be installed and configured, etc. The scope of this chapter, however, is the virtualization of an *existing* installation.

Your currently running machine is already installed and configured, so, all the installation steps are already done! **This is the inherent advantage to virtualizing a currently running machine**. However, virtualizing existing Windows installations has some other complications as compared to creating a new virtual instance. The primary complication comes from the fact that Windows has a concept of the underlying hardware and virtualization will change the hardware that the operating system is expecting to see—Windows interprets these changes as errors. There are two primary forms these errors take. One is licensing—if Windows sees too many changes to what it perceives as the underlying hardware, it will assume that it has been copied and generate an error; Windows will boot, but it will require reactivation. The other error is that if the underlying hardware changes, Windows will assume it has been corrupted and will not successfully boot. Both of these issues will have to be addressed. The final complication to address here is that of virtualization peripherals—the cards and ports on the computer that the applications need to use. The serial ports, the network interfaces, etc. all need to be made available to the virtual computer. There is one final error that can occur, the original computer may not have appropriate drivers installed to match what is presented by the virtual hardware; this tends to be rare for XP virtualizations.

As an example, I will be using a 2007 install of an LTQ-Orbitrap with a CTC-Leap autosampler and an Eksigent 2D LC system. The control computer is running Windows XP SP3. This configuration presents some nice challenges in that it requires the use of 2 network interface cards as well as several RS-232 serial connections. I have used very similar steps to virtualize spectrophotometers, DNA spotters, and gel documentation systems. While the presented process focuses on the virtualization of Microsoft Windows based installations (primarily XP) using VirtualBox virtualization software, the essential steps remain unchanged for different operating systems and are as follows:

1. Prepare the live machine to be virtualized.
2. Cleanly shutdown the live machine.
3. Duplicate the raw hard drive data to an image file contains the complete image of the hard drive.
    (a) This will usually be done by booting an alternate operating system on the live machine, attaching an external hard drive and executing a duplication job. Sometimes it will be required (or desired) to remove the current live hard drive and duplicate it on a different machine.
4. Prepare the host computer to run the VirtualBox Software
    (a) Adjust BIOS settings.
    (b) Install VirtualBox.
5. Convert the hard drive image file into an image readable by VirtualBox.
6. Create an initial hardware definition and attempt booting the virtualized machine.
7. Finish configuration and perform required post-virtualization processes.
    (a) Re-validate windows.
    (b) Configure peripherals.
    (c) Check application configurations.

2 Materials

1. Software—Within this chapter, there are many referenced tools that are to be downloaded from the Internet. Because of the uncontrollable volatility of Internet links, we have made available an uncontrollably volatile mirror site for these applications. The mirror link is:
   https://drive.google.com/folderview?id=0B41aLALlbwPWX2FQRjVqMjJoeFU
   Conveniently accessible by its volatile short link:
   https://goo.gl/3MDo5Y
   This mirror should only be used as a last resort when the official channels are unavailable.
2. Host computer.
   (a) Hardware—Most modern computers should be sufficiently powerful to virtualize a computer from a couple of years or more in the past. There are two important considerations. The first is that the available RAM should be at least 1–2 GB more than the RAM that is installed on the machine you are virtualizing. The second is that the computer should have free hard drive space at least the size of the hard drive of the computer you are virtualizing. For example, if I am virtualizing a machine with a 500 GB hard drive, I find a host computer with a 750 GB hard drive installed usually more than sufficient.
   (b) Operating system—My preference in virtualization software is VirtualBox from Oracle. Any operating system that can run VirtualBox will be sufficient, this includes Linux, Windows, and Mac OS.
   (c) Virtualization software—VirtualBox is available for free download from [1]; for Linux distributions there is often an easy install process, *see* **Note 1** for an Ubuntu example.
3. External hard drive—An external USB hard drive that is at least the size of the hard drive you wish to virtualize. Note, if you are virtualizing a 32-bit machine, it is probably best to use a 2 TB or smaller external USB drive.
4. External drive reader (optional)—For some people and for some significantly older computers, it is difficult to create the image of the hard drive from the machine that you wish to virtualize. This can happen for a number of reasons—the computer might not be able to boot external media, the computer might have severely limited resources, etc. In these cases, it is often easier to remove the hard drive and duplicate the contents on another machine. To do this, it is recommended to obtain a multi-format USB hard drive reader. I have had good

success with the SATA/IDE to USB converters made by Sabrent, however, other models should work similarly well. One that mounts SATA, PATA, and IDE will fulfill most needs.

5. External boot media (e.g., CD-R, USB flash drive)—If you are going to create your disk image on the existing physical computer, you will need to boot into an alternate operating system to allow for the duplication of hard drive contents without corruption. Depending on your physical computer, this can be either a blank CD-R or USB stick prepared with a bootable copy of an operating system suitable for making the image copy. Typically a 2 GB USB drive is more than sufficient.
6. External USB->Serial adapters (possibly required). It was very common for older devices to use RS-232 serial ports as their means of connection to a host computer. Sometimes the live computer would use "multiple port serial cards" to increase the available RS-232 ports. It is common for your new host machines to have less of these ports available. Further, it is sometimes difficult to virtualize the multiple-port RS-232 cards and even the onboard RS-232 port. These issues are readily solved with USB based RS-232 interfaces. I have had much success with the TRENDnet TU-S9 USB to RS-232 interfaces, however, other models should work similarly well.

3 Methods

1. Prepare the physical computer to be virtualized. Much of this section mirrors the unsupported protocols referenced from VirtualBox [2]:
   (a) The primary hardware incompatibility that stops successful Windows XP virtualization is that the Windows hard drive driver does a check of the attached hard drive controller; if the controller has changed, Windows will stop the boot process. The easiest remedy is to relax these checks before creating a hard drive image; VirtualBox provides a tool on their website (MergeIDE) that greatly simplifies this process. This is not a problem with Windows98.
   - Obtain MergeIDE (noting capitalization) from: https://www.virtualbox.org/attachment/wiki/Migrate_Windows/MergeIDE.zip
   - Extract the contents of MergeIDE.zip.
   - Execute MergeIDE.bat on the live computer ("double click" works fine for this. Be sure to double click "MergeIDB.bat" and not "MergeIDE.reg").

   (b) Cleanly shutdown the live computer.
2. Create a raw ("dd") copy (i.e., "image") of the live machine's hard drive. In this example, we will boot the existing live com-

puter with either a CD or USB containing a copy of OSFClone. This is the least invasive process and, in practice, perhaps the most useful technique but it does require the live computer to be able to boot off of secondary media such as a CD-ROM or a USB flash drive. While we give the OSFClone example here, this can also be performed from any live Linux distribution as described in **Note 2**. While least optimal, if the computer cannot boot from secondary media, it will be required to physically take the hard drive out of the live machine and then, on a different machine, use an external drive interface to create the image. This is also the required process if the live machine is having problems booting off of CD-R's and USB sticks. If needed, the process of removing a hard drive to image it is described in **Note 3**.

Below is the method for creating a full hard drive copy (i.e., creating a hard drive image) on a live system using OSFClone.

(a) Determine which bootable media type will work best on your current live computer. While more convenient, older machines often have difficulty booting from USB media and are often more successful at booting from CD-ROM.

(b) Go to http://www.osforensics.com/tools/create-disk-images.html and download either the.iso (to create a bootable CD-ROM) or the.zip (to create a bootable USB).

(c) Follow the instructions on the osforensics website to burn the image to your chosen media. For example, in Windows 7, for the.iso, right-click and select burn image, for the.zip, unpack it then run the ImageUSB tool. In Linux, one might chose to use "brasero" to write the.iso to a CD-R disk.

(d) Power off the live computer if not already.

(e) Attach the USB external hard drive you will use to hold your hard drive image.

(f) Boot OSFClone on your live computer using the CD-R or USB you created. For this to be successful, you may need to enter the BIOS to change the boot order (*see* Fig. 1 and **Note 4** for some further description of this).

(g) Figure 2 shows the boot process for OSFClone. You will be presented with a prompt "**boot:**" at which point you should press the enter key.

(h) You may be presented with a prompt to "**see video modes available**", it is ok to just press the space key here to continue.

(i) Once OSFClone is started, select the option to "**image**" a complete drive (type the number **2** then press the **enter** key; please note that this interface does not use the mouse or the arrow keys. You will need to type the values you are

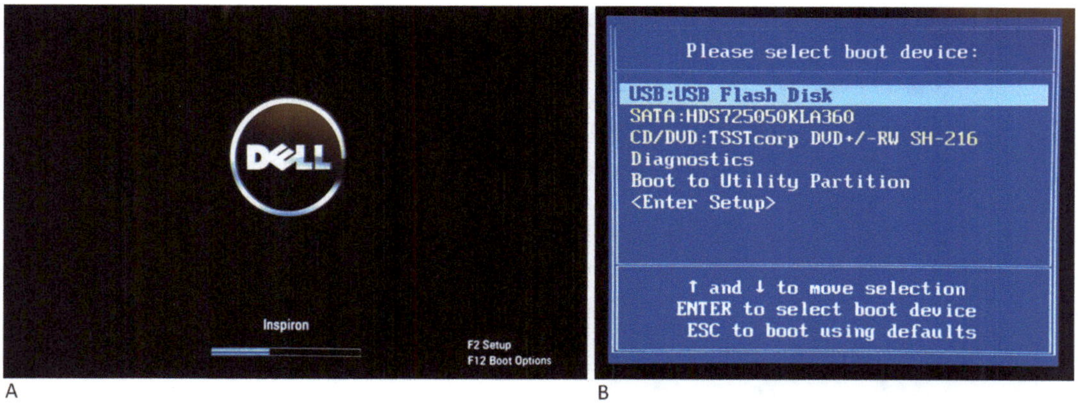

Fig. 1 Booting from secondary media. Panel (**a**) shows a representative example of a splash screen when a computer is powered on. Pressing F12, as prompted by this screen brings up the menu shown in panel (**b**). From this menu, the system can be booted from prepared USB flash drives, removable CD drives, or other media

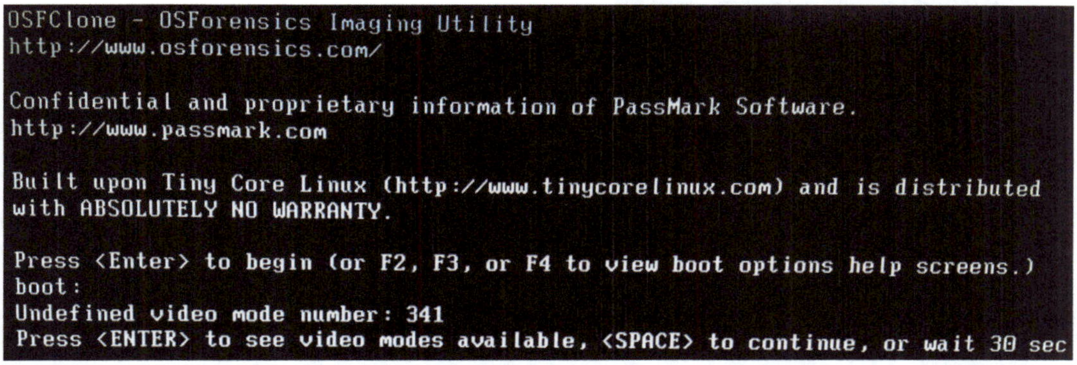

Fig. 2 Booting into OSFClone. OSFClone is built on an operating system with very rudimentary drivers. As such, it is entirely keyboard controlled. In this screen shot, the user was prompted to press enter at the "**boot:**" prompt and then asked to press the space bar to boot without further defining the video card

prompted for and press enter). Image will copy the entire hard drive into a file (in contrast to "clone" which would duplicate one device onto another device). You may be presented with options of which "**format you wish to use**" and you should select "**dd**" to perform this imaging operation (i.e., type **1** and press the **enter** key).

(j) You will now need to select a source and destination. *See* Fig. 3 as an example.

- Type **1** and press **enter** to select the source; most likely it will be "/dev/hda"—confirm this on your system; you will be able to confirm this based on the reported size (it will match the size of your "c:" drive). Return to the option menu.

- Type 2 and press enter to select the destination; most likely it will be "/dev/sda1"—confirm this on your system; you will be able to confim this based on the reported size (it will most likely be the largest size available). Note that for the destination of an image operation, the partition will be mounted and a file will be created in that partition.

A

B

C

Fig. 3 Setting Source and Destination in OSFClone. OSFClone uses a command line interface to set all the parameters. Initially, to set the source, type "1" and press enter (as shown in panel (**a**)). You will be presented with a screen similar to panel (**b**). Most likely you will select the option corresponding to /dev/hda—confirm this based on the device size. Then you will be back at panel (**a**) and you can select option 2 for the destination. Presented with a screen similar to Panel (**c**), you will probably select the option for /dev/sda1 or /dev/sdb1; confirm this by matching the size of your external hard drive to the option you wish to select

(k) Start the **dd**. After it is complete, shutdown the computer and remove the external drive containing your image.

3. Prepare the host computer to run the virtual computer
   (a) It may be required to enable virtualization extensions within the BIOS; this is often indicated by the error message "VT-x is not available." The process to enable Intel virtualization extensions is described in **Note 5**.
   (b) Boot the host machine and install VirtualBox which can be obtained from http://www.virtualbox.org/ (Ubuntu install described in **Note 1**).
   (c) If you have USB based peripherals, you will also want to download and install the VirtualBox extension pack from virtualbox.org; the downloaded file is then installed from within VirtualBox -- File->Preferences->Extensions. (**Note 1** also describes how to match the extension pack version in Ubuntu.)
   (d) Attach your external hard drive that contains the dd image file.
   (e) Convert the hard drive image file to a VirtualBox image. This is best done from the command line. In Windows, this process is described below (*see* **Note 6** for the Linux equivalent commands).
      - Press the "Start Button", type "**cmd**" and press enter to open a command line interface (Fig. 4a).
      - Execute the VBoxManage command to perform the conversion. The concept is:
      - *VBoxManage convertfromraw image.img out-drive.vdi*
      - where "VBoxManage" is the name of a command installed when you installed VirtualBox, "convertfromraw" is the action you want the command to perform "image.img" is the name of the image that you created and out-drive.vdi is the name of the VirtualBox compatible hard drive. All of the commands and file names will need to reference the full path to the folder where the commands or image files reside. It is easy to get the full path by navigating to the directory within the Windows file explorer and then right clicking and selecting "Select address as text" (Fig. 4b). Below is an example of the command as it might be typed. Note that the VBoxManage portion had to be in quotes because the path had a space in it. At the **C:\>** prompt type:
      - "C:\Program Files\Oracle\VirtualBox\VBoxManage" convertfromraw E:\diskcopy.img D:\HD.vdi

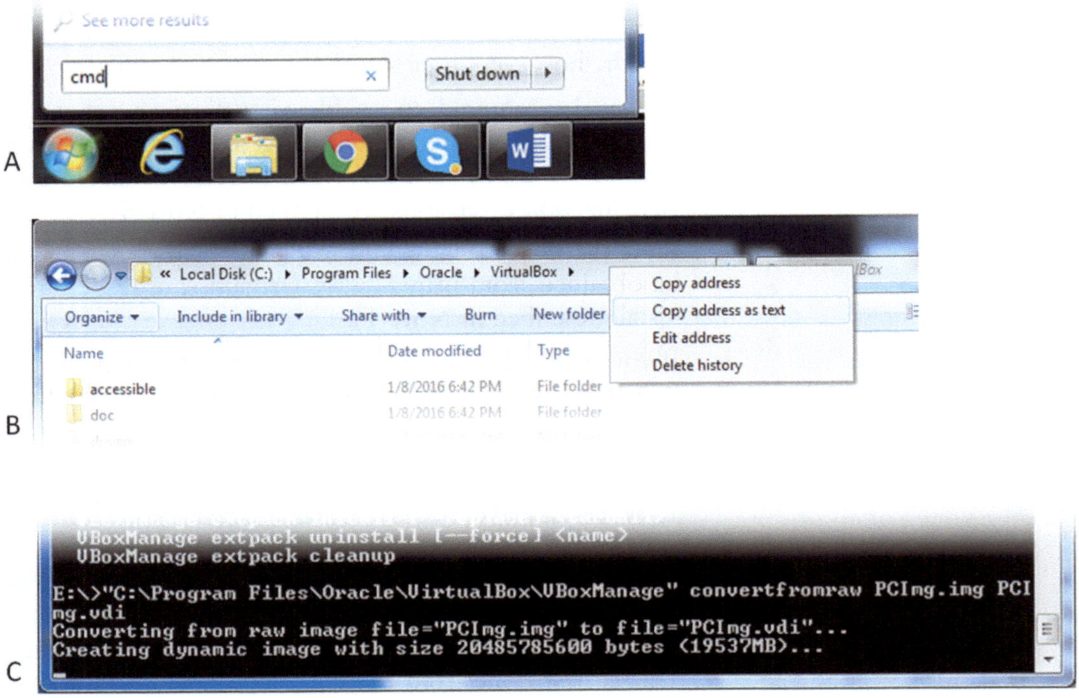

Fig. 4 Converting the dd image file into a VirtualBox image file. Panel (**a**) demonstrates how to open a command line to type commands. Press the "Start" button in the lower left, type "**cmd**" and press enter. Panel (**b**) demonstrates how to copy the complete path for the command line from within Windows Explorer. Panel (**c**) shows a sample of how the command window looks when executing the VBoxManage command under Windows 7

- replacing the paths with your paths; this assumes that the removable hard drive with your disk image was mounted as the E: drive and the path where you want to store the VirtualBox hard drive image is just D:\ and that VirtualBox was installed in C:\Program Files\Oracle (Fig. 4c).
- It is recommended that the.vdi image be placed on a physically installed hard drive, or, if it is to be on an external USB hard drive that USB 3.0 be used.

4. Start VirtualBox and select the option to create a new virtual machine. Select the appropriate type (e.g., "Microsoft Windows") and version (e.g., "Windows XP (32-bit)"). For memory, a reasonable starting setting is to match the setting on the machine you are virtualize—all things being equal, 2 or 3 GB is often a good starting point for XP machines. Note that for Windows 98, a maximum 512 MB memory setting avoids some compatibility issues. When asked for hard drive options, select the .vdi file that you created above.

5. Before starting the new virtual computer, connect all USB peripherals you want to control. In the VirtualBox list of virtual machines, select the machine you created, click on settings, click on USB. On the right hand side, click the icon of the USB connector with the "+ "symbol. Add all the devices that should be attached to that computer (e.g., USB based LC Pumps). If you are using USB to RS-232 interfaces, do not forget to connect these as well. You do not need to add mice or keyboards as they are handled in a separate fashion by VirtualBox.
6. Set up the network interfaces. For most machines, the default settings (e.g., one attached interface set as "Attached to: NAT") is sufficient. This will allow your virtual computer to share the host network connection in a similar fashion to connecting a computer to a router that is attached to your home internet feed. In our example, however, the Thermo-Fisher mass spectrometer uses a second network interface for control. To make this work properly in the virtual machine, select Network->Adapter 2 and then set it to be "Attached to: Bridged Adapter" and then select the network interface attached to the mass spectrometer router. Note that for Windows 98 installations, selecting the network interface type as "PCNet-PCI II" will usually work better than the default option.
7. Start the Virtual Machine; if Windows hangs on a black screen, power off the virtual machine and try enabling **I/O APIC**. In the VirtualBox list of virtual machines, select the machine you created, click on settings, I/O APIC will be found by navigating to System->Motherboard->Extended Features.
8. If Windows asks you to reactivate, use the license key that is associated with your installation. **Note 7** provides some pointers if Windows does require activation and your license key does not work.
9. Install the guest additions. The guest additions allow your virtual machine to more gracefully handle the graphics display as well as sharing keyboards, mice, and files with the host computer. Installation is performed on the running guest machine by choosing the "Devices" option from the menu bar, selecting to insert the "Guest Additions CD" (*see* Fig. 5) and then installing the software from within your virtual computer. You may be prompted to do this automatically after you "insert" the CD, or, you may have to navigate to the CD folder to execute the install script.
10. Check your system application settings to make sure they are interfacing to the appropriate network 'cards and RS-232 COM ports.

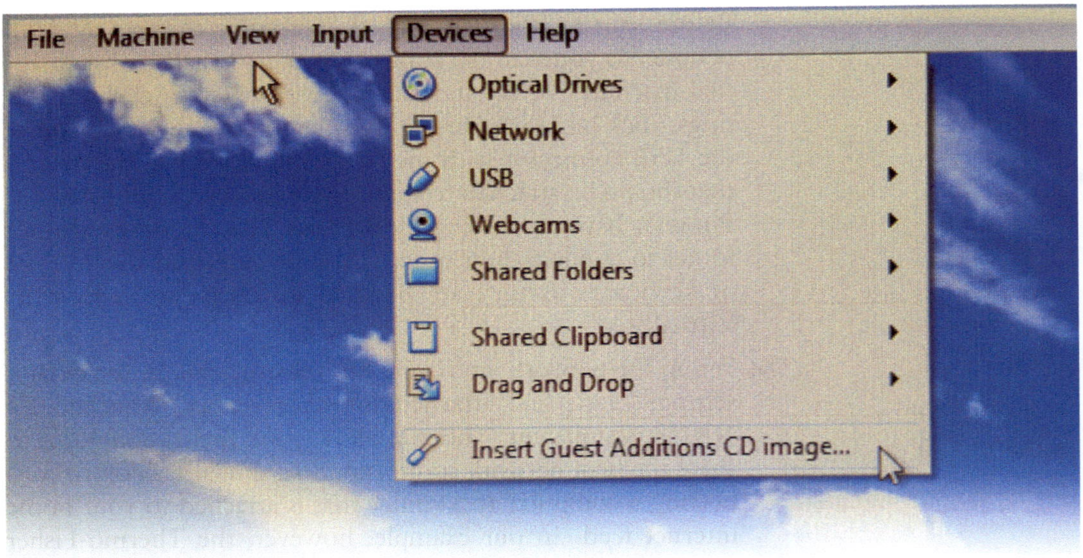

Fig. 5 Installing Guest Additions in Your Virtual Machine. Within the started virtual machine, select Devices->Insert Guest Additions CD. Then install the guest additions as prompted

11. Enjoy your easy to manage new virtual instrumentation control computer (**Note 8** describes how to export your new virtual machine suitable for rapid redeployment)

4 Notes

1. Ubuntu install of VirtualBox.
   (a) VirtualBox exists in the software repository for Ubuntu. While this facilitates installation, this is often not the most current copy of VirtualBox that is available; this will usually not present a problem, but, it does mean that you will have to look through the download archives to find the appropriately version-matched extension pack.
   (b) First, open a terminal window. Clicking on the top left icon to open the Dash, type **terminal** and click on the resulting icon. Or use the hotkey shortcut of "**ctrl+alt+t**".
   (c) Type "**sudo apt-get install virtualbox**" and enter your password when prompted.
   (d) Start up virtual box (*VirtualBox*), check the version, download the extensions pack from http://www.virtualbox.org/ that matches the version you installed from the repository.
2. Imaging a removed hard drive on a Linux computer.

(a) Remove the hard drive from the computer to be virtualized.

(b) Open a terminal window on your Linux machine (*see* **Note 1**(**b**)).

(c) Become "root" (e.g., type "**sudo /bin/bash**").

(d) Using your USB based external drive reader, attach the hard drive to the Linux machine.

(e) Figure out which device id was assigned (type **dmesg** and look for the last message, you will see something like "/dev/sde" referenced—sda is the first hard drive, sdb the second, and so forth. In this example, we will assume that this drive was assigned as /dev/sde.

- If you are booting Linux on the machine to be virtualized, this is often /dev/sda.

(f) Type **mount** and see if any partitions from the live drive were auto-mounted, if so, unmount them—e.g., **umount /dev/sde1** repeat for all mounted sde partitions.

(g) Connect the external USB drive that you acquired to hold the copy of the hard drive of the live computer.

(h) Figure out which device id was assigned (type **dmesg** and look for the last message, you will see something like "/dev/sdf" referenced. In this example, we will assume that this drive was assigned as /dev/sdf.

(i) Type **mount** and note the partitions from sdf (in this example) that have been mounted and where. Often this will take the form of /media/user/xxxx-xxxx-xxxx-xxxx/. If sdf has not automounted its partition, you can do this by hand: **mkdir /tmp/out; mount /dev/sdf1 /tmp/out** This assumes that the first partition on the external USB drive is the one you want to use and it was assigned as "sdf".

(j) This is dangerous. If you are unsure of the device and path names, you can accidentally overwrite data! Proceed with caution. From the root command prompt type the imaging command:

dd if=/dev/sde of=/tmp/out/file.dd bs=4M conv=noerror,sync

where "/tmp/out" should be replaced by the path identified in step 2.9. The "conv" options are meant to intelligently handle read errors if they were to arise; noerror means the copy will continue and sync means to replace error reads with the appropriate number of 0's to match the amount of data that was not correctly read.

3. Imaging a removed hard drive on a Windows computer.

(a) Remove the hard drive from the computer to be virtualized

(b) Using your USB based external drive reader, attach the hard drive to another Windows machine

(c) Connect the external USB drive that you acquired to hold the copy of the hard drive of the live computer

(d) Download and install the HDD raw copy tool from HDD Guru (http://hddguru.com/software/HDD-Raw-Copy-Tool/). I use the "portable Windows executable" that does not require installation.

(e) Run HDD Raw Copy, select the source hard drive—this will easily be findable because it will be a "USB" device and have the size you are expecting. For destination, select "FILE" and navigate where you want to store the image. **For file type, be sure to select "Raw Image (dd image)" versus compressed image.**

(f) Initiate the copy.

(g) Cleanly unmount the external USB drive and the mounted hard drive. If you are uncertain of how, you can just shut-down your computer.

(h) Remove the USB drive containing the hard drive image and remove the mounted hard drive. The drive with the disk image does not need to be removed if it is on the machine you are going to use as the host computer.

(i) If desired, reinstall the hard drive in the original physical computer.

(j) Windows does provide a tool to convert a physical device to a .vhd image file ("disk2vhd"), however, .vhd files are not the best to use with VirtualBox so conversion to. vdi-will still be required.

4. Configuring BIOS to boot from CD or USB before the hard drive.

   (a) For most instances, the example show in Fig. 1 will be sufficient—press the prompted for key and then select either the USB or CD as appropriate.

   (b) For some versions of the BIOS, you will need to find the options for the boot order and then set the CD or USB to boot before the hard drive.

5. Configuring BIOS settings to enable virtualization extensions

   (a) If you are running an AMD CPU, hardware virtualization is usually turned on by default. Intel machines tend to be a mix of default enabled and default disabled.

   (b) To check and change settings, you need to be in the BIOS configuration screen. Turn on the host computer and, on the initial splash screen, select the option to activate the "Setup"/BIOS menu. This is often the "F2", "DEL" or "ESC" key—look at the splash screen to see if they tell you the appropriate key; you can also perform an Internet search for the make and model of your computer. This can vary considerably, for example, for some Lenovo laptops, there is a little physical button that needs to be pressed during the boot process.

   (c) BIOS menus also vary considerably, you are looking to enable "Intel VT-x" aka "Intel Virtualization Extensions" aka "Intel Virtual Technology".

6. Converting a dd image into a vdi file in Linux.

   (a) Open a terminal window.

   (b) Become "root" (e.g., type **sudo /bin/bash**).

   (c) Type **VBoxManage convertfromraw /media/user/xx/image.dd ~user/out-drive.vdi** replacing the paths to the input and output file with the appropriate fully qualified pathnames.

7. Windows reactivation

   (a) Unfortunately, the new virtualized Windows will sometimes require reauthentication. This is complicated because Windows XP and earlier are no longer supported by Microsoft. Often the reauthentication can be accomplished by using the original license key. If, however, the original key will not validate Windows, there is no supported reactivation option. Many people have found success in a work-around that involves editing the Windows registry, resetting the countdown timer and then preventing the operating system from updating the countdown timer value. There are many sites on the internet that describe this process [3–5].

8. Comments about managing your new virtualized computer.

   (a) Some functions may work better if you install the VirtualBox guest additions. In the running virtualized computer select Devices->Insert Guest Additions CD Image. The image will be downloaded and inserted into the virtual CD drive. At this point, you will either be prompted to install the software, or, you can select the CD drive from within the virtualized computer.

   (b) Backups and redeployment are easy. With the virtual computer shut down, in the VirtualBox main console, select "File->Export Appliance". This exported appliance (i.e., a single file that contains everything required by the virtual image) is easily imported into other VirtualBox installations. The exported files can be archived and instantiated as needed.

(c) Recovery from configuration mistakes and from viruses is now trivial! After your machine is running, shut down the machine then, from the VirtualBox main console, select "Machine->Clone". To save disk space, select a "linked clone". Then, always use your new clone instance of your virtual machine. If a configuration gets corrupted, just start up the original instance and all the changes are undone.

Acknowledgments

I thank David B. Agus, The Lawrence J. Ellison Institute for Transformative Medicine at USC and the USC Center for Applied Molecular Medicine for generous institutional support. Also, I thank Lara Bideyan for her critical review of this document and the testing of the protocols described within.

References

1. VirtualBox download site: https://www.virtualbox.org/wiki/Download
2. https://www.virtualbox.org/wiki/Migrate_Windows
3. http://community.spiceworks.com/how_to/show/3381-how-to-fix-windows-xp-activation-after-a-windows-xp-repair
4. http://www.americancomputerenterprises.com/downloads/how_to_activate_windows_xp_witho.htm
5. http://www.technotrait.com/2012/02/27/activate-windows-xp-without-genuine-key/

Chapter 22

Statistical Assessment of QC Metrics on Raw LC-MS/MS Data

Xia Wang

Abstract

Data quality assessment is important for reproducibility of proteomics experiments and reusability of proteomics data. We describe a set of statistical tools to routinely visualize and examine the quality control (QC) metrics obtained for raw LC-MS/MS data on different instrument types and mass spectrometers. The QC metrics used here are the identification free QuaMeter metrics. Statistical assessments introduced include (a) principal component analysis, (b) dissimilarity measures, (c) T^2-chart for quality control, and (d) change point analysis. We demonstrate the workflow by a step-by-step assessment of a subset of Study 5 for the Clinical Proteomics Technology Assessment for Cancer (CPTAC) using our R functions.

Key words Abnormal experiment, Change point, Dissimilarity, Euclidean distance, MS/MS proteomics, Principal component analysis, Quality control

1 Introduction

Shotgun proteomics experiments are subject to multiple sources of variations, from the sample preparation, to the internal spectrometer operation and external environmental factors, to the data acquisition methods. Large variations inhibit the reproducibility of the experiments across labs and across time [1, 2]. Timely quality assessment is thus important in lab practice. Also, with massive data available in the field of proteomics, the data quality is an important problem to be addressed, especially for decision making by combing multiple data sets from different labs, such as the published data. When multiple labs have investigated the same condition, many challenges may hinder meta-analysis. The data must be scrutinized to determine whether they have been consistently collected and if outliers or batch effects exist. Both evaluation of experiment reproducibility and examination of data reusability help improve the effectiveness and accuracy of biological and clinical decisions based on the experimental proteomics data.

Lucio Comai et al. (eds.), *Proteomics: Methods and Protocols*, Methods in Molecular Biology, vol. 1550,
DOI 10.1007/978-1-4939-6747-6_22, © Springer Science+Business Media LLC 2017

This note describes a toolkit of multivariate statistical procedures and its R implementation we developed for quality assessment on raw data from a wide range of shotgun proteomics experiments. The quality assessment is based on the identification free, multidimensional quality metrics produced by QuaMeter [3]. We emphasize that the multidimensional features of a shotgun proteomics experiment should be kept integrated in its quality assessment and thus employ multivariate statistical procedures in all the assessments. The objective is to produce visualization of the multidimensional quality metrics, to detect potential abnormal experimental runs, to test for "out-of-control" experiments, and to examine batch effects.

The major statistical procedures include (a) principal component analysis (PCA). This facilitates visualization of multiple quality control (QC) metrics and generates the principal component (PC) scores for further quality assessment. (b) Dissimilarity measures. This measure is based on the Euclidean distance between the robust PCA coordinates for each LC-MS/MS experiment. It helps to detect potential abnormal experimental runs. (c) T^2-chart for quality control. This QC method provides a statistical test to detect those "out-of-control" experiments for further investigation with a cutoff significance level. The difference and connection between procedures (b) and (c) are discussed in Subheading 3. (d) Change point analysis. Change points are caused by sustained changes in the experimental conditions and thus are particularly importance in investigating longitudinal studies and examining batch effects. We include a step-by-step demonstration of the above analysis in R functions using CPTAC Study 5 data (*see* **Notes 1** and **3**).

2 Materials

Identification-free QC metrics on raw LC-MS/MS data are generated by QuaMeter. QuaMeter software [3] is a tool built on the ProteoWizard library [4] to produce quality metrics from LC-MS/MS data. It has a special "IDFree" mode to produce metrics that are independent of identification success rates for MS/MS scans (*see* **Note 2**). As only the raw data need to be accessed, the process time of QuaMeter is very short. It typically takes less than 5 min per file for Thermo Orbitrap Velos raw data files, with a large portion of time consumed in the extraction of ion chromatograms [5]. Thus, QuaMeter is well-positioned to give timely feedback in support of go/no-go decisions during experiments.

For each data file obtained from an LC-MS/MS experiment, QuaMeter IDFree generates 44 numerical measures in four categories: XIC (extracted ion chromatograms, 7 metrics), RT (retention times, 13 metrics), MS1 (mass spectrometry, 11 metrics), and MS2 (tandem mass spectrometry, 13 metrics). It aims to provide a full evaluation of multidimensional aspects of the experiment. Some of the 44 numerical metrics need to be omitted in the multivariate analysis due to low variance in the metric itself or high linear correlation with other metrics (*see* **Note 3** initial data examination). Besides these numerical measures, QuaMeter also records the name of the raw LC-MS/MS data file and the start time of the experiment, which are very useful for data file identification and for examining the temporal evolution of the experiments. Table 1 contains a full list of all 46 metrics along with their computations.

Table 1
46 metrics produced in the "IDFree" mode of QuaMeter

Filename	What is the name of the file from which the metrics were computed?
StartTimeStamp	At what time did acquisition begin for this experiment?
XIC-WideFrac	What fraction of precursor ions account for the top half of all peak width?
XIC-FWHM-Q1	What is the 25%ile of peak widths for the wide XICs?
XIC-FWHM-Q2	What is the 50%ile of peak widths for the wide XICs?
XIC-FWHM-Q3	What is the 75%ile of peak widths for the wide XICs?
XIC-Height-Q2	The log ratio for 50%ile of wide XIC heights over 25%ile of heights.
XIC-Height-Q3	The log ratio for 75%ile of wide XIC heights over 50%ile of heights.
XIC-Height-Q4	The log ratio for maximum of wide XIC heights over 75%ile of heights.
RT-Duration	What is the highest scan time observed minus the lowest scan time observed?
RT-TIC-Q1	The interval when the first 25 % of TIC accumulates divided by RT-Duration.
RT-TIC-Q2	The interval when the second 25 % of TIC accumulates divided by RT-Duration.
RT-TIC-Q3	The interval when the third 25 % of TIC accumulates divided by RT-Duration.
RT-TIC-Q4	The interval when the fourth 25 % of TIC accumulates divided by RT-Duration.
RT-MS-Q1	The interval for the first 25 % of all MS events divided by RT-Duration.
RT-MS-Q2	The interval for the second 25 % of all MS events divided by RT-Duration.
RT-MS-Q3	The interval for the third 25 % of all MS events divided by RT-Duration.
RT-MS-Q4	The interval for the fourth 25 % of all MS events divided by RT-Duration.
RT-MSMS-Q1	The interval for the first 25 % of all MS/MS events divided by RT-Duration.
RT-MSMS-Q2	The interval for the second 25 % of all MS/MS events divided by RT-Duration.

(continued)

Table 1 (continued)

RT-MSMS-Q3	The interval for the third 25% of all MS/MS events divided by RT-Duration.
RT-MSMS-Q4	The interval for the fourth 25% of all MS/MS events divided by RT-Duration.
MS1-TIC-Change-Q2	The log ratio for 50%ile of TIC changes over 25%ile of TIC changes.
MS1-TIC-Change-Q3	The log ratio for 75%ile of TIC changes over 50%ile of TIC changes.
MS1-TIC-Change-Q4	The log ratio for largest TIC change over 75%ile of TIC changes.
MS1-TIC-Q2	The log ratio for 50%ile of TIC over 25%ile of TIC.
MS1-TIC-Q3	The log ratio for 75%ile of TIC over 50%ile of TIC.
MS1-TIC-Q4	The log ratio for largest TIC over 75%ile TIC.
MS1-Count	How many MS scans were collected?
MS1-Freq-Max	What was the fastest frequency for MS collection in any minute? (Hz)
MS1-Density-Q1	What was the 25%ile of MS scan peak counts?
MS1-Density-Q2	What was the 50%ile of MS scan peak counts?
MS1-Density-Q3	What was the 75%ile of MS scan peak counts?
MS2-Count	How many MS/MS scans were collected?
MS2-Freq-Max	What was the fastest frequency for MS/MS collection in any minute? (Hz)
MS2-Density-Q1	What was the 25%ile of MS/MS scan peak counts?
MS2-Density-Q2	What was the 50%ile of MS/MS scan peak counts?
MS2-Density-Q3	What was the 75%ile of MS/MS scan peak counts?
MS2-PrecZ-1	What fraction of MS/MS precursors is singly charged?
MS2-PrecZ-2	What fraction of MS/MS precursors is doubly charged?
MS2-PrecZ-3	What fraction of MS/MS precursors is triply charged?
MS2-PrecZ-4	What fraction of MS/MS precursors is quadruply charged?
MS2-PrecZ-5	What fraction of MS/MS precursors is quintuply charged?
MS2-PrecZ-more	What fraction of MS/MS precursors is charged higher than +5?
MS2-PrecZ-likely-1	What fraction of MS/MS precursors lack known charge but look like +1 s?
MS2-PrecZ-likely-multi	What fraction of MS/MS precursors lack known charge but look like >+1 s?

3 Methods

All the statistical procedures are done using R software [6].

3.1 Initial Data Examination

A timeline plot of the data files, grouped by participating laboratories, individual instruments, sample types, or other features, helps visualize the data files in the order that they were produced in the study. The metric "StartTimeStamp" is used to extract starting time of the experiment for a raw data file.

Many of the QuaMeter metrics are based on quartiles of a quantity. For example, XIC-FWHM-Q1, XIC-FWHM-Q2, and XIC-FWHM-Q3 are the 25%ile, 50%ile, and 75%ile of peak widths for the wide XICs. This helps to provide a rough approximation of the distribution of the peak widths for the wide XICs. However, it also brings in potential high correlation among these measures. In practice, we usually remove any metric that shows correlation greater than 0.98 or less than −0.98 with the other metric(s). The analysis loses little information from this removal because one can predict the values for these removed variables almost perfectly from the retained information. Also, metrics that are related to directly configurable options, such as MS2-PrecZ-* and RT-Duration, are also omitted in the statistical analysis (*see* **Note 3** initial data examination).

3.2 Principal Component Analysis

Principal component analysis (PCA) helps to visualize the multidimensional quality control (QC) metrics in lower dimension. It makes the interpretation easier by transform correlated metrics to uncorrelated linear combinations of the metrics. To reduce impact by any extreme data files, we use robust PCA method [7]: the variance–covariance matrix of the QC metrics as well as the center of each metrics are estimated by cov.rob function in R and are used in the PCA analysis (cov.rob function in R). The PCA analysis is carried out by princomp R function using the QC metrics normalized by the individual metric's variance. The results contain PC scores matrix (linear combinations of original metrics), whose columns corresponding to the dimension of the QC metrics that are analyzed and whose rows are the newly transformed QC metrics for each data file (*see* **Note 3** principal component analysis for the example). The data structure can then be visually examined by systematically exploring two- or three-dimensional plots of different combinations of PC scores [8]. The one used most frequently is the plot based on the first two PCs, which account for the largest proportion of variance in the data. A more complete examination is needed for a comprehensive view of the data structure.

3.3 Dissimilarity Measures

This step is to exploratorily examine how similar two experiments are in terms of the QC metrics. If a spectrometer has been operating stably, it is expected that QC metrics obtained from all the experiments on this spectrometer should be randomly distributed around a "true" stable status, and thus the distance between each other is approximately equal. If an experiment is "far away" from the rest of the experiments based on the QC metrics, it reflects the overall difference in the QC metrics and thus implies this experiment may have been operated under quite different conditions compared to others. We call an experiment that is "far away" from the other experiments "an abnormal experimental run."

To detect if an experiment is "far away," we design the dissimilarity measure. Suppose the *p*-dimensional PC scores for the two experiments are x_1 and x_2, the dissimilarity between these two *p*-dimensional coordinates is defined as

$$\sqrt{\left(x_{11}-x_{21}\right)^2+\left(x_{12}-x_{22}\right)^2+\cdots\left(x_{1p}-x_{2p}\right)^2}\ .$$

Thus, this measure is based on the Euclidean distance between the robust PCA coordinates for each LC-MS/MS experiment. The larger the dissimilarity values, the lower the similarity between the two experimental data files. Compared to T^2-chart in Subheading 3.4, this measure has a wider application as it does not require that most of experiments are similar to each other. Actually, since these is no benchmark profile required in comparison, this distance measure is designed to be automatically outlier-proof and can be used in almost any situation. An abnormal experiments can be easily identified by its large distance (higher dissimilarity) from a few other experiments.

3.4 T^2-Chart for Quality Control

For experiment *i* with the *p*-dimensional PC scores xi, a T^2 statistic is computed as

$$T_i^2=\left(x_{i1}-\overline{x}_1\right)^2/\lambda_1+\left(x_{i2}-\overline{x}_2\right)^2/\lambda_2+\cdots+\left(x_{ip}-\overline{x}_p\right)^2/\lambda_p,$$

where the *p*-dimensional vector $\overline{x}$ is the median of the *p* QC metrics and λj is variance of the *j*th PC component. The statistic Ti^2 is a normalized Euclidean distance. Since xi's are PC scores and the variance–covariance matrix is a diagonal matrix with λj, $j = 1, \ldots, p$, as the diagonal element, it is assumed that Ti^2 approximately follows a χ^2-distribution with *p* degrees of freedom [9]. The upper and lower control limits can be obtained at a given significance level α as cutoff between in-control and out-of-control experiments. While the comparison based on the dissimilarity measure is more exploratory, this statistic provides a more rigorous evaluation of the abnormal experiments with a cutoff significance level. It may be impacted, however, by data files from extreme experiments.

3.5 Change Point Analysis

Change point analysis is a tool to study the potential batch effects in experiments. The multidimensional QC metrics is summarized in the normalized *p*-dimensional Euclidean distance (Ti^2). The batch effects may be caused by the changes in the mean of Ti^2, or in the variability of Ti^2, or both. With a smaller number of experimental runs on each spectrometer, we only assume changes in the mean. The R function cpt.mean in the R library changepoint is used in studying the batch effects in the mean level [10]. When variance or mean and variance changes are suspected, cpt.var or cpt.meanvar functions can be used.

4 Notes

1. Overview of CPTAC Study 5. Study 5 for the Clinical Proteomics Technology Assessment for Cancer (CPTAC) had multiple-site participants with standard operating procedures (SOPs) implementation [11]. It focused on Thermo LTQ and Orbitrap instruments. The six spectrometers included are LTQ73, LTQ295, LTQc65, Orbi86, OrbiP65, and OrbiW56, where the first three to five letters and number denote the instrument type and the last two numbers are the anonymous code for the laboratory that participated. The samples included the digested NCI-20 (labeled "1B"), the yeast reference material ("3A"), and the yeast reference material with bovine serum albumin spiked at 10 fmol/μL in 60 ng/μL yeast lysate ("3B"). The run order in Study 5 repeated the following block twice 1B, 3A, 3A, 3A, 1B, 3B, 3B, 3B, followed by an additional 1B. This leads to a total of 103 experimental runs, including 30 on Sample 1B (five runs for each spectrometer). Here we only study the data collected from experiments on Sample 1B. Table 1 in the Supporting Information of the CPTAC Repeatability article [6] lists the commercial sources of the proteins in the NCI-20 mixture and the detailed description on the creation of NCI-20 reference materials can be found in [4]. Data were collected on three LTQ and three Orbitrap mass spectrometers between October of 2007 and January of 2008. Raw data files for Study 5 can be found at the CPTAC Public Portal: https://cptac-data-portal.georgetown.edu/cptacPublic/. QuaMeter metrics for Study 5 data and the R functions are available on the author's website (http://homepages.uc.edu/~wang2x7/).
2. QuaMeter IDFree execution. The QuaMeter software was executed with this command line:

```
quameter.exe *.raw -MetricsType idfree
-OutputFilepath metrics.tsv
```

The quameter.cfg files used for each type of instrument included the following:

Orbitrap configuration:

```
ChromatogramMzLowerOffset = "0.01mz"
ChromatogramMzUpperOffset = "0.01mz"
Instrument = "orbi"
```

Ion Trap configuration:

```
ChromatogramMzLowerOffset = "1.5mz"
```

```
ChromatogramMzUpperOffset = "1.5mz"
Instrument = "LTQ"
```
QqTOF configuration:
```
ChromatogramMzLowerOffset = "0.1mz"
ChromatogramMzUpperOffset = "0.1mz"
Instrument = "orbi"   #Here, "Orbi" means
"can resolve isotopes."
```

3. Demonstration using CPTAC Study 5.

   Read in the QuaMeter data. There were a total six .tsv files for QuaMeter metrics with the 46 metrics. The sample type, the spectrometer instrument type, and a label for the spectrometer number were added. Thus the "QuaMeter.csv" file had a total of 50 columns with the first column as the index number. In this demonstration, we only examined 30 runs from Sample 1B, 5 runs for each spectrometer.

   Initial data examination. Figure 1 shows the timeline plot for all the data files in CPTAC Study 5. We first removed RT-Duration and MS2.PrecZ.* as they are operating conditions set by the operator. We then used cor R function to detect and removed metrics with high correlations with other metrics. At the end, 27 numeric metrics, excluding Filename and StartTimeStamp, were used in the following statistical analysis. These 27 numeric metrics were:

   XIC (5): XIC.WideFrac, XIC.FWHM.Q2, XIC.Height.Q2, XIC.Height.Q3, XIC.Height.Q4

   RT (11): RT.TIC.Q1, RT.TIC.Q2, RT.TIC.Q3, RT.TIC.Q4, RT.MS.Q1, RT.MS.Q2, RT.MS.Q3, RT.MS.Q4, RT.MSMS.Q1, RT.MSMS.Q2, RT.MSMS.Q3

   MS1 (9): MS1.TIC.Change.Q2, MS1.TIC.Change.Q3, MS1.TIC.Change.Q4, MS1.TIC.Q2, MS1.TIC.Q3, MS1.TIC.Q4, MS1.Count, MS1.Freq.Max, MS1.Density.Q1, MS1.Density.Q2, MS1.Density.Q3

   MS2 (2): MS2.Freq.Max, MS2.Density.Q1

   Principal component analysis. The R object metrics.pca contained the 27 metrics for the 30 experiment runs. The following R codes carried out robust PCA analysis.

```
metrics.pca<-metrics.1[,5:(ncol(metrics.1)-3)]
robust.cov<-cov.rob(metrics.pca)
robust.cor<-cov2cor(robust.cov$cov)
robust.cor.1<-robust.cov
robust.cor.1$cov<-robust.cor
metrics.pca.1<-metrics.pca
for ( i in 1:ncol(metrics.pca)){
```

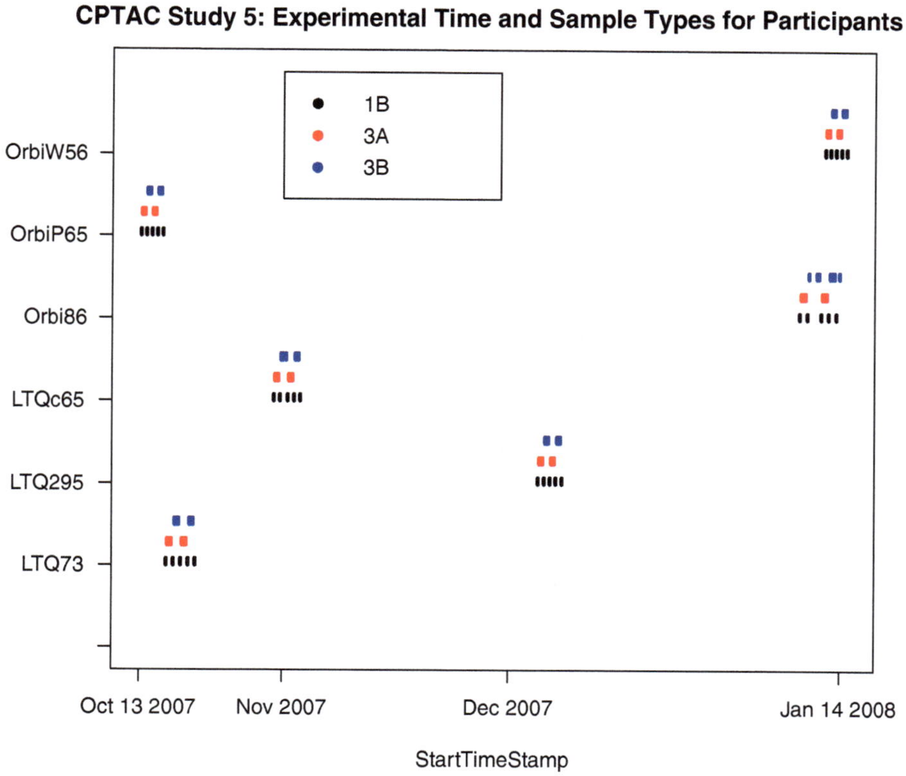

Fig. 1 Timeline for experiments in CPTAC Study 5. It illustrates the run times for each LC-MS/MS experiment in the course of Study 5 across six spectrometers. The run order was prescribed by the SOPs under which Study 5 was conducted. Sample 1B: digested NCI-20 mixture; Samples 3A: yeast; 3B: yeast + BSA

```
robust.cor.1$center[i]<-robust.cov$center[i]/sqrt(robust.cov$cov[i,i])
metrics.pca.1[,i]<-metrics.pca[,i]/sqrt(robust.cov$cov[i,i])
}
print(summary(pc.cr <- princomp(metrics.pca.1, covmat = robust.cor.1, scores=T)))
```

Figure 2 shows the PC plot for the first two principal components. The experimental runs from the same spectrometer tended to cluster together, and LTQ and Orbi instruments tended to cluster together, along with some potential abnormal experimental runs, such as the one in LTQ73.

Dissimilarity measures. For pairwise comparison, we used run1 to represent the p-dimensional PC coordinate for one experiment and run2 to represent that of the other experiment. The dissimilarity measures were then calculated in R as distance <-sqrt(sum((run1-run2)^2)). Figure 3 shows all the values of dissimilarity measure grouped by spectrometers. Clearly, there was a cluster of large distances on LTQ73, which were all

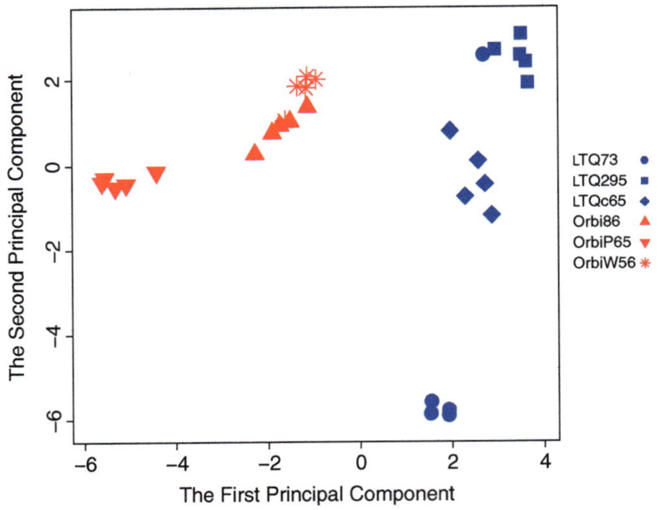

Fig. 2 PCA plots based on the first two principal components of quality metrics for experiments on Sample 1B on the six spectrometers

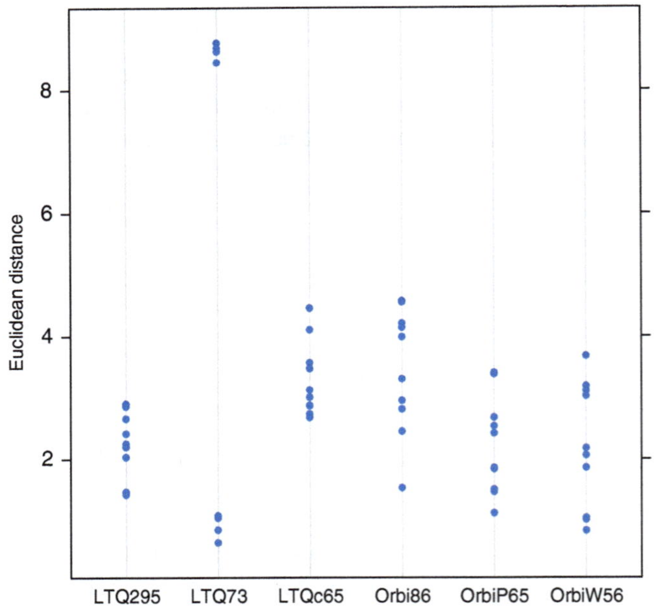

Fig. 3 Dissimilarity measures for experiments on Sample 1B on the six spectrometers

distances of the first run compared to the other runs on LTQ73. This implied that the first experiment on LTQ73 was potentially abnormal/different compared to the other runs and needed further investigation. When there is a large number of experiments to compare, the number of dissimilarity measures increases fast (in the order of n^2). A boxplot of the dissimilarity measures is a helpful visualization tool.

T^2-chart for quality control. For each spectrometer, the T^2 statistics for each data file was calculated using the following R codes:

```
temp.dist$T2[j] <- t(temp)%*%solve(pca.var)%*%temp,
```

where temp.dist was the R object containing the PC scores on experiment runs for the given spectrometer, temp was the metrics vector for the experiment centralized by the median vector and pca.var was a diagonal matrix with the diagonal elements equal to the variance of each PC component. The median vector of a given spectrometer was calculated by R function l1median and saved as L1median. The T^2's were assumed following a χ^2-distribution with 27 degrees of freedom. We used an overall familywise error rate $\alpha = 0.01$ and used Bonferroni correction for the multiple testing on each spectrometer (five files per spectrometer).

Figure 4 shows the T^2 statistics along the experiment starting time, grouped by the six spectrometers. The blue dots represent normal experiment runs, while the pink dot represents the "out-of-control" one detected by the χ^2 test and the "*" symbol represents the experiment with large dissimilarity measures. As shown in Fig. 4, the first experiment in LTQ73 (labeled as "*") was an experiment quite different from the rest of experiments on LTQ73 and the last experiment in OrbiW56 (pink dot) was detected as an out-of-control experiment. Further investigation was needed to identify the cause of the abnormality.

Change point analysis. We first removed any isolated out-of-control experiments detected by the T^2 statistic because change point analysis studies sustained changes. Change point detection was then carried out upon the remaining experiment runs. In Fig. 4, the last experiment in OrbiW56 was removed from change point analysis as it is an isolated out-of-control experiment (black circled). Except LTQc65, the other five spectrometers all had one change point, which divided the experiments on Sample 1B into two batches (first batch: red squares; second batch: green diamonds). Particularly, it turned out that the first run in LTQ73 was different from the rest of the runs and was detected as a change point location.

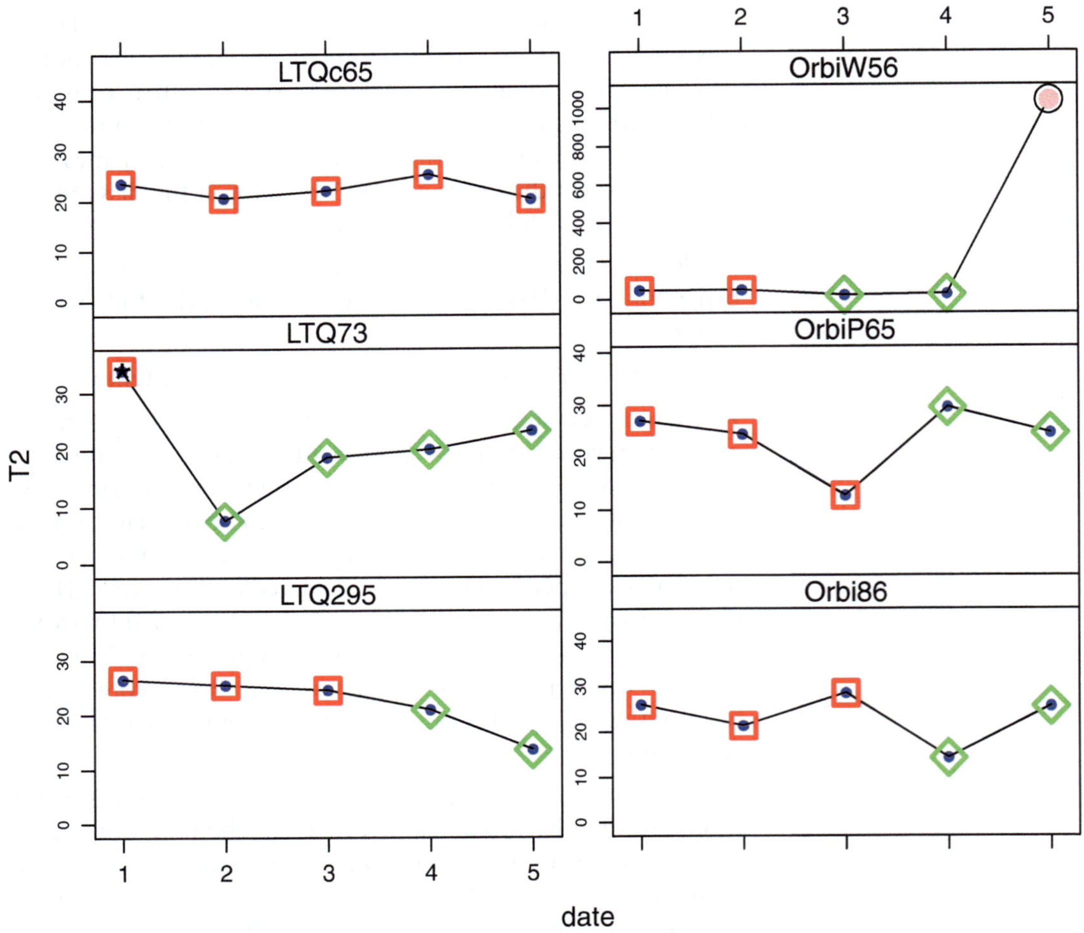

Fig. 4 T^2 statistic and change point analysis results for the six spectrometers. The *dots* represent the values of T^2 statistic with *blue dots* for "in-control" experiments, *pink dots* for "out-of-control" experiments, and "*" for those with large dissimilarity measures. If a *pink dot* is *circled* by a *black circle*, then it is not included in the change point study (as the one for OrbiW56). The *red squares* and *green diamonds* distinguish the two batches of the experiments runs, if there is a change point

References

1. Bell AW, Deutsch EW, Au CE et al (2009) A HUPO test sample study reveals common problems in mass spectrometry-based proteomics. Nat Methods 6:423–430
2. Mann M (2009) Comparative analysis to guide quality improvements in proteomics. Nat Methods 6:717–719
3. Ma ZQ, Polzin KO, Dasari S (2012) QuaMeter: multivendor performance metrics for LC–MS/MS proteomics instrumentation. Anal Chem 84(14):5845–5850
4. Chambers MC, Maclean B, Burke R et al (2012) A cross-platform toolkit for mass spectrometry and proteomics. Nat Biotechnol 30:918–920
5. Wang X, Chambers MC, Vega-Montoto LJ et al (2014) QC metrics from CPTA raw LC-MS/MS data interpreted through multivariate statistics. Anal Chem 86(5):2497–2509
6. R Core Team (2015) R: a language and environment for statistical computing. R Foundation for Statistical Computing, Vienna, Austria. http://www.R-project.org/

7. Rousseeuw PJ, van Zomeren BC (1990) Unmasking multivariate outliers and leverage points. J Am Stat Assoc 85:633–639
8. Ringnér M (2008) What is principal component analysis? Nat Biotechnol 26(3):303–304
9. Johnson RA, Wichern DW (2007) Applied multivariate statistical analysis. Pearson Prentice Hall, Upper Saddle River
10. Killick R, Eckley IA (2014) changepoint: an R package for changepoint analysis. J Stat Softw 58(3):1–19
11. Tabb DL, Vega-Montoto L, Rudnick PA et al (2010) Repeatability and reproducibility in proteomic identifications by liquid chromatography-tandem mass spectrometry. J Proteome Res 9:761–776

Chapter 23

Data Conversion with ProteoWizard msConvert

Ravali Adusumilli and Parag Mallick

Abstract

Recent advances in proteome informatics have led to an explosion in tools to analyze mass spectrometry data. These tools operate across the analysis pipeline doing everything from assessing quality control to matching peptides to spectra to quantification. Unfortunately, the vast majority of these tools are not able to operate directly on the proprietary formats generated by the diverse mass spectrometers. Consequently, the first step in many protocols is the conversion of data from vendor-specific binary files to open-format files. This protocol details the use of ProteoWizard's msConvert and msConvertGUI software for this conversion, taking format features, coding options, and vendor particularities into account. We specifically describe the various options available when doing conversions and the implications of each option.

Key words Proteomics, Proteome informatics, Data conversion, Open formats, mzML, mzXML, ProteoWizard

1 Introduction

Mass-spectrometry-based proteomics has become an important component of biological research. Numerous proteomics methods have been developed to identify and quantify the proteins in biological and clinical samples, identify pathways affected by endogenous and exogenous perturbations, and characterize protein complexes. However, before it is possible to explore any of these important biological questions, a battery of computational tools is required to convert mass spectrometry data into knowledge. Each vendor has developed its own ecosystem of tools, such as ProteomeDiscoverer (Thermo) and Analyst (SCIEX). to help interpret the complex data produced by their instruments. However, the proteome informatics research community has developed thousands of alternate tools with a wide range of capabilities. Unfortunately, the vast majority of these tools are not able to directly read vendors' binary formats and instead are dependent upon "open" formats such as mzML and mzXML. Translating from closed binary formats to open formats requires specialized

Lucio Comai et al. (eds.), *Proteomics: Methods and Protocols*, Methods in Molecular Biology, vol. 1550,
DOI 10.1007/978-1-4939-6747-6_23, © Springer Science+Business Media LLC 2017

software. Initially a set of tools that specifically converted data from one vendor format into an open format (e.g., ReAdW) emerged. More recently, the ProteoWizard toolkit [1, 2] has provided a swiss-army knife tool in msConvert that is able to convert all popular vendor formats into diverse open formats.

Beyond performing a direct translation of file contents, from a closed binary format to an openly readable format, conversion is also a data processing step. In this step, researchers may perform lossy data compressions (such as centroiding) or fundamentally alter the values in files through recalibration (such as in mzRefinery). Here we describe a variety of protocols for data conversion and particularly detail all the diverse options available to users during conversion.

The msConvert tool described below is available as user-friendly graphical user interface (GUI) program (msConvertGUI) for Windows and as cross-platform (Windows, Mac, Linux) command line tool (msConvert). We note that there are more options available in the command line tool than in the GUI. MS Convert offers the flexibility of processing multiple input file formats with options for several filters and transformations. The details of the filters and transformations available for msConvertGUI and msconvert command line will be further discussed below. In addition to converting from vendor formats to open formats, msConvert and msConvertGUI can also convert among open formats, such as from mzXML to mzML. Conversion from vendor formats is only achievable on Windows. However, conversion among open formats can be performed on any platform.

2 Materials

2.1 Installation of ProteoWizard (Includes Both msConvert, msConvertGUI and msPicture):

1. Make sure you have installed Microsoft .NET Framework **3.5** SP1 and **4.0**.
   - (a) MS .NET Framework 3.5 SP1 can be found at: http://www.microsoft.com/en-us/download/details.aspx?id=22
   - (b) MS .NET Framework 4 can be found at: http://www.microsoft.com/en-us/download/details.aspx?id=17851
2. Download the latest version of proteoWizard from: http://proteowizard.sourceforge.net/
   - (a) We typically recommend the Windows 64-bit installer for most users.
   - (b) Upon download, double-click the .msi file to install.
   - (c) proteoWizard will typically be installed into a folder 'C:\Program Files\ProteoWizard\ProteoWizard 3.0.DDDD'

3. (For Command Line Use) Add "proteoWizard" folder to the path variable:
   (a) Go to control panel > System and security > System > Advanced settings > Environment Variables
   (b) You should see a table/list of variables and values. Double click on "Path".
   (c) You should see a bunch of values separated by ";" in the Variable value field.
   (d) Add your proteowizard location to the list separating with a semicolon.

   For example: "C:\Windows; C:\Program Files\ProteoWizard 3.0.8708\"

   (e) Restart Windows.

3 Methods

3.1 Using msConvertGUI to Convert the File Format of Data Files

1. Navigate to the proteowizard directory (Fig. 1) (For ex: C:\Program Files\ProteoWizard\ProteoWizard 3.0.8708)
2. Start the msConvertGUI application

   Double-click on "msConvertGUI" application to open. This should open the msconvert GUI application. Visually, msconvertGUI is divided into four boxes (Fig. 2), which are described in greater detail below.

   (a) Select input files and output files directory location (**Top left**).

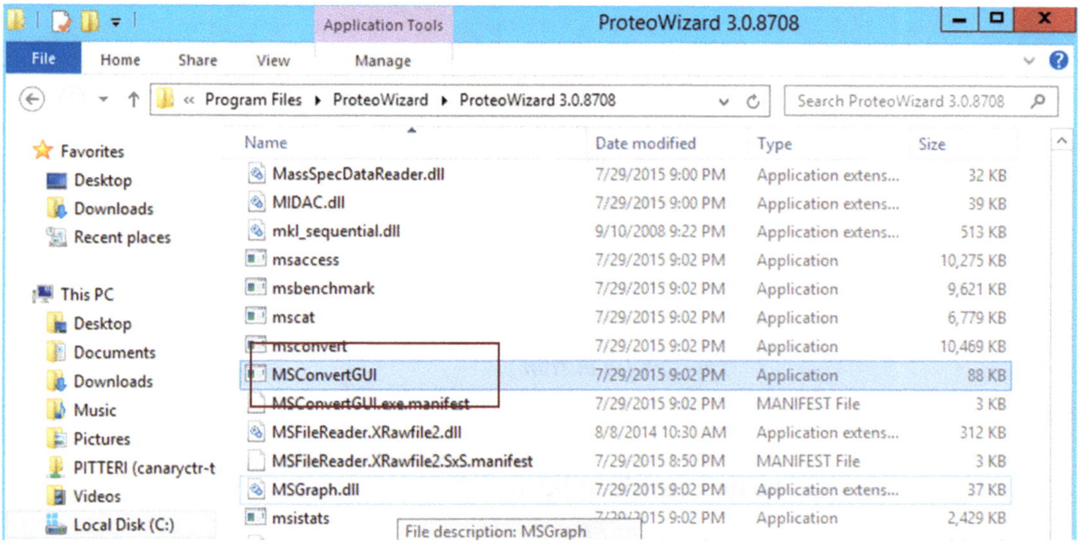

Fig. 1 Open from the folder

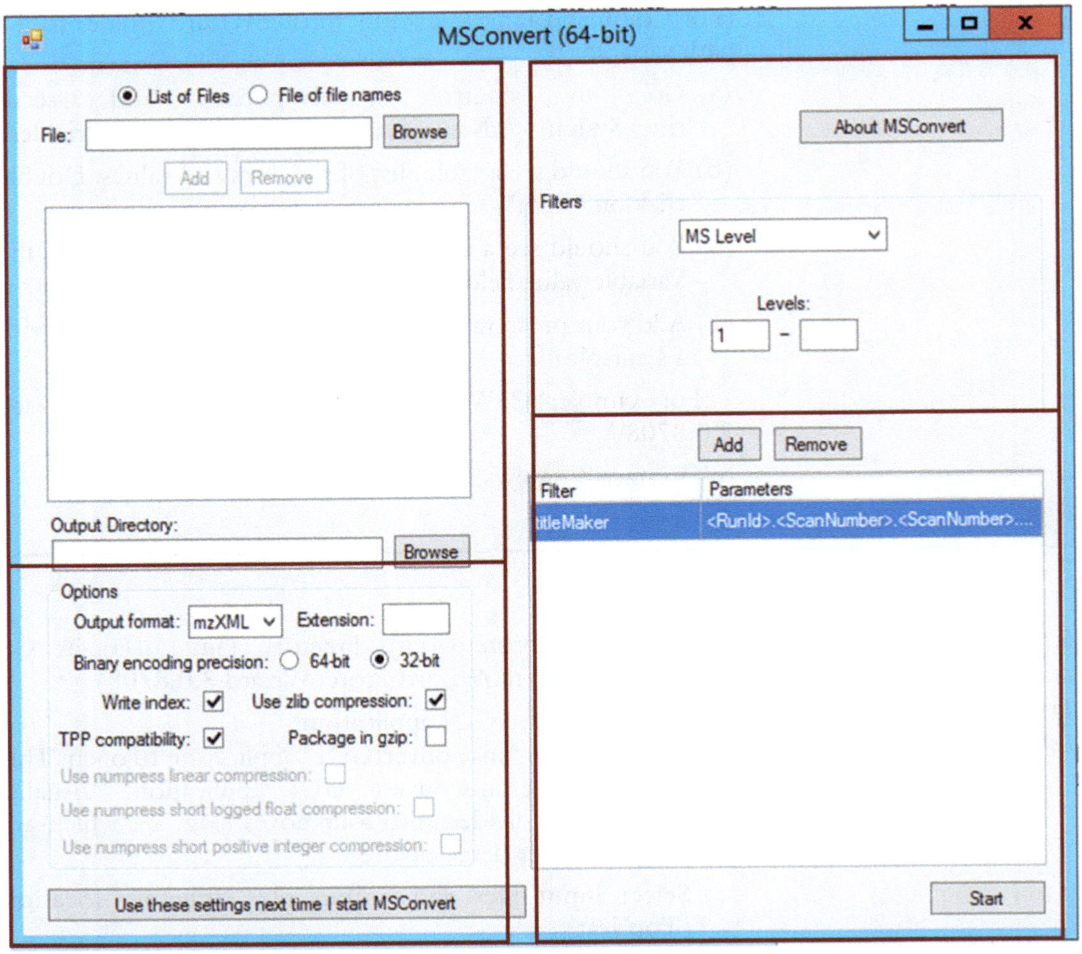

Fig. 2 MSConvertGUI

(b) **Bottom left**: Options—to select the output format, extension, binary encoding precision, write index, use zlib compression, TPP compatibility, package in gzip, numpress compression types (linear, short logged float, short positive integer).

(c) **Top Right**: Filters—Type of filters and filtering levels/options to add or remove.

(d) **Bottom Right**: Box to view the selected filters in **step c**.

3. Choosing files to convert (top left, Figs. 3 and 4):

Type: radio button

Default: List of Files

Select the file import type (list of files or file of filenames)—further explained below.

(a) *List of Files:*

Type: Button ("Browse")

Additional types: Button ("Add" and "Remove")

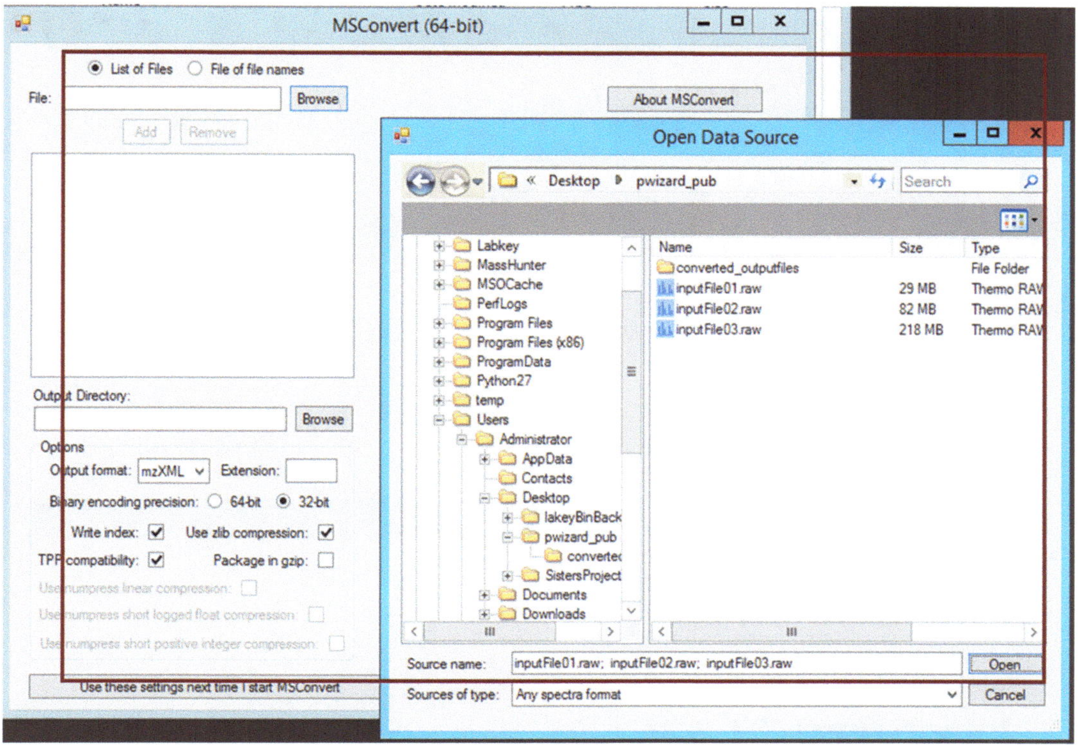

Fig. 3 Click on Browse and select input files from MSConvertGUI

i. Select the "List of Files" option on top-left of the msconvert GUI.

ii. Click on the "Browse" button-this should open a desktop browser window to select the input files.

iii. Go to the directory with the input formatted (*.RAW files) from the browser window in **step ii**.

iv. Select one input file or multiple files by holding down the shift/control button on the keyboard.

v. Click "open"—the selected files should now be visible in the msconvert GUI (in the box right below the "List of Files" and "File of filenames" options.

vi. To remove some files after adding them to the GUI, select the file/files to remove in the box from **step v**. Click "Remove" right above the box with the list of files.

(b) *File of file names:*

Type: Button ("Browse")

Additional types: Button ("Add" and "Remove")

Input files to be converted can be imported from a text file with input file names and locations.

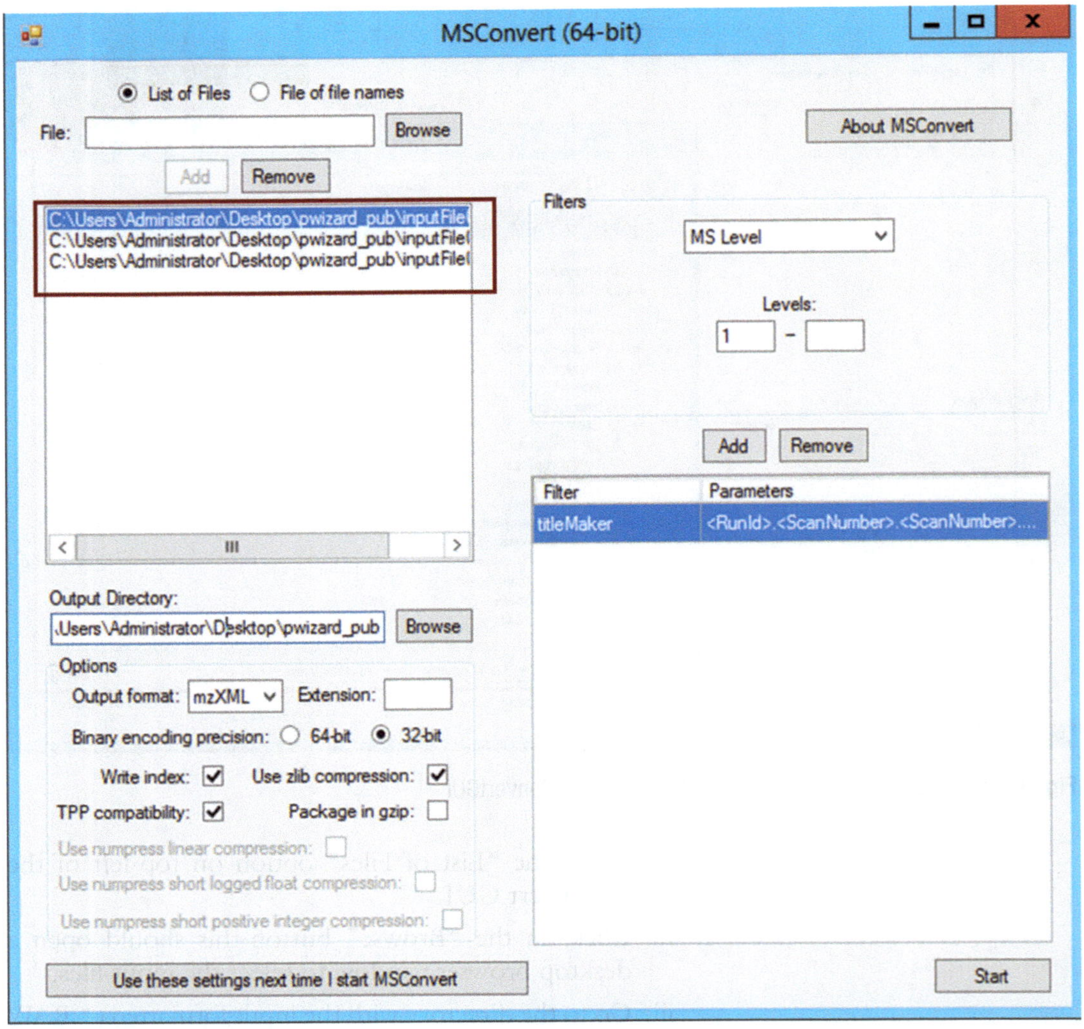

Fig. 4 MSConvertGUI list of selected files

i. Select the "File of File names" option on top-left of the msconvert GUI.

ii. Click on the "Browse" button-this should open a desktop browser window to select the input files.

iii. Go to the folder containing the text file with "input filename and paths" to be converted.

iv. Select the text file/files with "input filename and paths" to be converted.

v. Click "open".

The files to be imported (selected in either **step a** or **b**) can be viewed in **step 2**.

4. Select an Output Directory *(top left):*

   Type: Button ("Browse")

   Default: Folder path with the input files

   (a) The default output location is set to the directory with the input formatted files selected. If input files are selected across multiple directories, the directory containing the first selected input file is selected.
   (b) To choose an output location other than the default, click on "Browse". This should open a browser window.
   (c) Select an already existing folder from the desktop or create a new folder to save the output files.
   (d) To create new output folder, click on "Make New Folder"—this will create a new folder. Type in the output name for the folder (For Example: "Project1_output")
   (e) Click "OK" to select the folder or "Cancel" to cancel the folder selection.

5. **Select Conversion Options (Fig.** 5**):**

   (a) Output format:

   Type: Drop down list

   Default: mzXML

   Select the format of the converted output file. The following output formats can be selected:

   i. mzXML (default)
   ii. mz5
   iii. mgf (Mascot generic format)
   iv. text (Proteowizard internal text format)
   v. ms1
   vi. cms1
   vii. ms2
   viii. cms2

   (b) Extension:

   Type: Text box

   Default: Blank

   To add an extension to the output formatted file other than the option chosen above. For example: mzXML can be saved as .fred.mzXML if needed

   (c) Binary encoding precision:

   Type: radio button

   Default: 32-bit

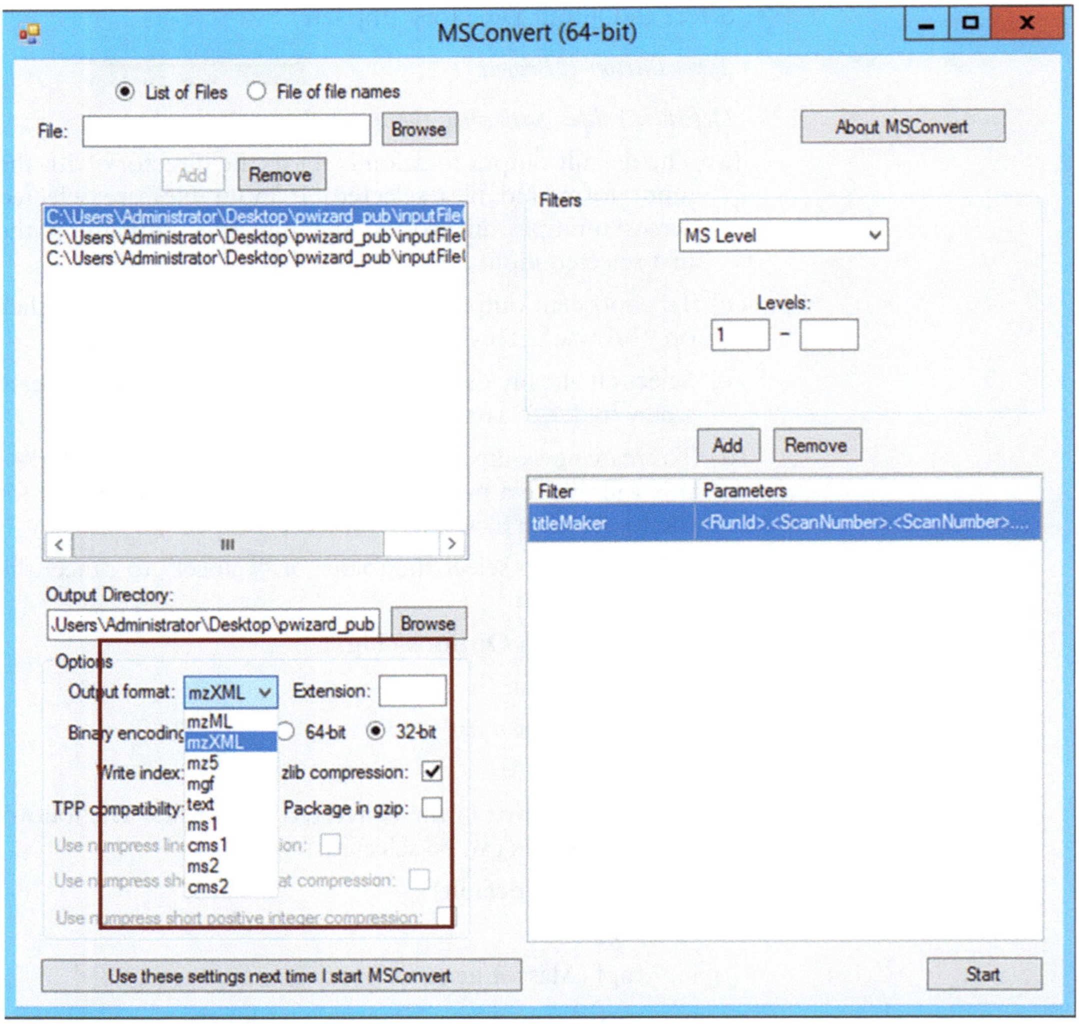

Fig. 5 MSConvertGUI output format options

Select the output file encoding type. The output file can be encoded as 32-bit (default—smaller file size) or 64-bit.

(d) Write index:

Type: Checkbox

Default: Checked

Check this box to write index to the output file (*see* **Note 1**).

(e) Use zlib compression:

Type: Checkbox

Default: Checked

Check this box to "zlib" compress the output file (*see* **Note 2**).

(f) TPP compatibility:

Type: Checkbox

Default: Checked

Check this box to make the output file "TPP compatible."

(g) Package in gzip:

Type: Checkbox

Default: NOT Checked

Check this box to "gzip" the output file (*see* **Note 3**).

6. **Select Filter Options (Figs.** 5, 6 **and** 7) (*see* **Note 4**)

Filters:

Type: Dropdown list

Default: MS Level

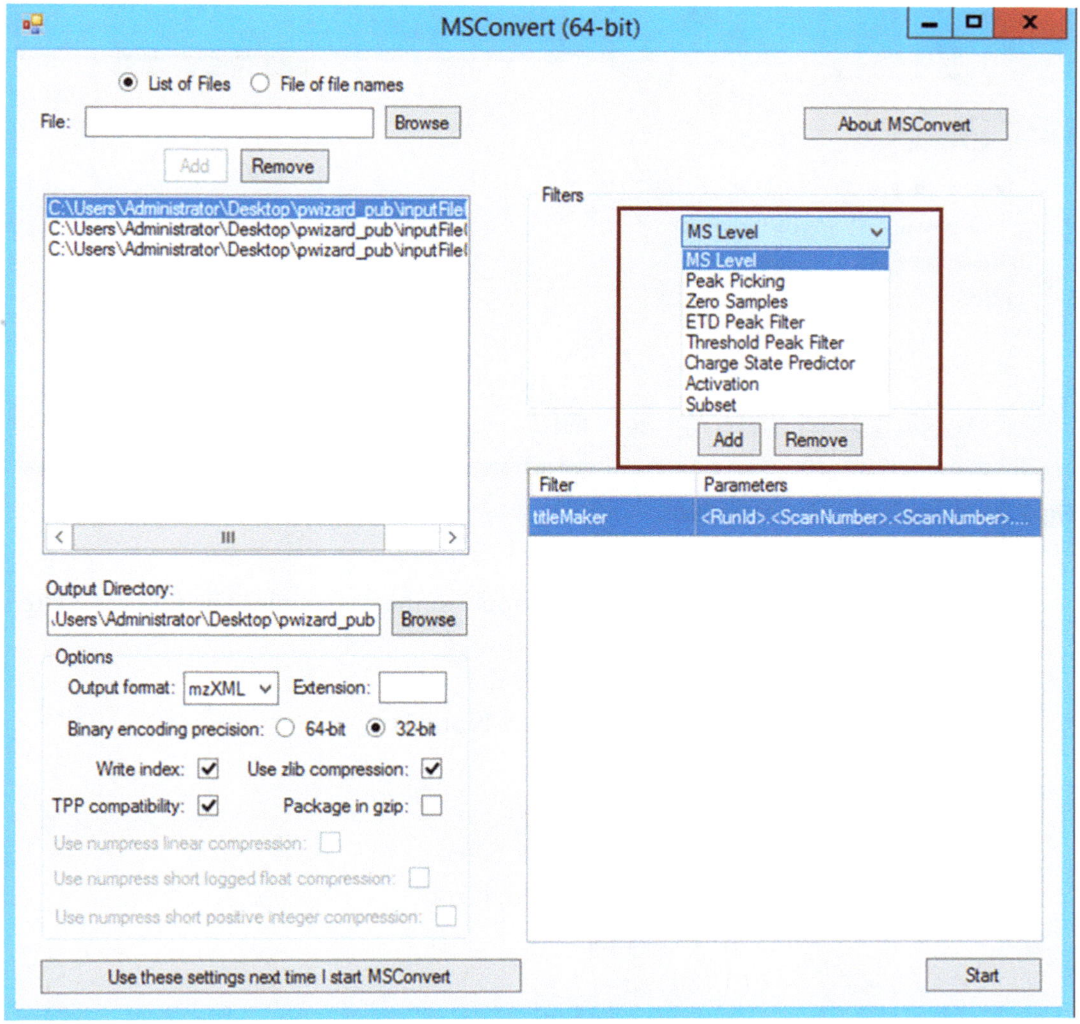

Fig. 6 MSConvertGUI filtering options

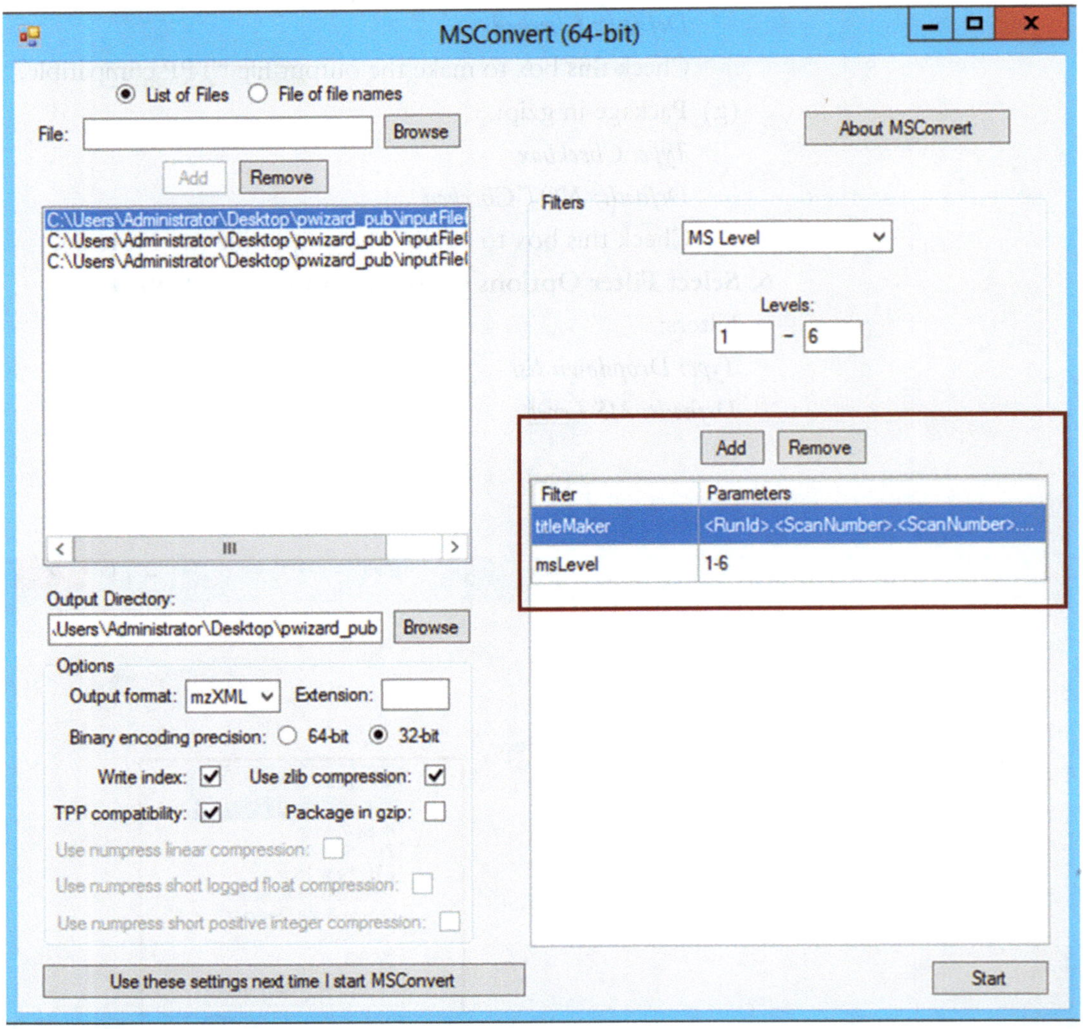

Fig. 7 MSConvertGUI list of selected filters

Select the filter to apply to the output file. The following filters can be selected:

(a) MS Level (Default)

(b) Peak Picking (*see* **Note 5**)

(c) Zero Samples

(d) ETD Peak Filter

(e) Threshold Peak Filter

(f) Charge State Predictor

(g) Activation

(h) Subset

Click "Add" to add the settings selected from **step a** and/or **b**. Alternately, remove the settings added by clicking "Remove". View the selected filters and parameters in the box.

7. **Save Settings.** Click "Use these settings next time I start MSConvert" (bottom left) to set the settings from steps 1 to 10 as defaults so they do not need to be set the next time you run the program.
8. Double-check all the settings. Once all options have been filled click "Start." A new window will appear with a list of files to be converted and current progress. Details on the conversion of the currently selected file will be shown in the text box at the bottom of the new menu. Once all files have finished converting it is safe to close the progress window and use the resulting files

3.2 Using the Command Line Version of msConvert to Convert the File Format of Data Files

1. Make sure the latest version of proteowizard is downloaded and setup as instructed above.
2. Go to command prompt/windows power shell (for windows 8+). If the path variable is set under "**SETUP PROTEOWIZARD**" instructions, all the proteowizard tools can be called without typing the absolute path. For example: *msconvert.exe* can be called directly if path variable is set. If not, *C:\Program Files\ProteoWizard 3.0.8708\msconvert.exe* will be required to call the tool.

   For this protocol, the assumption is that path has been set and so all the tools will be called directly without the absolute path.
3. To view all options under the msConvert commandline, type *msconvert.exe* (or *C:\Program Files\ProteoWizard 3.0.8708\msconvert.exe*) and hit ENTER/return.
   (a) This should show all the options available under msConvert commandline (screenshot attached).
   (b) All Command line usage terms for msconvert can be summarized into four sections:

1. **Usage:** Generic command format to run the msconvert tool
2. **Options:** List of all available options to run msconvert (
   - For example: -f for "text file with input filenames" (like File of file names *in msconvertGUI)*
     1. -o for "output directory "
     2. -c for configuration file
3. **FILTER OPTIONS:** This section lists all the options to filter the data in the output file
   - For example: index (similar to write index on the GUI);

     1. msLevel to select the level of ms data (same as MS Levels in the GUI)

     4. **Examples:** This section list few examples with explanations to run msconvert commandline options.

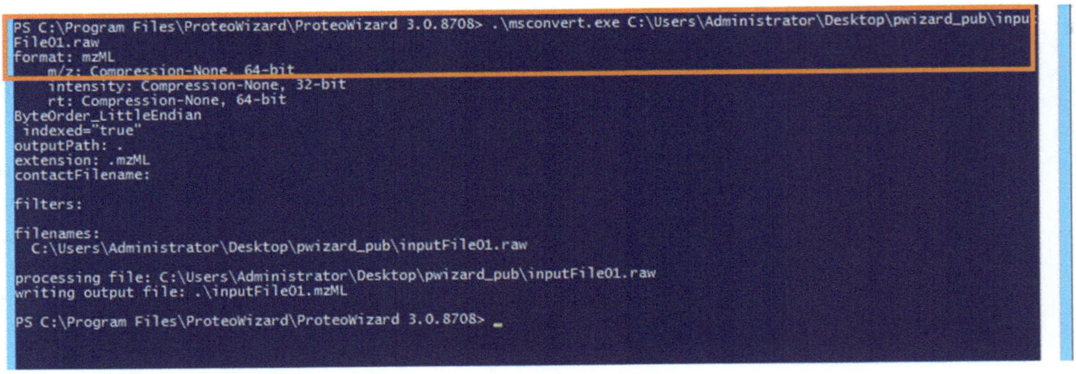

Fig. 8 msconvert—default RAW (to mzML) file conversion from commandline

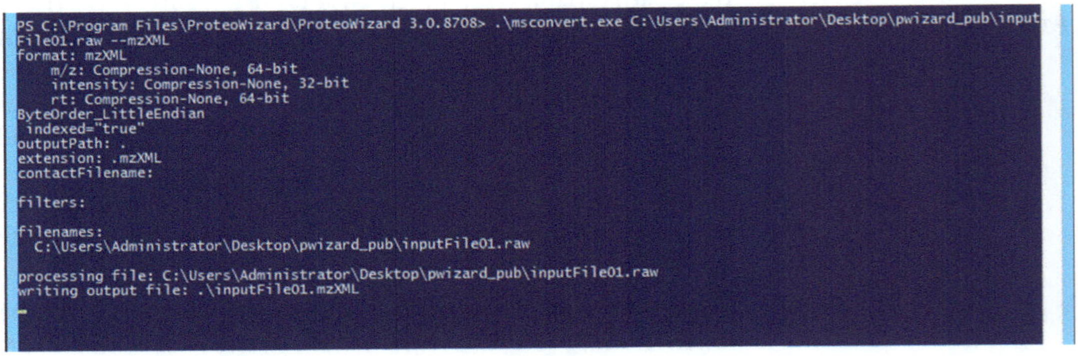

Fig. 9 msconvert—RAW to mzXML file conversion from commandline

4. Execute msConvert (Figs. 8, 9 and 10)

   The simplest way to run msConvert is as:

   msconvert C:\Users\Administrator\Desktop\pwizard_pub\inputFile01.raw

   The above is the most basic command to run a single file on one input *.RAW file. The "*inputFile01.RAW*" in the above command is converted to *"inputFile01.mzML"* formatted file by default. The *"inputFile01.mzML"* file will be located in the current working directory (i.e., the directory from which the command is executed).

   However, msconvert provides the user with flexible options to choose the output conversion format, set the output directory location, set file format extensions, provide configuration file choice to set all the conversion properties together. These options are elaborated in **Note 6**.

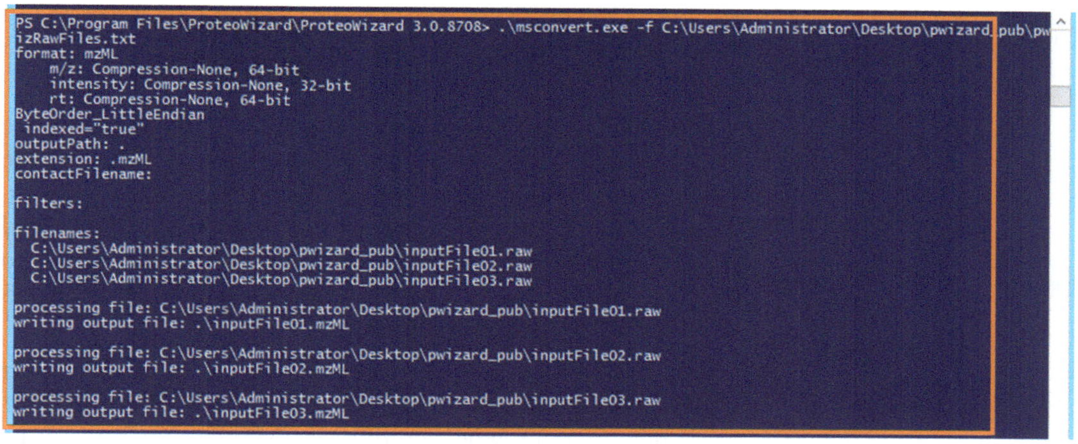

Fig. 10 msconvert commandline using -f (or --filelist) to convert multiple files together

5. Execute msConvert with variations in parameters and filters.
   (a) For each filter to be included (*see* **Note 7**), add text structured like the following: --filter "peakPicking true 1-". The first piece signals that a new data filter is being added. The section in double quotes supplies the name of the filter (see below) and the configuration options for it.
   (b) Specify the output format options, such as specifying the overall format by one of the following flags: "--mzML", "--mzXML", "--mz5", "--mgf". If subsequent software allows it, specify "zip" encoding for data within the files through use of the "-z" option. The full list of these options can be found by running the msConvert executable without any parameters.
   (c) Specify which files are to be converted. Both absolute and relative paths are permitted, and wildcard characters such as '*' and '?' can be used to process multiple files in a single pass.
   (d) Combine these elements into a single command line, such as the following:

```
msconvert.exe --filter "peakPicking true 1-" --mz5 -z *.raw
```

4 Notes

1. A file index allows for more rapid scan level access of the file. The index itself makes files slightly larger. However, instead of having to read a file from beginning to end, it can now be read from random positions.
2. zlib compression is a seemingly invisible change to the file. Essentially, the ms spectra themselves are zlib compressed prior

to writing the file. This greatly compresses the file, but also can slow down its reading. Furthermore, some readers are unable to handle zlib compressed files.

3. Unlike zlib compression, gzip actually fully gzips the entire file. This shrink files by about ~20–30%. Once gzipped, the files are no longer directly human readable without being ungzipped first. ProteWizard and some other tools are able to read files, even when gzipped. However, they are no longer able to provide random access. The vast majority of open source tools are not able to read gzipped files directly.
4. Filters are operations that are performed on spectra to alter them prior to writing out a converted file. While some filters are simple (e.g., only writing out MS/MS data), others can be quite complex and substantially alter your data. The commandLine msConvert has access to many more filters than are available in the GUI. Please note, that as filters can fundamentally change your data, they may substantially impact downstream operations, such as quantification.
5. Peak Picking, is the msConvert nomenclature for "centroiding" which is a very common procedure for shrinking file sizes. However, this is an irreversible operation that discards a substantial amount of data. It should be used with caution.
6. msConvert has a large number of available options (Fig. 8):
   All the available options are listed on running the command "msconvert" from the command prompt.

Parameter Flag	Explanation
-f [--filelist] arg	Specify text file containing filenames
-o [--outdir] arg (=.)	Set output directory ('-' for stdout) [.]
-c [--config] arg	Configuration file (optionName = value)
--outfile arg	Override the name of output file
-e [--ext] arg	Set extension for output files [mzML|mzXML|mgf|txt|mz5]
--mzML	Write mzML format [default]
--mzXML	Write mzXML format
--mz5	Write mz5 format
--mgf	Write Mascot generic format
--text	Write ProteoWizard internal text format
--ms1	Write MS1 format
--cms1	Write CMS1 format
--ms2	Write MS2 format
--cms2	Write CMS2 format

(continued)

Parameter Flag	Explanation
-v [--verbose]	Display detailed progress information
--64	Set default binary encoding to 64-bit precision [default]
--32	Set default binary encoding to 32-bit precision
--mz64	Encode *m/z* values in 64-bit precision [default]
--mz32	Encode *m/z* values in 32-bit precision
--inten64	Encode intensity values in 64-bit precision
--inten32	Encode intensity values in 32-bit precision [default]
--noindex	Do not write index
-i [--contactInfo] arg	Filename for contact info
-z [--zlib]	Use zlib compression for binary data
--numpressLinear [=arg(=2e-009)]	Use numpress linear prediction compression for binary mz and rt data (relative accuracy loss will not exceed given tolerance arg, unless set to 0)
--numpressPic intensities	Use numpress positive integer compression for binary intensities (absolute accuracy loss will not exceed 0.5)
--numpressSlof [=arg(=0.0002)]	Use numpress short logged float compression for binary intensities (relative accuracy loss will not exceed given tolerance arg, unless set to 0)
-n [--numpressAll]	Same as --numpressLinear --numpressSlof (see https://github.com/fickludd/ms-numpress for more info)
-g [--gzip]	gzip entire output file (adds .gz to filename)
--filter arg	Add a spectrum list filter
--merge level	Create a single output file from multiple input files by merging file-level metadata and concatenating spectrum lists
--simAsSpectra	Write selected ion monitoring as spectra, not chromatograms
--srmAsSpectra chromatograms	Write selected reaction monitoring as spectra, not chromatograms
--combineIonMobilitySpectra	Write all drift bins/scans in a frame/block as one spectrum instead of individual spectra
--acceptZeroLengthSpectra	Some vendor readers have an efficient way of filtering out empty spectra, but ittakes more time to open the file
--ignoreUnknownInstrumentError	If true, if an instrument cannot be determined from a vendor file, it will not be an error
--help	Show this message, with extra detail on filter options

-f or --filelist (Fig. 10)

- The -f or --filelist flag is used in case the user would like to convert multiple input files together.
- For instance, consider a text file containing the filenames of three *.RAW files which need to be converted to *. mzML format.
- Sample text file (pwizRawFiles.txt) with list of filenames and locations:

  C:\Users\Administrator\Desktop\pwizard_pub\input-File01.raw

  C:\Users\Administrator\Desktop\pwizard_pub\input-File02.raw

  C:\Users\Administrator\Desktop\pwizard_pub\input-File03.raw
- Using the *pwizRawFiles.txt* file above with the flag -f, the command should look like:
- msconvert -f C:\Users\Administrator\Desktop\pwizard_pub\pwizRawFiles.txt
- Additional flags can be added to convert the files to different formats as discussed further.
- NOTE: One line inside the *pwizRawFiles.txt* file should have only one input filename. Each input filename in the list should be present in a new line

-o or --outdir

- The flags "-o" or "--outdir" can be used to specify the directory to write the output file
- For example:
- mconvert -o C:\Users\Administrator\Desktop\pwizard_pub\
- Default: Current working directory

-c or --config

- This option provides the user to enter all file conversion properties together on one or more files.
- Sample configuration (msconvert.config) file:
- mzXML=true
- zlib=true
- filter="index [3,7]"
- filter="precursorRecalculation"
- IMPORTANT: DO NOT use spaces before and after "=" inside the configuration file

--outfile

- The --outfile argument lets the user set the name of the output file. If missing, the output filename will be based on the given input filename.

-e or --ext

- This argument can be used to select the extension of the output filename.
- Options available: mzML, mzXML, mgf, txt, mz5
- If none selected, the extension will be based on the output file conversion format (For example: using argument --mzXML with msconvert will result in the <output_filename> .mzXML).
- If no output file conversion is selected, then the default as mentioned earlier will be "mzML"

--mzML

- The output converted file will be ".mzML" format.
- This is the default if no other format is selected

--mzXML

- Very similar to the above "--mzML" format
- The output file will be created as <output_filename> . mzXML

--mz5

- Similar to --mzML and --mzXML options
- Writes the output file to "mz5" format

--64

- The "--64" option is used to set the binary encoding to 64-bit precision for the converted file (same as 64bit precision option on the msconvertGUI)
- Default if no other encoding flags are selected

--32

- Like above, this is used to set the binary encoding.
- As the flag suggests, this option lets us set the output file conversion to 32-bit encoding precision
- The *m/z* and intensity values can also be encoded independently using the --mz23 and --inten64 flags.

--mz64

- The "mz" prefix to the flag "--mz64" refers to *m/z* values This flag along with the following "--mz32" flag below offers the flexibility of setting the *m/z* values only encoding to 64-bit precision in case the user would like to specify the "intensity" coding separately.

- --mz64 is the default for the *m/z* encoding.

--mz32

- Similar to the "--mz64" option above.
- In case of mz and intensity independent encoding, We can select "--mz32" for 32-bit encoding of "*m/z*" values

--inten64

- The "inten" prefix to the flag "--inten64" refers to intensity values.
- This and the below given "--inten32" flags can be used for encoding intensity values independently.
- --inten64 is the default for encoding the intensity values.

--noindex

- Setting this "--noindex" flag will tell the msconvert command line program NOT to write index to the output file.

-i or --contactInfo

- Provide the msconvert program with contact information in a filename and path by using this flag.

-z or --zlib

- Use zlib compression for binary data by setting this flag. This option is used for data compression.

--numpressLinear

- Use numpress linear prediction lossy compression for binary mz and rt data (relative error guaranteed less than given tolerance)

--numpressPic

- Use numpress positive integer lossy compression for binary intensities (maximum 0.5 absolute error guaranteed)

--numpressSlof

- Use numpress short logged float lossy compression for binary intensities (relative error guaranteed less than given tolerance)

-n or --numpressAll

- Same as --numpressLinear --numpressSlof
- See https://github.com/fickludd/ms-numpress for more info

-g or --gzip

- Use "-g" or "--gzip" option to gzip the entire output file.
- Selecting this option reduces the file size and adds ".gz" to the output filename

--filter

- IMPORTANT: Adding this argument provides the option to add/set filters before writing the data to output file.
- Several options are available for filtering the data. All the available filter types/options are explained in greater detail under the next section C (FILTER OPTIONS).
- Each type of filter mentioned in section C should be preceded by the "--filter" flag.
- For Example:
  - msconvert C:\Users\Administrator\Desktop\pwizard_pub\inputFile01.raw
  - --filter "peakPicking true 1-" --filter "threshold bpi-relative .5 most-intense"

--merge

- IMPORTANT: This argument/flag provides us the option to merge the output of multiple input files into a single output files. This option can be very useful if needed to merge fractions with each fraction having an individual *.RAW file.

--simAsSpectra

- "sim" refers to selected ion-monitoring. This is a useful option available in case the selected ion-monitoring is required as spectra and NOT as chromatograms. This is often used to represent MRM data in mzXML files, which do not natively support chromatogram data.

--srmAsSpectra

- "srm" refers to selected reaction monitoring. Like the "--simAsSpectra" option, the "--srmAsSpectra" flag can be used to write the selected reaction monitoring as sepctra and NOT chromatograms. This is often used to represent MRM data in mzXML files, which do not natively support chromatogram data.

--combineIonMobilitySpectra

- Usually, ion mobility spectra are written individually for each scan. This option can be used to write all drift bins/scans in a frame/block as one spectrum instead of individual spectra.

--acceptZeroLengthSpectra

- This option lets you accept zero length spectra in case the file contains empty spectra.
- Some (but not all) vendor readers have an efficient way of filtering out empty spectra.
- NOTE: Takes more time to open the file.

--ignoreUnknownInstrumentError

- The instrument information should be present in the vendor file usually but in case it is missing, this option can be useful.
- If true, if an instrument cannot be determined from a vendor file, it will not be an error

--help

- Use this option to display all filtering options in detailed

7. The msconvert commandline has additional filtering options when compared to the msConvertGUI.

   For example, msLevel is present in both commandline and msConvertGUI (as MS Levels on GUI). This filter selects only spectra with indicated mslevels (ms level 1). Similarly, the analyzer option in the commandline allows the user to retain the spectra with the specified mass analyzer ("FTMS" or "ITMS") type only.

   Note: Filters are applied sequentially in the order that you list them, and the sequence order can make a large difference in your output. In particular, the peakPicking filter must be first in line if you wish to use the vendor-supplied centroiding algorithms since these use the vendor DLLs, which only operate on raw untransformed data.

Filter Flag	Explanation
index	<index_value_set>
msLevel	<mslevels>
chargeState	<charge_states>
precursorRecalculation	
mzRefiner	mzRefinerinput1
precursorRefine	
peakPicking	[<PickerType>[snr=<minimumsignal-to-noiseratio>] [peakSpace=<minimumpeakspacing>][msLevel=<ms_levels>]]
scanNumber	<scan_numbers>

(continued)

(continued)

Filter Flag	Explanation
scanEvent	<scan_event_set>
scanTime	<scan_time_range>
sortByScanTime	
stripIT	
metadataFixer	
titleMaker	<format_string>
threshold	<type><threshold><orientation>[<mslevels>]
mzWindow	<mzrange>
mzPrecursors	<precursor_mz_list>
defaultArrayLength	<peak_count_range>
zeroSamples	<mode>[<MS_levels>]
mzPresent	<tolerance><type><threshold><orientation> <mz_list>[<include_or_exclude>]
scanSumming	[precursorTol=<precursortolerance>][scanTimeTol=<scantimetolerance>]
MS2Denoise	[<peaks_in_window>[<window_width_Da>[multicharge_fragment_ relaxation]]]
MS2Deisotope	[hi_res[mzTol=<mzTol>]][Poisson[minCharge=<minCharge>] [maxCharge=<maxCharge>]]
ETDFilter	[<removePrecursor>[<removeChargeReduced>[<removeNeutralLoss>[<bl anketRemoval>[<matchingTolerance>]]]]]
chargeStatePredictor	[overrideExistingCharge=<true|false(false)>] [maxMultipleCharge=<int(3)>][minMultipleCharge=<int(2)>][singleC hargeFractionTIC=<real(0
turbocharger	[minCharge=<minCharge>][maxCharge=<maxCharge>] [precursorsBefore=<before>][precursorsAfter = <after>] [halfIsoWidth=<half-widthofisolationwindow>][defaultMinCharge=<def aultMinCharge>][defaultMaxCharge=<defaultMaxCharge>][useVendor Peaks=<useVendorPeaks>]
activation	<precursor_activation_type>
analyzer	<analyzer>
analyzerType	<analyzer>
polarity	<polarity>

- *msLevel*

  This filter selects only spectra with the indicated <mslevels>.

  - The value set to this argument needs to be an integer.
  - For Example:

    # extract MS1 scans only

    msconvert C:\Users\Administrator\Desktop\pwizard_pub\ inputFile01.raw

    --filter "msLevel 1"

    # Alternatively, inside the configuration file from the msconvert.config file (input as in section 3.2, **step 2**), add the filter for "msLevel 1" as:

    mzXML=true

    zlib=true

    filter="index [3,7]"

    filter="msLevel 1"   #For ms1 scans only

- *chargeState*
  - This filter keeps spectra that match the listed charge state(s).
  - The value set to this argument needs to be an integer.
  - Both known/single and possible/multiple charge states are tested. Use 0 to include spectra with no charge state at all.

- *precursorRecalculation*
  - This filter recalculates the precursor *m/z* and charge for MS2 spectra.
  - Looks at the prior MS1 scan to better infer the parent mass.
  - Works only on orbitrap and FT data and does not use any third party (vendor DLL) code.
  - ONLY need to use as a flag. For example:

    # Add this line if using config file

    filter="precursorRecalculation"

    # If using as a flag to the command line parameter, add the following to the command:

    --filter "precursorRecalculation"

- *mzRefiner*
  - This filter recalculates the *m/z* and charges, adjusting precursors for MS2 spectra and spectra masses for MS1 spectra.
  - It uses an ident file with a threshold field and value to calculate the error and will then choose a shifting mechanism to correct masses throughout the file. It only works on Orbitrap, FT, and TOF data. It is designed to work on mzML files.

- *precursorRefine*
  - Similar to *precursorRecalculation*
  - This filter refines the precursor *m/z* and charge for MS2 spectra. It looks at the prior MS1 scan to better infer the parent mass. It only works on orbitrap, FT, and TOF data. It does not use any third party (vendor DLL) code.
- *peakPicking*
  - This filter performs centroiding on spectrawith the selected ms_level argument

    peakPicking [<PickerType>[snr=<minimum signal-to-noise ratio>] [peakSpace=<minimum peak spacing>]

    [msLevel=<ms_levels>]]
- *scanNumber*
  - This filter takes input as an integer and lets the user select spectra by scan number
  - Scan numbers are 1-based and not contiguous
- *scanEvent*
  - scanEvent filters spectra by scan event and takes input in the form of a set.
  - This filter offers the flexibility of excluding the selected scan events as well.
  - For Example:

    # All scan event EXCEPT 5 can be included using-

    --filter "scanEvent 1-4 6- "
- *scanTime*
  - Spectra can be selected using a specific time range using this filter.
- *sortByScanTime*
  - As the name suggests, the spectra can be a sorted in ascending order by scan start time.
- *stripIT*
  - "IT" in the filter name refers to "ion trap".
  - This filter removes out the ion trap data spectra with MS level 1.
- *metadataFixer*
  - As the name suggests, this filter can be used to "fix" metadata—meaning, it adds/replaces a spectra's TIC/BPI metadata.
  - This filter traverses the *m/z* intensity arrays to find the sum and max.

  - Usually used after peak-picking, for Example:

    --filter "peakPicking true 1-" --filter metadataFixer

- *titleMaker*
  - Uses a string format as the input
  - For Example:

    --filter "titleMaker<RunId>.<ScanNumber>.<ScanNumber>.<ChargeState>"

- *Threshold*
  - This filter can be used to set several thresholds.
  - Data in the input file meeting all the thresholds is retained and everything else is filtered out.
  - Threshold filter can be set to "*most or least*" (i.e., either upper or lower limit can be set) for all give threshold types. This can be set using the orientation in the threshold format as given below.
  - Format for threshold-

    threshold<type><threshold><orientation>[<mslevels>]
  - Various threshold types with their input format and explanation are given below:

Threshold type	Threshold format	Explanation
count	Integer (*n*)	Retains intensity values for the selected number of "*n*" data points. Data points where intensity is equal to *n*th intense data point are removed
count-after-ties	Integer (*n*)	Same as the count but retains data points where intensity = *n*th data point
absolute	Flag (no argument)	Keep data whose absolute intensity matches threshold
bpi-relative	Flag (no argument)	Data is retained if the intensity matches the percentage of base-peak intensity. 0.75 = 75 %
tic-relative	Flag (no argument)	Similar to above but for retains data for individual intensities greater than or less than the percentage of total ion current for the scan. 1 = 100 %
tic-cutoff	Flag (no argument)	As the name suggests, this is a cutoff. Data is retained UPTO the percentage of the total ion current for the scan
ms_levels	Integer (*n*)	OPTIONAL. If selected, only scans with MS level = *n* will be selected

- *mzWindow*
  - As the name suggest, this filter is used to provide a range to select the *m/z* within the selected window.
  - The *m/z* values falling ONLY within the selected range will be retained.
  - Input type: range [mzLow, mzHigh]
  - For Example:

    --filter "mzWindow [100.1,307.5]"
- *mzPrecursors*
  - This filter allows us the flexibility of selecting spectra within a given list of precursor *m/z* values.
  - The input type is a list of *m/z* precursor value
  - For Example:

    --filter "mzPrecursors [123.4,567.8]"
- *defaultArrayLength*
  - For this filter, the default array length refers to the range of peak counts.
  - The name is derived from mzML format file where "defaultArrayLength" refers to peak list.
  - This can be specified as range of integers
  - For Example:

    # Retain only peakcounts >=100:

    --filter "defaultArrayLength 100-"
- *zeroSamples*
  - Usage: zeroSamples <mode> [<MS_levels>]
  - This is a useful filter to deal with zero values in the spectrum. This can be used one of two ways—either remove spectra with zero values or add the sero value incase the spectra is missing.
  - Mode options-

    i. *removeExtra: consecutive zero intensity peaks are removed from spectra*

    "100.1,1000 100.2,0 100.3,0 100.4,0 100.5,0 100.6,1030"

    would become

    "100.1,1000 100.2,0 100.5,0 100.6,1030"

    and a peak list

    "100.1,0 100.2,0 100.3,0 100.4,0 100.5,0 100.6,1030 100.7,0 100.8,1020 100.9,0 101.0,0"

    would become

    "100.5,0 100.6,1030 100.7,0 100.8,1020 100.9,0"

ii. *addMissing: When <mode> is "addMissing", each spectrum's sample rate is automatically determined and flanking zeros are inserted around non-zero data points. The optional [=<flankingZeroCount>] value can be used to limit the number of flanking zeros, otherwise the spectrum is completely populated between nonzero points.*

For example, to make sure spectra have at least five flanking zeros around runs on nonzero points, use filter-

"addMissing = 5".

– The <MS levels> is optional but when used, retains spectra with only the ms levels specified using the option. Format: integer set (For example: 1–5)

- *mzPresent*
  - Usage: mzPresent <tolerance> <type> <threshold> <orientation> <mz_list> [<include_or_exclude>]
  - This filter is very similar to threshold filter

    i. tolerance: Specified as a number and units (PPM or MZ).

    For example, "5 PPM" or "2.1 MZ".

    ii. <type>, <threshold>, and <orientation> operate as in the "threshold" filter (refer to *step 14*).

    iii. <mz_list>: List of mz - [mz1,mz2, … mzn]

    #Data points within <tolerance> of any of these values will be kept. For example,

    "[100, 300, 405.6]"

    iv. <include_or_exclude>: as the terms suggest, using *include* (default) retains all the values that match the criteria where as *exclude* drops data points matching the criteria.

- *scanSumming*
  - Usage: scanSumming [precursorTol=<precursor tolerance>] [scanTimeTol=<scan time tolerance>]
  - Used to sum "*MS2201D* sub-scans whose precursors are within the given <precursor tolerance> and <scan time tolerance>
  - Defaults:

    i. Precursor tolerance: 0.05

    ii. Scan time tolerance: 10 s
  - Used where sub-scans need to be summed to increase the signal-to-noise ratio.
  - IMPORTANT: Only tested on waters data so far.

- *MS2Denoise*
  - Usage: MS2Denoise [<peaks_in_window> [<window_width_Da> [multicharge_fragment_relaxation]]]
  - Use to denoise, i.e., remove peaks with noise for spectra with precursor ions.
    - i. <peaks_in_window>: the number peaks to select in moving window, default is 6.
    - ii. <window_width_Da>: the width of the window in Da, default is 30.
    - iii. <multicharge_fragment_relaxation>—if "true" (the default), allows more data below multiply charged precursors.
- *MS2Deisotope*
  - MS2Deisotope [hi_res [mzTol=<mzTol>]] [Poisson [minCharge=<minCharge>][maxCharge=<maxCharge>]]
  - Uses the Markey method or a Poisson model to deisotope the ms2 spectra.
    - i. hi_res: set to "false" (default) or "true"
      - *mzTol: Input format is a decimal value. Sets the m/z tolerances .Regular default is 0.5, high resolution default is 0.01*
    - ii. Poisson: Used to define the search range for the charge. Default is 1
      1. minCharge: Default is 1.
      2. maxCharge: Default is 3.
  - For Example:

    --filter "MS2Deisotope true mzTol=0.4 1 minCharge=1 maxCharge=3"
- *ETDFilter*
  - Usage: ETDFilter [<removePrecursor> [<removeChargeReduced> [<removeNeutralLoss> [<blanketRemoval> [<matchingTolerance>]]]]]
    - i. <removePrecursor>—specify "true" to remove unreacted precursor (default is "false")
    - ii. <removeChargeReduced>—specify "true" to remove charge reduced precursor (default is "false")
    - iii. <removeNeutralLoss>—specify "true" to remove neutral loss species from charge reduced precursor (default is "false")
    - iv. <matchingTolerance>—specify matching tolerance in MZ or PPM (examples: "3.1 MZ" (the default) or "2.2 PPM")

- *chargeStatePredictor*
  - Usage: chargeStatePredictor [overrideExistingCharge=<true|false (false)>] [maxMultipleCharge=<int (3)>] [minMultipleCharge=<int(2)>][singleChargeFractionTIC=<real (0.9)>] [maxKnownCharge=<int (0)>] [makeMS2=<true|false (false)>]
    - i. <overrideExistingCharge> : always override existing charge information (default:"false")
    - ii. <maxMultipleCharge> (default 3) and <minMultipleCharge> (default 2): range of values to add to the spectrum's existing "MS_possible_charge_state" values.If these are the same values, the spectrum's MS_possible_charge_state values are removed and replaced with this single value.
    - iii. <singleChargeFractionTIC>: is a percentage expressed as a value between 0 and 1 (the default is 0.9, or 90%). This is the value used as the previously mentioned ratio of intensity above and below the precursor *m/z*.
    - iv. <maxKnownCharge> (default is 0, meaning no maximum): the maximum charge allowed for "known" charges even if override existing charge is false. This allows overriding junk charge calls like +15 peptides.
    - v. <algorithmMakeMS2>: default is "false", when set to "true" the "makeMS2" algorithm is used instead of the one described above.
- *turbocharger*
  - turbocharger [minCharge=<minCharge>] [maxCharge=<maxCharge>] [precursorsBefore=<before>] [precursors After=<after>] [halfIsoWidth=<half-width of isolation window>] [defaultMinCharge=<defaultMinCharge>] [defaultMaxCharge=<defaultMaxCharge>] [useVendorPe aks=<useVendorPeaks>]
  - <maxCharge> (default: 8) and <minCharge> (default 1): defines range of possible precursor charge states.
  - <before> (default: 2) and <after> (default 0): number of survey (MS1) scans to check for precursor isotopes, before and after a MS/MS in retention time.
  - <half-width of isolation window> (default: 1.25): half-width of the isolation window from which precursor is derived. Window is centered at target *m/z* with a total size of ± the value entered.
  - <defaultMinCharge> (default: 0) and <defaultMaxCharge> (default: 0): in the event that no isotope is found in the iso-

lation window, a range of charges between these two values will be assigned to the spectrum. If both values are left at zero, no charge will be assigned to the spectrum.

- *activation*
  - Usage: activation <precursor_activation_type>
  - Filter to retain only spectra of a specified type of activation. DOES NOT effect non-MS spectra.
  - Input options:
    i. ETD
    ii. CID
    iii. SA
    iv. HCD
    v. HECID
    vi. BIRD
    vii. ECD
    viii. IRMPD
    ix. PD
    x. PSD
    xi. PQD
    xii. SID
    xiii. SORI
- *analyzer*
  - Usage: analyzer <analyzer>
  - Filter to retain only spectra of a specified type of mass analyzer
  - Four options for input:
    i. "quad"
    ii. "orbi"
    iii. "FT"
    iv. "IT"
    v. "TOF"
- *analyzerType*
  - Currently deprecated but accepted two options are: "FTMS" or "ITMS"
- *Polarity*
  - This filter allows the user to retain spectra with scan of the selected polarity.

- Available Options:
  i. "negative"
  ii. "positive"
  iii. "+"
  iv. "–"

References

1. Chambers MC, Maclean B, Burke R, Amodei D, Ruderman DL, Neumann S, Gatto L, Fischer B, Pratt B, Egertson J, Hoff K, Kessner D, Tasman N, Shulman N, Frewen B, Baker TA, Brusniak MY, Paulse C, Creasy D, Flashner L, Kani K, Moulding C, Seymour SL, Nuwaysir LM, Lefebvre B, Kuhlmann F, Roark J, Rainer P, Detlev S, Hemenway T, Huhmer A, Langridge J, Connolly B, Chadick T, Holly K, Eckels J, Deutsch EW, Moritz RL, Katz JE, Agus DB, MacCoss M, Tabb DL, Mallick P (2012) A cross-platform toolkit for mass spectrometry and proteomics. Nat Biotechnol 30(10):918–920. doi:10.1038/nbt.2377
2. Kessner D, Chambers M, Burke R, Agus D, Mallick P (2008) ProteoWizard: open source software for rapid proteomics tools development. Bioinformatics 24(21):2534–2536. doi:10.1093/bioinformatics/btn323

INDEX

Lucio Comai et al. (eds.), *Proteomics: Methods and Protocols*, Methods in Molecular Biology, vol. 1550, DOI 10.1007/978-1-4939-6747-6, © Springer Science+Business Media LLC 2017

MIX
Papier aus verantwortungsvollen Quellen
Paper from responsible sources
FSC® C105338

If you have any concerns about our products,
you can contact us on
ProductSafety@springernature.com

In case Publisher is established outside the EU,
the EU authorized representative is:
Springer Nature Customer Service Center GmbH
Europaplatz 3, 69115 Heidelberg, Germany

Printed by Libri Plureos GmbH
in Hamburg, Germany